CELL AND MOLECULAR BIOLOGY OF VERTEBRATE HARD TISSUES

The Ciba Foundation is an international scientific and educational charity. It was established in 1947 by the Swiss chemical and pharmaceutical company of CIBA Limited—now CIBA-GEIGY Limited. The Foundation operates independently in London under English trust law.

The Ciba Foundation exists to promote international cooperation in biological, medical and chemical research. It organizes about eight international multidisciplinary symposia each year on topics that seem ready for discussion by a small group of research workers. The papers and discussions are published in the Ciba Foundation symposium series. The Foundation also holds many shorter meetings (not published), organized by the Foundation itself or by outside scientific organizations. The staff always welcome suggestions for future meetings.

The Foundation's house at 41 Portland Place, London, W1N 4BN, provides facilities for meetings of all kinds. Its Media Resource Service supplies information to journalists on all scientific and technological topics. The library, open seven days a week to any graduate in science or medicine, also provides information on scientific meetings throughout the world and answers general enquiries on biomedical and chemical subjects. Scientists from any part of the world may stay in the house during working visits to London.

Ciba Foundation Symposium 136

CELL AND MOLECULAR BIOLOGY OF VERTEBRATE HARD TISSUES

A Wiley – Interscience Publication

1988

JOHN WILEY & SONS

Chichester · New York · Brisbane · Toronto · Singapore

© Ciba Foundation 1988

Published in 1988 by John Wiley & Sons Ltd, Chichester, UK.

Suggested series entry for library catalogues:
Ciba Foundation Symposia

Ciba Foundation Symposium 136
× + 307 pages, 51 figures, 21 tables

Library of Congress Cataloging-in-Publication Data

Cell and molecular biology of vertebrate hard tissues.
      p. cm. — (Ciba Foundation symposium ; 136)
    Proceedings of a symposium held at the Ciba Foundation, London,
Oct. 13–15, 1987.
    Edited by David Evered and Sara Harnett.
    "A Wiley–Interscience publication."
    Includes indexes.
    ISBN 0 471 91885 7
    1. Bones — Cytology — Congresses. 2. Bones — Differentiation —
Congresses. 3. Bones — Growth — Congresses. I. Evered, David.
II. Harnett, Sara. III. Ciba Foundation. IV. Series.
    [DNLM: 1. Cell Differentiation — congresses. 2. Gene Expression
Regulation — congresses. 3. Osteoblasts — congresses.
4. Osteoclasts — congresses. 5. Osteogenesis — congresses. W3 C161F
v. 136 / WE 200 C393 1987]
QP88.2.C45 1988
596'.01852 — dc19
DNLM/DLC
for Library of Congress

88–10776
CIP

British Library Cataloguing in Publication Data

Cell and molecular biology of vertebrate
  hard tissues
  — (Ciba Foundation symposium ; 136).
  1. Vertebrates. Calcified tissues. Cells
  & molecules. Structures & properties
  I. Series
  596'.0471

  ISBN 0 471 91885 7

Typeset by Inforum Ltd, Portsmouth
Printed and bound in Great Britain at The Bath Press, Avon

Contents

Symposium on Cell and Molecular Biology of Vertebrate Hard Tissues, held at the Ciba Foundation, London, 13–15 October 1987
The topic for this symposium was proposed by Professor A.I. Caplan

Editors: David Evered (Organizer) and Sara Harnett

G.A. Rodan Introduction 1

A.I. Caplan Bone development 3
Discussion 18

H.C. Slavkin, M.L. Snead, M. Zeichner-David, M. MacDougall, A. Fincham, E.C. Lau, W. Luo, M. Nakamura, P. Oliver and **J. Evans** Factors influencing the expression of dental extracellular matrix biomineralization 22
Discussion 35

M.E. Owen and **A.J. Friedenstein** Stromal stem cells: marrow-derived osteogenic precursors 42
Discussion 53

P.J. Nijweide, A. van der Plas and **A.A. Olthof** Osteoblastic differentiation 61
Discussion 72

G.A. Rodan, J.K. Heath, K. Yoon, M. Noda and **S.B. Rodan** Diversity of the osteoblastic phenotype 78
Discussion 85

T.J. Chambers The regulation of osteoclastic development and function 92
Discussion 100

P. Osdoby, M.J. Oursler, T. Salino-Hugg and **M. Krukowski** Osteoclast development: the cell surface and the bone environment 108
Discussion 122

General discussion I   Osteoclast activity   125

M.D. Pierschbacher, S. Dedhar, E. Ruoslahti, S. Argraves and **S. Suzuki**   An adhesion variant of the MG-63 osteosarcoma cell line displays an osteoblast-like phenotype   131
Discussion   136

D.J. Prockop, K.E. Kadler, Y. Hojima, C.D. Constantinou, K.E. Dombrowski, H. Kuivaniemi, G. Tromp and **B. Vogel**   Expression of type I procollagen genes   142
Discussion   156

A. Veis   Phosphoproteins from teeth and bone   161
Discussion   172

J.D. Termine   Non-collagen proteins in bone   178
Discussion   191

General discussion II   Osteopontin: structure and biological activity 203

P.V. Hauschka, T.L. Chen and **A.E. Mavrakos**   Polypeptide growth factors in bone matrix   207
Discussion   220

L.G. Raisz   Hormonal regulation of bone growth and remodelling   226
Discussion   234

S.M. Krane, M.B. Goldring and **S.R. Goldring**   Cytokines   239
Discussion   252

N.G. Testa, T.D. Allen, G. Molineux, B.I. Lord and **D. Onions**   Haemopoietic growth factors: their relevance in osteoclast formation and function   257
Discussion   269

Final general discussion: Clinical implications   275
                                 Future perspectives   288

G.A. Rodan   Chairman's summary   297

Index of contributors   298

Subject index   300

Participants

R. Baron   Department of Cell Biology, Yale University School of Medicine, Sterling Hall of Medicine, PO Box 3333, New Haven, Connecticut 06510-8002, USA

W.T. Butler   Dental Branch, University of Texas Health Science Center, PO Box 10068, Texas 77025, Houston, USA

E. Canalis   Department of Medicine, University of Connecticut, School of Medicine, St Francis Hospital & Medical Center, 114 Woodland Street, Hartford, Connecticut 06105, USA

A.I. Caplan   Department of Biology, Case Western Reserve University, Cleveland, Ohio 44106, USA

T.J. Chambers   Department of Histopathology, St George's Hospital Medical School, Cranmer Terrace, London SW17 0RE, UK

H. Fleisch   Pathophysiologisches Institut, Universität Bern, Murtenstrasse 35, CH-3010 Bern, Switzerland

D. Gazit (*Ciba Foundation Bursar*)   Division of Oral Pathology, Faculty of Dental Medicine, The Hebrew University, Hadassah School of Dental Medicine, PO Box 1172, Jerusalem 91010, Israel

M.J. Glimcher   Laboratory for the Study of Skeletal Disorders & Rehabilitation, Harvard Medical School, The Children's Hospital, 300 Longwood Avenue, Boston, Massachusetts 02115, USA

J. Glowacki   Harvard Medical School, Surgical Research, Children's Hospital, 300 Longwood Avenue, Boston, Massachusetts 02115, USA

H. Hanaoka   Department of Orthopaedic Surgery, Keio University School of Medicine, 35 Shinanomachi, Shinjuku-Ku, Tokyo 160, Japan

P.V. Hauschka   Harvard School of Dental Medicine, The Children's Hospital/Enders-12, 300 Longwood Avenue, Boston, Massachusetts 02115, USA

S.M. Krane   Arthritis Unit, Harvard Medical School, Massachusetts General Hospital, Boston, Massachusetts 02114, USA

T.J. Martin   Department of Medicine, University of Melbourne, Repatriation General Hospital, West Heidelberg, Victoria 3081, Australia

P.J. Nijweide   Laboratorium voor Celbiologie en Histologie, der Rijksuniversiteit Leiden, Rijnsburgerweg 10, 2333 AA Leiden, The Netherlands

P. Osdoby   Washington University School of Dental Medicine, Washington University Medical Center, 4559 Scott Avenue, St Louis, Missouri 63110, USA

M.E. Owen   MRC Bone Research Laboratory, Nuffield Orthopaedic Centre, Headington, Oxford OX3 7LD, UK

M.D. Pierschbacher   La Jolla Cancer Research Foundation, Cancer Research Center (NCI), 10901 North Torrey Pines Road, La Jolla, California 92037, USA

D.J. Prockop   Department of Biochemistry & Molecular Biology, Jefferson Institute of Molecular Medicine, Jefferson Medical College, Room 490, 1020 Locust Street, Philadelphia, Pennsylvania 19107, USA

L.G. Raisz   Division of Endocrinology & Metabolism, University of Connecticut Health Center, School of Medicine, Farmington, Connecticut 06032, USA

G.A. Rodan (*Chairman*)   Department of Bone Biology & Osteoporosis Research, Merck Sharp & Dohme Research Laboratories, Division of Merck & Co Inc, West Point, Pennsylvania 19486, USA

J.V. Ruch   Faculté de Médecine, Institut de Biologe Médicale, Université Louis Pasteur, 11 rue Humann, F-67085 Strasbourg Cedex, France

H.C. Slavkin   Laboratory for Developmental Biology, Department of Basic Sciences, School of Dentistry, Andrus Gerontology Center, University Park-MC 0191, University of Southern California, Los Angeles, California 90089-0191, USA

J.D. Termine   Bone Research Branch, National Institute of Dental Research, National Institutes of Health, Building 30, Room 106, Bethesda, Maryland 20892, USA

N.G. Testa   Paterson Institute for Cancer Research, Christie Hospital & Holt Radium Institute, Wilmslow Road, Manchester M20 9BX, UK

M.R. Urist   Department of Orthopedics, UCLA Bone Research Laboratory, Rehabilitation Center, RM A3-34, 1000 Veteran Avenue, Los Angeles, California 90024, USA

A. Veis   Department of Oral Biology, Connective Tissue Research Laboratories, Northwestern University Dental School, 303 E Chicago Avenue, Chicago, Illinois 60611, USA

S. Weiner   Isotope Department, The Weizmann Institute of Science, IL-76100 Rehovot, Israel

R.J.P. Williams   Inorganic Chemistry Laboratory, University of Oxford, South Parks Road, Oxford OX1 3QR, UK

Introduction

G.A. Rodan

Department of Bone Biology and Osteoporosis Research, Merck Sharp & Dohme Research Laboratories, Division Merck & Co Inc., West Point, Pennsylvania 19486, USA

We are a very diverse group of people with interests in many aspects of mineralized tissue, bones and teeth. A unifying theme for this symposium may be the question of to what extent recent advances in biological sciences, which have been spectacular in the last few years, can be translated into medically applicable information. This is a general question which underlies all of biomedical research. In other words, how can we apply molecular information obtained from systems approached in a reductionist way to the whole animal?

In our discussion of the cells of bones and teeth we should address their characteristics, their origin, their life cycle, the interactions between various cells in the tissue, and the integration of the tissue within the organism in the context of developmental biology, physiology and pathology.

In hard tissues the extracellular matrix plays an important role. In the last few years we have seen considerable advances in identifying molecules that are part of this matrix. Our discussion will focus on their biochemical characterization, their structure, the interaction of these molecules with cells and with other similar molecules, and the regulation of their production and degradation. Discussion of these issues may give us some insight into the function of the extracellular matrix molecules, which remains unclear.

I hope that during the symposium our discussions will focus on finding consensus and, if necessary, defining clearly areas in which controversy remains. If this meeting is successful the result will be a 'road map' for further research in this area.

1988 Cell and molecular biology of vertebrate hard tissue. Wiley, Chichester (Ciba Foundation Symposium 136) p 1

Bone development

Arnold I. Caplan

Skeletal Research Center, Department of Biology, Case Western Reserve University, Cleveland, Ohio 44106, USA

Abstract. The sequential cellular and molecular details of the initial embryonic formation of bone can be used to gain insight into the control of this process and subsequent bone physiology and repair. The functioning of osteogenic cells is governed by a complex balance between the intrinsic capacities of these cells in the context of extrinsic information and signalling. As with other mesenchymal tissues, the balance of intrinsic versus extrinsic capacities and influences is central to understanding both the sequence and consequence of bone development. It has been suggested that the cartilaginous model which forms at the centre of limbs is responsible for, and provides the scaffolding for, subsequent bone formation. Our recent studies of the embryonic chick tibia indicate that osteogenic progenitor cells are observed *before* the formation of the chondrogenic core. In particular, a layer of four to six cells, referred to as Stacked Cells, forms around a prechondrogenic core of undifferentiated cells. These osteoprogenitor cells give rise to all of the newly forming bone. Importantly, this newly forming bone arises outside and away from the chondrogenic core in a manner similar to the intramembranous bone formation seen in calvariae. Indeed, the cartilaginous core is replaced not by bone but by vascular and marrow tissues. The interplay between the osteogenic collar and the chondrogenic core provides an environment which stimulates the further differentiation of the cartilage core into hypertrophic cartilage and eventually renders this core replaceable by vascular and marrow tissue. There is an intimate relationship between the osteogenic cells and the vasculature which is obligatory for active bone formation. Bone formation in long bones, such as the tibia, as well as in the calvaria seems to proceed in a similar manner, with vascular tissue interaction being the most important aspect of successful osteogenesis, as opposed to the presence or interaction of cartilage. Our studies have focused on the development of long bones in aves, but detailed study of mouse and man indicates that many of the general fetaures observed for birds apply to bone development in mammals. It is our current thesis that the general rules governing embryonic formation of long bones also apply to the formation of ectopic bone and are related to aspects of fracture repair.

1988 Cell and molecular biology of vertebrate hard tissues. Wiley, Chichester (Ciba Foundation Symposium 136) p 3–21

What are the rules by which bone does business? How does it keep itself going? How does it keep itself fit? And how does it react to crisis? These questions require an understanding of aspects of bone growth, physiology and repair. The answers are related to aspects of the structure, both cellular and molecular, of the living tissue and, in part, to how that structure arose and the identities of the cells responsible for its formation and current maintenance. Using the natural formation processes of embryology to obtain clues about the eventual physiology of bone from a cellular and molecular perspective provides the strategy for attempting to delineate a detailed description of the development of bone.

The process of development, maturation and ageing is a continuum of sequential cellular and molecular *replacement* events wherein tissues, cells and molecules are replaced by structures of progressively reduced developmental potential and increased degree of specialization (Caplan et al 1984). Some of the new structures in this sequence of replacement events are variants, or isoforms, of their predecessors. Some of the new structures are unique to that particular site: in this case the process of their fabrication is referred to as differentiation. Three fundamental variables control both the developmental and maintenance processes:

(1) The *genomic repertoire* of the organism sets the limits to the developmental and maturation possibilities. For example, birds cannot form teeth although their evolutionary ancestors had these structures. Moreover, the size, shape and responsiveness of a particular tissue is directly related to the response capabilities genetically programmed into the cells which make up that tissue.

(2) The *previous developmental decisions* also set the limits since the process of development is progressive and irreversible. There is no 'de-differentiation' to an earlier developmental state in repair or crisis situations (Caplan & Ordahl 1978). Once committed to a particular phenotypic pathway, cells may cease to synthesize specialized molecules and proliferate, but the descendants are nonetheless committed to the parental lineage. Thus, in order for the repair of particular structures to be effected, undifferentiated stem cells must be brought into the location to provide a source for the necessary phenotypic diversity for the regeneration of complex multicellular tissues or body parts.

(3) The local molecular and cellular environment provides cues which affect the rate and extent of developmental, maturational and repair events. These cues can be collectively referred to as *positional information* and can profoundly influence the process of cellular decision and expression. For example, a local site at which a mesenchymal stem cell finds itself can determine whether this cell will become an osteoblast or a chondrocyte.

With the above in mind, it should be obvious that the sequential maturation or regeneration/repair process cannot be an exact recapitulation of embryonic development, although the two may be very similar. The process of tissue maturation and repair takes place on the fabric of existing *adult* structures in an ambiance of adult physiology. Thus, while a detailed description of embryology can help in the understanding of how a tissue arises and how the main cellular contributors originate, it cannot predict how the maturation will continue or the exact sequence and emphasis of repair events, even though the repair and embryological processes must be very similar.

It is probable that adult tissues are endowed with a very sparse population of their own repair cells — in some cases, mesenchymal stem cells. An important example of this principle is skeletal muscle which is equipped with its own repair cells. These are the muscle satellite cells, which are formative cells that are committed to the myogenic lineage and are situated within the myofibrillar basement membrane so that repair processes can take place within a well-defined and limited region (Lipton & Schultz 1979). This repair potential is controlled by the genomic repertoire of those satellite cells, their previous decisional history, which has already brought them into a committed myogenic lineage, and local cueing which will activate and modulate the satellite cells' activity with regard to repair. Clearly, as individuals interested in clinical medicine, we cannot yet manipulate the genomic repertoire or the previous decisional history of adult human tissue; however, we have considerable experience of, and the opportunity to affect, the environment or local cueing in which maturation and repair take place. As a short-term goal, we might consider manipulation of the hosts' local repair cells. For cartilage and bone, these are the mesenchymal stem cells, which have not made major developmental commitments and can be directed to maturation and repair sites.

With the concept of *replacement* as the backdrop, the following description of the embryology and development of bone focuses on the pluripotential mesenchymal stem cell and how it gives rise to the two major skeletal phenotypes, cartilage and bone, and their interaction and interdependence.

Make no bones about it

The bones of the body are of different lengths, shapes, and, in some cases, cellular origins. For example, the cranial/facial bones and the tibia have different physical shapes and are derived from cells of distinct embryonic lineages: ectomesenchyme (neural crest) for cranial/facial bones and lateral plate mesenchyme for the tibia (Hall 1978). However, these bones are populated by the same general classes of cells: osteoblasts, osteocytes, osteoclasts, vascular cells and, in some cases, marrow cells. The unresolved question can be immediately stated: 'Are the cells at these two different anatomical sites the same or different?' It is clear that osteoclasts are of the same origin (systemic monocytes) and are probably identical at the two locations (Walker 1975). This strong statement cannot be made for osteoblasts or osteocytes and this indeed provides a basis for future experimentation. Moreover, the details of the formation of bones in the cranium and the tibia were thought to be completely different — intramembranous versus endochondral processes — but, as will be discussed below, it now appears that these two formative processes are in many ways identical.

The replacement events which precede the formation of an adult structure provide the anatomical boundaries of the tissue and the means by which tissue interactions are established. The replacement phenomena are complicated by the fact that embryonic tissues are still expanding, 'growing', while complex morphologies are being generated. In this regard, the example of tissue replacement often used is that of the replacement of embryonic tibial cartilage by bone. As detailed below, in the embryonic development of the tibia, cartilage is *not* replaced by bone; cartilage is replaced by marrow and vascular tissue.

An analysis of the formation, maturation and maintenance of any tissue, bone included, should concentrate on the cells responsible for giving this tissue its distinctive properties, in terms of both shape and physiological responsiveness. The properties of individual cells are governed by a complex balance between the *intrinsic* capacities and properties of these cells in the context of the *extrinsic* information and signalling provided to them. This balance of intrinsic and extrinsic signals is the key to describing the functioning of any tissue or group of cells. For example, the intrinsic response machinery of a particular cell must be in place for extrinsic cueing to be effective. Thus, for a cell to be affected by parathyroid hormone (PTH), it must have an appropriate PTH receptor and its response to receptor occupancy must be part of its intrinsic machinery.

Steps in long bone formation

The individual steps involved in *first bone* formation in the embryonic chick limb are outlined in Table 1 and diagrammatically summarized in Fig. 1. The osteogenic progenitor cells are observed as a distinct grouping of four to six cells in a *Stacked* conformation around a prechondrogenic core of cells. Importantly, this critical mass of cells at the centre of the newly forming and expanding limb bud is completely avascular. Thus, both the Stacked Cells and the central core that will differentiate into a rod of cartilage, referred to as a cartilage 'model', have, as a preliminary step to the developmental process, excluded vascular tissue. The capillary network has been segregated to a position outside and surrounding the central core region, the outer boundary to which is the Stacked Cell layer (Caplan & Pechak 1987, Pechak et al 1986a, b).

Eventually the Stacked Cell layer produces the osteogenic progenitor cells which differentiate into osteoblasts as a sheet or layer of cells facing the cartilage model. These osteoblasts fabricate a unique matrix, type I collagen-rich osteoid, at the mid-diaphysis as a continuous, circumferential collar around and outside the cartilage rod. This newly formed osteoid eventually

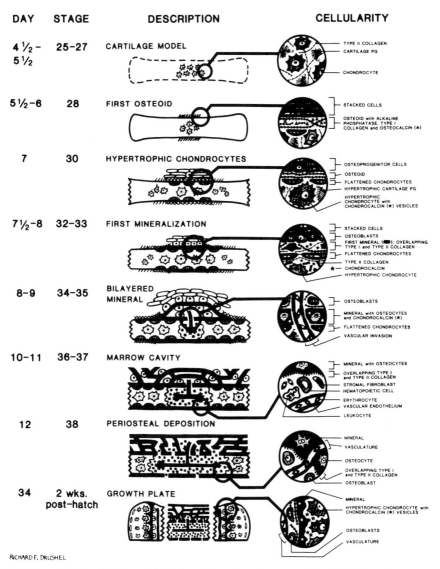

RICHARD F. DRUSHEL

FIG. 1.   Cellular details of chick long bone development. Simplified sketches of the major events associated with long bone development are shown from stage 25 to two weeks post-hatching.

acts as a physical restraint and, we speculate, as a barrier for nutrients and other vascular-derived molecules which are diffusing to the cells of the cartilage core. This restricted access may result in the initiation of hypertrophy of the core chondrocytes. The osteoid collar then becomes mineralized — that is, it becomes bone. Further chondrocyte hypertrophy is observed at the

TABLE 1   The embryology of bone formation

Morphology	Days of development		
	Chick	Mouse	Human
1. Formation of limb buds	3		
2. Commitment of mesenchymal cells to osteogenic lineage	4	12	40
3. Commitment of mesenchymal cells to chondrogenic lineage	4	13	40
4. Expression of phenotypic characteristics	4.5	14	40
5. Formation of cartilage core	4.5–7	14	40
6. Osteoprogenitor cells of the Stacked Cell layer	4.5	15	40
7. Production of mid-diaphyseal osteoid	6	15	50
8. Phase boundary between osteoid and cartilage core	6.5	15	50
9. Initiation of hypertrophy in cartilage core	6.5	15	50
10. Progressive proximal and distal spreading of osteoid layer	7.0–16	15–16	50
11. Mineralization of osteoid	7.5	15	50
12. Vascular invasion onto the mineralized collar	8.0	15	50
13. Cartilage hypertrophy culmination (cessation of synthesis of anti-angiogenesis factors)	9.0	14–15	50–55
14. Formation of vertical struts between capillaries	8.5	16	56–57
15. Initiation of second layer of trabecular osteoid	9.0	16	57–58
16. Marrow elements associated with collar vasculature	8.5	16	60
17. Mid-diaphyseal invasion of first bone by osteoclasts	9.0	—	56
18. Vascular penetration and erosion of cartilage	9.0	15	56
19. Cartilage replaced by vasculature and marrow	9.0–14.0	16–17	60
20. Continued sequential formation of 12 more layers of trabecular bone	9.0–19.0	—	—
21. Dissolution of the first layer of bone by marrow elements	11.0	—	—

mid-diaphysis coincident with this mineralization and in an expanding manner as the osteoid layer is continuously fabricated in both a proximal and distal direction from the mid-diaphyseal region.

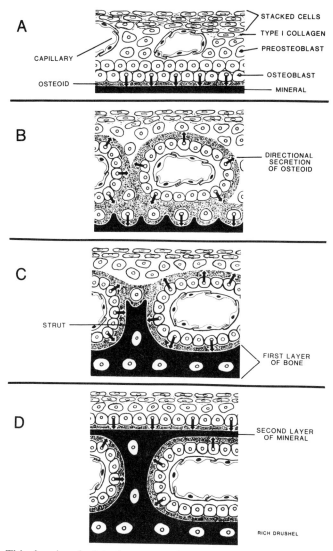

FIG. 2. This drawing depicts the progressive re-positioning of the vasculature from outside the Stacked Cell layer to a position close to the first layer of osteoblasts responsible for the formation of the first bony collar. The orientation of the osteoblast secretion of osteoid is with the osteoblast's back towards the invading capillary and the osteoblast's front secreting osteoid towards the cartilage core. In this model, osteoblasts secrete osteoid in a direction away from vasculature (B), causing the formation of a strut (C) and eventually the second layer of mineral (D). In this way, the vascular channels determine the trabecular layout within bone and the channels between the trabeculae are interconnected.

The next (and most important) phase of subsequent development is the invasion and penetration of the Stacked Cell layer by vascular elements which are positioned just outside this central region of newly formed bone (Pechak et al 1986a, b). The capillaries invade from outside the osteoprogenitor layer to lie as a web on top of the first layer of newly mineralized bone. Between individual capillaries and perpendicular to this first bony layer, a series of osteoid struts is fabricated and then mineralized. This is followed by the deposition of a second layer of bone parallel to the first layer, so that the individual capillaries are now completely surrounded and bony 'trabeculae' are formed. It is during this process of capillary invasion, strut formation and formation of the second layer of bone that the unique interrelationship between osteoblasts and capillaries can be directly observed. This relationship involves the directional secretion of osteoid by osteoblasts in an orientation away from the neighbouring capillary. As shown in Fig. 2, the osteoblasts have their backs to the capillaries and are secreting, from their opposite face, osteoid which eventually becomes mineralized. Thus, the osteoblast (like a gut epithelial cell) is a secretory cell with a highly structured and controlled directionality to its biosynthetic processes. Nutrients and essential ingredients enter the back face of osteoblasts with the secretion face opposite this aspect of the cell, in a manner comparable to gut epithelial or secretory cells which have a distinct directionality. Moreover, the osteoblasts are connected as a continuous monolayer of cells held together in close-packed conformation with their side faces completely in contact with their neighbours. This cobblestone arrangement allows continuous sheets of osteoid to be fabricated; we would predict that this arrangement may play a role in the production of unusually large diameter collagen fibrils indigenous and specific to osteoid.

Clearly, this intimate interaction between vasculature and newly forming bone is obligatory. *Bone cannot form in the absence of vasculature.* When, for example, a bone breaks, the establishment of vascular continuity across the bone break will determine whether the mesenchymal repair blastema will differentiate into cartilage or trabecular bone (Page & Ashhurst 1987). In a mechanically unstable situation, vascular continuity cannot be established across the break and an avascular repair blastema results. In this case, a cartilaginous plug will form to span this bone-break gap. If, however, there is no mechanical instability, only interconnecting trabecular (i.e. vascular) bone, not cartilage, will form. Previously it was thought that this intimate interaction between vasculature and newly forming bone was directly related to local nutrient supply and the high oxygen tension necessary for osteogenesis. It is possible that vascular endothelial cells may directly influence osteoblasts by providing unique endothelial cell-specific molecular cueing which is sensed by the osteoblasts and whose response is intrinsic to this unique mesenchymal phenotype. The experimental exploration of this speculation will provide an active area for future research.

The cartilage model

While the first and second layers of new mid-diaphyseal bone are forming in a direction away from the cartilaginous model, the chondrocytes are also undergoing developmental change. As mentioned above, the osteoid probably provides a diffusion barrier which impedes nutrient movement from the capillary network residing outside the Stacked Cell layer or on top of the newly mineralized first bony collar. We speculate that this diffusion barrier provides the signal, by virtue of nutrient deprivation, for the continued change of the chondrocyte into the hypertrophic chondrocyte phenotype. Hypertrophic chondrocytes are not dying cells but represent an active cell state in which unique differentiated products first appear, such as type X collagen (Schmid & Linsenmayer 1985) and a large chondroitin sulphate proteoglycan indigenous to hypertrophic chondrocytes (Carrino et al 1985). We believe that the mineralization of the first osteoid layer further exacerbates the inhibition of nutrient diffusion and eventually causes the starvation and demise of the hypertrophic chondrocytes. These hypertrophic chondrocytes (Caplan & Pechak 1987) can be rescued and placed into tissue culture where they will survive for months in an active and specific biosynthetic state (Syftestad et al 1985). Thus, these cells are not programmed to die but do so as a direct consequence of environmental conditions probably controlled by the newly forming circumferential bone.

It is clear that the surrounding osteoid/bone collar provides an absolute physical restraint to the further and continuous expansion of the cartilage model. The expansion of the cartilage model is due to the slow division of core chondrocytes as well as the production of voluminous cartilage extracellular matrix. Once the collar of first osteoid is formed, it restrains the further physical expansion of this cartilaginous model and limits the diameter of that model at the mid-diaphysis. This physical limitation on the still continuing expansion of the cartilage model is the cause of the flattening of usually round chondrocytes at the bone–cartilage border with a higher amount of extracellular matrix molecules on the cartilaginous side of this phase interface. This interface with its now distinct morphology has been mistakenly referred to as a perichondrial layer and, furthermore, these flattened chondrocytes were thought to be different from the more central, round chondrocytes. Our view is that this is not the case.

The replacement of cartilage by marrow/vascular tissue

While the first bone collar makes the second layer of bone, a variety of single cells appear in contact with the capillary web. These cells appear to be marrow and marrow derivatives. Thus, the bony collar becomes populated with marrow cells before any vascular elements have intruded into the

12

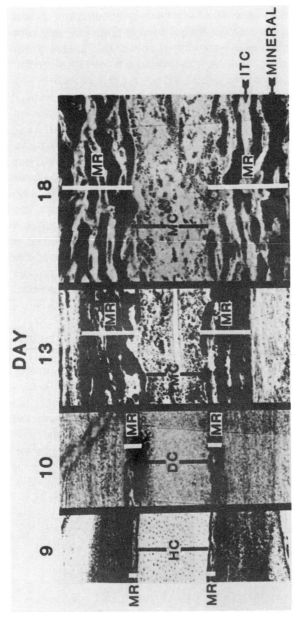

FIG. 3. A collage of longitudinal sections of the central mid-diaphyseal regions of 9, 10, 13 and 18-day-old embryonic chick tibiae. Sections on different days were photographed and printed at identical magnification. A comparison of the width of the Day 9 hypertrophic cartilage (HC) zone with the width of the Day 10 degenerating cartilage (DC) zone and the width of the Day 13 and Day 18 marrow cavity (MC) reveals that the size of the marrow cavity is determined by the width of the hypertrophic cartilage model. Also, note the several-fold increase in the thickness of the mineralized region (MR) during development. ITC refers to the intertrabecular channels which are substantial in chicken (so that the bone is never as dense as in mammals).

cartilaginous core. The next major phase of development involves the dissolution of aspects of the first bony collar with the invasion of vascular elements into the hypertrophic cartilaginous core. This core is systematically replaced by vascular and marrow elements, as is documented by Fig. 3, which compares mid-diaphyseal sections of Day 9, 10, 13, and 18 embryonic tibiae.

In the developing chick tibia, as in many other chick bones, the hypertrophic cartilage is not calcified before it is replaced. In contrast, in mammals the hypertrophic cartilage is calcified before its replacement by marrow and vasculature. Also, in certain circumstances in mammals some areas of this hypertrophic cartilage become encrusted in a layer of newly formed bone. Eventually, however, both this bone and the calcified cartilage are replaced by vascular and marrow elements. Thus, in chick, man and mouse, the embryonic cartilage model is completely replaced by vasculature and marrow, not by bone (Caplan & Pechak 1987). Moreover, the speculation by others that cartilage provides the scaffolding for new bone formation is not correct. Cartilage may indeed provide the scaffolding for vasculature and marrow, but not for bone. Bone forms external to, and separated from, the cartilaginous model. In the chick embryo, the formation of bone continues for 10–15 layers in addition to the formation of the two layers described above.

It therefore seems clear that bone forms as a direct result of the conversion of the osteoprogenitor cells in the Stacked Cell layer into osteoblasts which fabricate bone independently of contributions by cartilage. In this regard, structural bone of the tibia and intramembranous bone formation in calvariae have the same or similar formative sequence. Both the cellular and molecular aspects of these osteogenic events depend on an intimate interplay between vasculature and mesenchymal cells and/or mesenchymal derivatives (i.e. osteoblasts). Additionally, this Stacked Cell layer provides the organizational layer for the anchorage of tendons and ligaments which coordinate the insertion of muscles to bone and the protection for outer joint formation. Furthermore, the cartilage core is completely replaced by, and exactly defines the limits of, the initial marrow cavity. Endochondral bone formation may thus be defined as cartilage (which becomes calcified in mammals) which is subsequently eroded by vascular and marrow elements. The remnant cartilage that is not eroded may be covered by a layer of bony tissue. However, the resultant composite tissue is usually totally resorbed by osteoclasts and eventually replaced by vasculature and/or marrow elements.

Ectopic bone formation

Adult bone contains bioactive factors which can regulate and control the formation of ectopic bone. The experiments of M.R. Urist, A.H. Reddi and others indicate that, when an extract of adult bone is placed in a subcutaneous or intramuscular site, mesenchymal stem cells are attracted to that site and factors in the extract are capable of stimulating the formation of cartilage which is then associated with subsequent bone formation (Urist et al 1983). One can view these events as a 'classical endochondral' sequence. Experimentally, demineralized bone chips are implanted subcutaneously and followed temporally. The first phase involves the chemotactic attraction of mesenchymal cells to the wound site and positioning in proximity to the bone chips. Next, a mitogenic phase of 1–3 days involves the expansion of these mesenchymal stem cells which is followed by the specific differentiation of cartilage that surrounds each of the bone chips. Eventually, this cartilage becomes calcified and bone is seen to form. The formation of marrow in this tissue has also been documented (Reddi & Kuettner 1981).

We have studied histological preparations of this sequence of events and interpret the sequence in a way that is complementary to that described above for embryonic first bone formation. The mesenchymal cells which differentiate into cartilage eventually mature into hypertrophic chondrocytes and are replaced by vasculature and marrow elements. Bone forms from a layer of osteoblasts which are seen as sheets on the surfaces of the bone chips and which secrete new osteoid that becomes mineralized onto the fabric of the existing chips. Importantly, the osteoblasts have their backs to vasculature which has invaded the degenerating hypertrophic cartilage and the majority of the newly formed bone is observed to be built on the fabric of the demineralized bone chips rather than on scaffolding provided by cartilage. A detailed analysis of this ectopic bone chip system is complicated by the absence of fixed positions of the bone chips. The advantage of the embryonic long bone system analysed above is that there are fixed boundaries which can be used as morphological and temporal guides for understanding the sequence of events involved in the formation of bone. Although samples of the bone chip system can be obtained at different times in the process, the lack of definitive morphological boundaries prohibits a detailed analysis comparable to the analysis of embryonic bone formation.

The sequence of events is similar to that seen in massive bone fracture, when a space is created between the two broken ends of the bone followed by the filling of this gap with a cartilaginous callus. Again, the intimate involvement of vasculature with osteoblasts seems to be a prerequisite for the eventual bony repair and replacement of this callus.

Of mice and man (and chicken)

A number of fundamentally similar rules govern the genesis of bone in aves and mammals. A detailed study is now under way and preliminary observations have been published (Caplan & Pechak 1987). Common to all these species is the early appearance of an osteoprogenitor or Stacked Cell layer. This osteoprogenitor layer acts as a physical barrier against further expansion of the cartilaginous core. Osteoblasts develop from a common internal aspect of the Stacked Cell layer and eventually orient with respect to vasculature to secrete osteoid away from vascular sources. The cartilaginous model which has formed inside this Stacked Cell layer does not provide the scaffolding onto which bone is built; rather, the cartilage is replaced by vascular and marrow elements. The interplay between the osteogenic and vascular elements determines the directionality and the sequence of bone deposition in embryonic bone formation, in ectopic bone formation and also in repair. The exquisite timing and diversity of cell types and macromolecules which are deposited and then replaced in the developing animal is an orchestration whose harmony and tempo are as beautiful and as complex as any symphony. The challenge now is to understand the genetic and phenomenological controls of this orchestration as well as to identify all the players and instruments involved in the performance — namely, the manufacture and maintenance of skeletal elements. The ultimate challenge will be the description of the complex regulatory phenomena governing the processes of skeletal morphogenesis, maturation and ageing.

Acknowledgements

My thanks to my colleagues who have collaborated with me these last ten years on aspects of bone and mineral research. A special thanks to Dr David Carrino who continually monitors all my writing and helps me communicate with precision and proper form. These studies were supported by grants from the National Institutes of Health.

References

Caplan AI, Ordahl CP 1978 Irreversible gene repression model for control of development. Science (Wash DC) 201:120–130
Caplan AI, Pechak DG 1987 The cellular and molecular embryology of bone formation. In: Peck WA (ed) Bone and mineral research. Elsevier, New York, Vol 5:117–184
Caplan AI, Fiszman MY, Eppenberger HM 1984 Molecular and cell isoforms during development. Science (Wash DC) 221:921–927
Carrino DA, Weitzhandler M, Caplan AI 1985 Proteoglycans synthesized during the cartilage to bone transition. In: Butler WT (ed) The chemistry and biology of mineralized tissues. EBSCO Media, Birmingham, Alabama, p 197–208
Hall BK 1978 Developmental and cellular skeletal biology. Academic Press, New York
Lipton BH, Schultz E 1979 Developmental fate of skeletal muscle satellite cells. Science (Wash DC) 205:1292–1294
Page M, Ashhurst D 1987 The effects of mechanical stability on the macromolecules of the connective tissue matrices produced during fracture healing. II. The glycosaminoglycans. Histochem J 19:39–61
Pechak DG, Kujawa MJ, Caplan AI 1986a Morphological and histochemical events during first bone formation in embryonic chick limbs. Bone (NY) 7:441–458

Pechak DG, Kujawa MJ, Caplan AI 1986b Morphology of bone development and
    bone remodeling in embryonic chick limbs. Bone (NY) 7:459–472
Reddi AH, Kuettner KE 1981 Vascular invasion of cartilage: correlation of morpholo-
    gy with lysozyme, glycosaminoglycans, protease, and protease-inhibitory activity
    during endochondral bone development. Dev Biol 82:217–223
Schmid TM, Linsenmayer TF 1985 Immunohistochemical localization of short chain
    cartilage collagen (type X) in avian tissues. J Cell Biol 100:598–605
Syftestad GT, Weitzhandler M, Caplan AI 1985 Isolation and characterization of
    osteogenic cells derived from first bone of the embryonic tibia. Dev Biol 110:275–
    283
Urist MR, DeLange R, Finerman GAM 1983 Bone cell differentiation and growth
    factors. Science (Wash DC) 220:680–686
Walker DG 1975 Bone resorption restored in osteopetrotic mice by transplantation of
    normal bone and spleen cells. Science (Wash DC) 190:784–786

DISCUSSION

Krane: How does the growth plate fit into your scheme of bone development?

Caplan: We have tried to analyse the growth plate systematically in the way
we analysed the embryonic tibia. It is hard to describe the dynamics of the
process in the same way that is possible in embryology. My interpretation is
that the hypertrophic cartilage is calcified; this is followed by vascular/chon-
droclast erosion of a substantial portion of that tissue. There is actually a layer
of bone that forms on the remaining hypertrophic cartilage, but the entire
composite structure is replaced by vasculature. I believe that the structural
bone forms underneath the preceding wave of vasculature.

Krane: But the rate at which the chondrocytes in the growth plate proliferate
and the amount of matrix they produce determine the ultimate length of the
long bones.

Caplan: That is the present dogma. We know that in vitamin D deficiency the
normal sequence of replacement phenomena is not completed. Evidently 1,25-
and/or 24,25-$(OH)_2$ vitamin D plays some role in the conversion of the bottom
portion of the growth plate into bone. I anticipate that this will be found to be
connected to chondroclast and vascular activity and eventual vascular invasion.
Without the vascular replacement of that part of the growth plate, bone
formation is absent. I can't prove this unequivocally, but rather than thinking
about it as a cartilage-driven phenomenon, I would see vascular replacement as
a key event in bone formation at the growth plate. This view changes the way
one designs experiments, and also eliminates the need to say that calvarial
osteoblasts differ from growth plate osteoblasts.

Raisz: What is the role of the cartilage if it is not a 'dress pattern' for bone? Is
there bone which forms in the absence of cartilage at the embryonic level?

Caplan: Under normal circumstances one never sees cartilage in the calvaria.

After the first layer of bone, all the bone that forms in the chick tibia is away from cartilage and has nothing to do with it.

Raisz: I want to discuss the first layer!

Caplan: That first layer of bone at the cartilage interface is a real phase change with no mixing of ingredients. Because the rod of cartilage is observed first, we previously believed that bone was formed around it. However, what is observed first in freeze-fracture studies of early limb buds is the stacked cell layer. The differentiation of this layer is apparent before other specialization of the mesenchyme is observed in the chick, mouse and human limb. Therefore, I believe that the morphology of the cartilage model is controlled by the surrounding stacked cell layer. One of the mysteries that I have struggled with for 15 years is what gives that cartilage rod its shape. Now I think we know: its girdle determines its shape.

Glimcher: It is often thought that, excluding its contribution in the suture lines, cartilage is never seen in calvariae. It is therefore assumed that bone tissue in calvariae develops solely by primary, intramembranous bone formation without any contribution from cartilage via the endochondral bone sequence. In fact, a significant amount of the periorbital bone of the early developing chick calvaria consists of cartilage as a *tissue* which then undergoes endochondral ossification. We have demonstrated this by serial coronal and cross sections of this portion of the cranium during the development of chick calvariae using light and electron microscopy and by biochemical analyses.

Caplan: The question is which comes first—is cartilage obligatory for bone formation? This logic has driven many people for a long time. I am saying that one need not invoke the obligatory predifferentiation of cartilage in order to get osteogenesis. We know that the C-propeptide for type II collagen finds its way into first bone formation and that there may be some molecular communication between products of the cartilage and the new bone formation. We don't know all the dynamics of that kind of interchange.

Nijweide: I always understood that the hypertrophic area had an influence on periosteal bone formation. Why does that bone formation always start at exactly the same place where the hypertrophic area will be? Why doesn't it occur elsewhere?

Caplan: We never see hypertrophy of chondrocytes in the chick embryo tibia until there is this girdle of osteoid which then becomes mineralized. We have taken the hypertrophic chondrocytes underneath the collar of first bone and the collar osteoblasts and studied them separately in cell culture. It is clear that the hypertrophic chondrocytes, which are thought to be programmed to die, can survive in culture for months. These cells are vital, active, and highly specialized. They make type X collagen and a proteoglycan which is unique to hypertrophic chondrocytes and differs from the normal cartilage proteoglycan. This synthesis continues for months in cell culture. We think that the chondrocytes in the core die because they are nutritionally choked, because the osteoid, and eventually mineral of first bone, prevents nutrients diffusing from the vasculature. This contrasts with articular cartilage which receives its nutrition via load bearing and washing in and out of nutrients. This never happens at the cartilage in the core. I am not excluding an influence of cartilage on osteogenesis, but I don't think it's obligatory. I suggest that what's driving the osteogenesis is actually driving the morphogenesis of that limb part. This is different from the previously held view in which the cartilage model drives the subsequent morphogenesis of the bone.

Testa: I would like to comment on the persistence of chondrocytes in culture and what it means physiologically. In haemopoiesis we have examples of cells which are not really programmed to persist *in vivo* and we can certainly keep them alive in culture for a long time. In that sense, a population can be manipulated. We have to use those extrapolations to *in vivo* situations with care.

Glimcher: Dr Caplan, I think we are dealing with two independent phenomena. Whether cartilage or bone cells are the first formed in the initial anlage of a bone does not answer whether formation of cartilage cells and tissue is obligatory for the formation of bone as an organ.

I don't think you should question the obligatory role of cartilage in osteogenesis without specifying to which bone you refer. Unlike long bones, most calvarial bones appear to be formed morphologically without being preceded by, or first developed as, organs with cartilage as the tissue, or indeed without cartilage or chondrogenic cells, or cartilage tissue, ever being present. On the other hand, long bones clearly consist of cartilage as a tissue before the cartilage tissue in the bone is replaced by bone tissue. I see no reason to doubt your data that the first outer layer of cells in the anlage are osteogenic and not chondroblasts or chondrogenic stem cells, but I don't see any evidence which conclusively proves whether or not the synthesis of this cartilage tissue is obligatory for the formation of long bones, whereas we know that it is not obligatory for formation of specific bones in calvariae.

Caplan: We can take chick limb mesenchymal cells (Osdoby & Caplan 1981a,b) and put them into culture under conditions where no cartilage is formed and we see osteoblasts form in culture.

Glimcher: We can look at the problem in another way: have you ever seen the development of a normal limb bone without the organ (tibia, femur etc.) being first formed and composed of cartilage cells and cartilage tissue?

Caplan: No.

Hanaoka: You stressed the importance of the vascular invasion when the bone marrow is formed. Have you seen the osteoclasts before the vascular invasion? The chick embryo may be different, but in the mouse embryo the osteoclast invades first, before the vascular invasion, in the same manner as the chondroclast erodes the calcified cartilage of the growth plate.

Caplan: Vascular invasion is preceded by an erosion of the first bony plate by multinucleated osteoclast-like cells, which can be seen in section. The vasculature is not erosive. Some other group of cells erodes away this first layer of bone in the embryonic chick and mouse tibia.

Rodan: My sense of the discussion so far is that the generalization of Dr Caplan's model to other bones and other species is being questioned.

Caplan: The commonly accepted features of my model are: there is a preosteogenic layer, which we refer to as the stacked cell layer; there is an intimate relationship between vascular invasion and bone formation; and this relationship is actually orienting in terms of the secretion of osteoid. What is in question is the intimate relationship and necessity for cartilage in this process.

Glowacki: Your ideas about cartilage providing the geometry of the space for the vascular element are consistent with what is seen in the model of induced endochondral ossification after the implantation of demineralized bone powder. These events take place in a four-dimensional frame of reference. I am having difficulty with your ideas about the very early events. You suggest that the preosteoblast layer is not interacting with the chondrocytes and the matrix that is being resorbed. Can you be more precise about the timing of the appearance of those two populations of cells?

Caplan: We don't know what stimulates the genesis of osteoblasts. After the first layer of bone is formed, all the subsequent layers are away from that area. I think there is no direct association of cartilage with this later osteogenic phenomenon.

Glowacki: But was the cartilage there first?

Caplan: Cartilage as a differentiated tissue is seen first. The stacked cell layer, however, is seen before that; it may be that the cartilage rod affects the orientation of the first layer of osteoprogenitor cells and may also affect the orientation of the first thick filaments of type I collagen between these two tissues.

Glowacki: What about alkaline phosphatase?

Caplan: Alkaline phosphatase is not observed by light microscopic histochemistry until long after a cartilage model is seen. The problem is that markers for the earliest osteoblasts do not exist. We lack any way of marking *bona fide* osteoprogenitor cells.

Glowacki: If we accept alkaline phosphatase as a reasonable marker for preosteoblasts, the induced endochondral model may be revealing to us that, at the very early stages, the demineralized bone promotes the formation of chondrocytes. At a distance from those chondrocytes one sees alkaline phosphatase-positive cells but they are always adjacent to induced cartilage. The cartilage itself could be inducing the differentiation of preosteoblasts from progenitor cells, rather than there being two pathways of differentiation, into chondrocytes or osteoblasts. This could be tested experimentally, the hypothesis being that cartilage matrix factors or soluble factors produced by the chondrocytes actually promote the differentiation of the preosteoblasts.

Caplan: In chick and mouse limb mesenchymal cells in culture, osteoblasts differentiate without prior differentiation of cartilage. In diffusion chambers which are broached by vasculature or arranged in a particular way to allow vasculature inside, bone is formed without any preceding cartilage formation. These cases indicate that the differentiation of osteoblasts proceeds without the predifferentiation or direct involvement of cartilage.

Glowacki: Those are not *in vivo* models. They may be different.

Owen: Dr Caplan, your observations fit with the hypothesis that cartilage arises in an avascular situation and bone in a vascular situation. I find your suggestion that marrow replaces cartilage intriguing. Which cells do you think give rise to the stroma of marrow in this replacement?

Caplan: I'm convinced that the stromal cell lineage is a cousin to the same lineage of mesenchymal cells that gives rise to stacked cells and cartilage cells. The data indicate that there is a mesenchymal cell lineage completely separate from the haemopoietic cell lineage in marrow which gives rise to these two different tissues.

I think the mesenchymal cells of the marrow lineage give rise to trabecular bone on the inside cavity of newly forming bone. In the mid-diaphyseal region of mouse and man you see two layers of this trabecular bone, but eventually it becomes cortical bone. In man, in the samples we have studied, the cortical bone is vascularized on both the marrow side and the stacked cell side; the cortical bone thickens by the production of osteoid on both sides of this bone. Starting with one and a half or two layers of trabecular bone, the trabeculae eventually fill in to form cortical bone with the vasculature facing both sides of that bone during the thickening at the mid-diaphyseal region. This is a very different mechanism from that seen in chick, where there are ten or twelve layers of trabecular bone at the mid-diaphysis.

Slavkin: You described the position of the stacked cells relative to the chondrogenic mass. One would assume that if you changed the polarity of either the stacked cells or the progenitor chondrogenic cells, the cartilage and bone phenotypes wouldn't emerge. Have you revisited Bob Trelstad's studies (Trelstad 1977, Holmes & Trelstad 1977) which used silver staining to identify the orientation of the Golgi, to try to identify the positions of the initial progenitor cell phenotypes, before alkaline phosphatase expression? Have you looked earlier to examine cell adhesion and/or substrate adhesion molecules (CAMs and SAMs) at the interface, and have you tried to perturb the interface?

Caplan: The orientation of Golgi, and therefore the secretion of specialized molecules by chondrocytes, may be influenced by the vasculature which resides outside the stacked cells. Likewise, the osteoblasts seem to be oriented by the vasculature. The chondrogenic cells in those two studies of the cartilage model in mouse and chick embryos seem to be oriented towards the major blood vessels surrounding this avascular core. Therefore, orientation of the cartilage cells seems to be dictated by the vascular network. It is not clear what orients the stacked cell layer.

Termine: Dr Caplan, how do you explain mantellar calcification, which is the formation of bone in Meckel's cartilage of the mandible?

Caplan: I have not studied that system, so I can't really comment.

Osdoby: In chick bone, at least, where alkaline phosphatase activity is first seen there is usually an associated large blood vessel. We know that bone responds to ionic influences. There is usually a neuronal network associated with the vasculature. Might this be playing a role in the differentiation?

Caplan: I don't believe there is any neuronal component to the early first tibial bone in the chick embryo. The vasculature is not invaded by a neural network of sensor or sympathetic nerves at the stages that we are discussing. The major neural involvement in the limb is motor, but this does not seem to correlate with osteogenic events.

Canalis: I find it difficult to believe that the cartilage could be avascular and survive.

Caplan: The cartilage is unequivocally avascular in the limb model.

Canalis: How does it survive? Most such tissues become necrotic.

Caplan: We are dealing with small distances: the simple diffusion of nutrients is possible. When that first layer of osteoid appears, it may act as a diffusion barrier between the vasculature and the cartilaginous core. This barrier to diffusion may be used as a signal to initiate hypertrophy of the cartilage.

References

Holmes LB, Trelstad RL 1977 Patterns of cell polarity in the developing mouse limb. Dev Biol 59:164–173

Osdoby P, Caplan AI 1981a First bone formation in embryonic chick limbs. Dev Biol 86:147–156

Osdoby P, Caplan AI 1981b Characterization of bone specific alkaline phosphatase in cultures of chick limb mesenchymal cells. Dev Biol 86:136–146

Trelstad RL 1977 Mesenchymal cell polarity and morphogenesis of chick cartilage. Dev Biol 59:153–163

Factors influencing the expression of dental extracellular matrix biomineralization

Harold C. Slavkin, Malcolm L. Snead, Margarita Zeichner-David, Mary MacDougall, Alan Fincham, Eduardo C. Lau, Wen Luo, Masanori Nakamura, Peter Oliver and John Evans

Laboratory For Developmental Biology, Department of Basic Sciences (Biochemistry), School of Dentistry, University of Southern California, Los Angeles, California 90089–0181, USA

Abstract. The forming tooth organ provides a number of opportunities to investigate the cellular and molecular biology of cell-mediated extracellular matrix (ECM) biomineralization. Regulatory processes associated with tooth formation are being investigated by identifying when and where cell adhesion molecules (CAMs), substrate adhesion molecules (SAMs), dentine phosphoprotein and enamel gene products are expressed during sequential developmental stages. *In vitro* organotypic culture studies in serumless, chemically-defined medium, have shown that instructive and permissive signalling are required for both morphogenesis and cytodifferentiation. Intrinsic developmental instructions (autocrine and paracrine factors) act independently of long-range hormonal or exogenous growth factors and mediate morphogenesis from the initiation of the dental lamina to the crown stages of tooth development. This review summarizes the results of studies using experimental embryology, recombinant DNA technology and immunocytology to elucidate mechanisms responsive to instructive epithelial–mesenchymal interactions associated with ameloblast differentiation, odontoblast differentiation, and dentine and enamel ECM biomineralization.

1988 Cell and molecular biology of vertebrate hard tissues. Wiley, Chichester (Ciba Foundation Symposium 136) p 22–41

One of the most important questions in developmental biology is how tissue-specific gene expression is regionally specified during embryonic, fetal and postnatal morphogenesis. The developing tooth organ provides a model with which to investigate how the one-dimensional genetic code generates the three-dimensional complexity of tooth morphogenesis, cytodifferentiation and tissue-specific extracellular matrix (ECM) biomineralization. Evidence is presented which supports the hypothesis that dentine and enamel ECM biomineralization are regulated by the sequential expression of an intrinsic genetic programme.

Model system for the development of mandibular first molar tooth organ

During the determination of mouse tooth morphogenetic positional values, branchial arch-derived epithelia appear to provide positional information for subsequent tooth development. Recent evidence indicates that hetero-typic tissue recombinations, formed between first branchial arch epithelia and second branchial arch ectomesenchyme, demonstrate that early mandibular arch epithelia (before Theiler stage 20, 12 days of gestation) possess odon-togenic potential and can induce neural crest-derived ectomesenchyme of the second branchial arch to aggregate into a dental morphogenetic pattern (Lumsden 1987). The nature of the instructions and their mode of transmission are not known. Mandibular ectomesenchyme appears to be required to interact with mandibular epithelia and to enable incisor and molar tooth determination (Lumsden 1987). Subsequent tooth morphogenesis (e.g. dental lamina, bud stage, cap stage, bell stage) and epithelial cytodifferentiation to ameloblasts, which synthesize the enamel extracellular matrix, are regulated by regional ectomesenchyme-specific instructions (see reviews by Kollar 1983, Ruch 1985, Slavkin et al 1984, Thesleff & Hurmerinta 1981).

However, the significant developmental problem remains: when, where and how do epithelia and/or ectomesenchyme signal during instructive and permissive tissue interactions, and how are these sequential processes related to tissue-specific ECM biomineralization?

Cell adhesion and substrate adhesion molecules expressed during tooth formation

Cell-to-cell and cell-to-ECM interactions participate in morphogenetic patterning during embryogenesis (see discussion by Bissell et al 1982). The regulation required to coordinate the timing and position of epithelia and mesenchyme orignates in the sequential expression of cell transmembrane linkage molecules (e.g. integrin), cell surface adhesion molecules (CAMs, e.g. L-CAM, N-CAM) and substrate adhesion molecules (SAMs, e.g. tena-scin, cytotactin, laminin, type IV, V and VII collagens, fibronectin, and basement membrane heparan sulphate proteoglycans) (see review by Edel-man 1985). This type of regulatory scheme involves the assumption that CAMs and/or SAMs are expressed at defined times and locations during tooth organ development (see summary in Table 1).

A number of reports suggest that the ECM and the cytoskeleton of cells are intimately related (Bissell et al 1982). For example, the fibronectin (FN) cell attachment domain binds to an 140 kDa FN receptor complex that co-distributes with intracellular microfilament bundles at the cell–substrate adhesion sites. The domain defined by an amino acid sequence of R-G-D (Arg-Gly-Asp) within the FN attachment domain, as well as within a number

TABLE 1 Molecular determinants expressed during tooth development

Theiler stages	Developmental processes	Molecular determinants
15	Mesencephalic & rhombencephalic neural crest migrations	CAMs & SAMs
16	Mandibular processes merge and midline established	CAMs & SAMs
18	Bilateral positions for molariform teeth determined (stereoisomerism)	CAMs & SAMs
20	Dental lamina	
22–23	Bud stage (inner (IEE) & outer enamel epithelia delineated)	CAMs & SAMs (e.g. tenascin)
23–25	Cap stage Odontoblast differentiation (dentinogenesis initiated) IEE differentiation (amelogenesis initiated) Morphogenesis (5 cusps, 3 buccal & 2 lingual)	DPP, Gla-protein Enamel proteins (EP) (anionic)
26	Bell stage ECM biomineralization initiated	Enamel proteins (anionic)
27	Birth Dentine/enamel junction Prismatic enamel formation Dentine/predentine junction	Anionic EP and 26 kDa amelogenin DPP in ECM
—	Postnatal development 0–6 days Crown completed 7–12 days Enamel maturation 5–28 days Root formation/tooth eruption/articulation with opposing maxillary molar tooth organ	EP Post-translational processing/bio-mineralization

of other SAM molecules, appears to provide the attachment biological activity of the molecule (E. Ruoslahti, unpublished paper, European Developmental Biology Congress, Helsinki, 16 June 1987; see also Pierschbacher et al, this volume). A biomechanical connection between ECM constituents and intracellular cytoskeletal constituents would seem a reasonable model for epigenetic signalling between heterotypic tissues during tooth organ development. Another interpretation of this scheme is that sequential expression and distribution of transmembrane linkage molecules, CAMs, SAMs and ECM molecules, may represent developmental regulation associated with regional specification of morphogenetic patterns and tissue-specific cytodifferentiation (see review by Slavkin et al 1988a).

Major changes which occur during enamel organ epithelial morphogenesis do so in the cap and bell stages of embryonic mouse mandibular first molar (M_1) development. For example, tenascin (SAM) immunostaining was inversely related to odontoblast differentiation (Chiquet-Ehrismann et al 1986). Cranial neural crest-derived dental ectomesenchyme cells that were stained for tenascin lost this staining as they differentiated into the odontoblast lineage (Chiquet-Ehrismann et al 1986, I. Thesleff, unpublished paper, III International Tooth Morphogenesis and Differentiation Workshop, Alund, 12 June 1987). During ectomesenchyme differentiation into odontoblasts in cap stage M_1, fibronectin and type III collagen were no longer detected (Thesleff & Hurmerinta 1981). Type I collagen was identified as characteristic of odontoblasts and as their major gene product. It is the major organic constituent of the dentine ECM.

Dentine phosphoprotein expression by odontoblasts

Dentine phosphoprotein (DPP) is a tissue-specific biochemical marker which characterizes the odontoblast cell lineage. During tooth development, odontoblasts express type I, V and type I trimer collagens, several proteoglycans, glycoproteins, glycosaminoglycans, γ-carboxyglutamate-containing proteins (Gla proteins or dentine osteocalcin) and dentine phosphoprotein which form the dentine extracellular matrix.

DPP is a highly phosphorylated and acidic protein (isoelectric point ≈ 1.0); serine and aspartic acid amino acid residues constitute approximately 70–88% of the total amino acids. So far, only limited primary structure data is available. A partial amino terminal sequence for rat DPP is Asp-Asp-Asp-Asn and Asp-Asp-Pro-Asn (Butler et al 1983). Polyclonal antibodies have been produced against mouse DPP (MacDougall et al 1985a), and these have been used to establish a site-restricted expression of DPP within odontoblasts (MacDougall et al 1985b). Preliminary characterization suggests that mouse DPP mRNA codes for a 43 kDa polypeptide. This contrasts with the 72 kDa mouse dentine phosphoprotein that is archetypal in mineralizing dentine ECM (M. MacDougall, unpublished results). We assume that this apparent discrepancy arises because DPP is a highly post-translationally modified dentine ECM constituent.

A biological function for DPP has been suggested based on its strong affinity for calcium ions. This calcium-binding property is useful for the isolation, purification and partial characterization of such proteins. Evidence for the function of DPP in the regulation of dentine calcium hydroxyapatite crystal formation comes from recent immunolocalization studies that report the distribution of mouse DPP within forming odontoblasts during early cap stage M_1 morphogenesis, well in advance of the initiation of ECM mineralization (MacDougall et al 1985b). DPP is localized within odontoblast cytoplasm

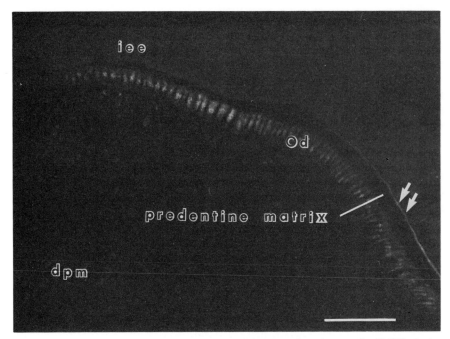

FIG. 1. Immunocytochemical localization of dentine phosphoprotein (DPP) during the process of ectomesenchyme differentiation into odontoblasts. The odontoblast phenotype shows positive immunostaining with anti-mouse DPP polyclonal antibody in differentiation zone II–III (Kallenbach 1971). Localization of DPP is restricted to the odontoblasts (od) and their cell processes until dentine extracellular biomineralization first appears; DPP is then identified at the mineralization front (arrows) but is not detected within the predentine regions of the forming mouse molar. iee, inner enamel epithelia; dpm, dental papilla ectomesenchyme. Barline = 50 μm.

concentrated in the Golgi region, odontoblast cell processes, and is subsequently secreted at the mineralizing front of the dentine ECM *after the initiation of ECM biomineralization* (M. MacDougall, unpublished results). No immunostaining is observed within the adjacent dental papilla ectomesenchyme cells or in the predentine (MacDougall et al 1985a, b) (Fig. 1). This pattern of site-restricted expression is also reported for osteocalcin and dentine Gla protein during rodent tooth development (DeVries et al 1987, Finkelman & Butler 1985). However, osteocalcin is not tissue-specific for odontoblasts: it is also found in forming bone appearing *after the initiation of osteogenic mineralization* (see discussion by DeVries et al 1987).

Enamel protein expression by inner enamel epithelia

The biochemical phenotype of the ameloblast is represented by two classes of enamel proteins — amelogenins and enamelins (Termine et al 1980). The

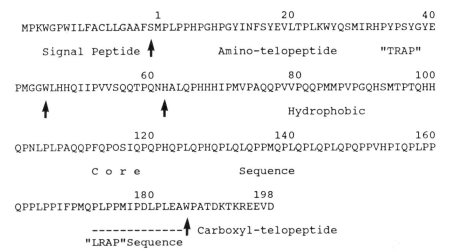

```
            1                   20                          40
MPKWGPWILFACLLGAAFSMPLPPHPGHPGYINFSYEVLTPLKWYQSMIRHPYPSYGYE

   Signal Peptide ↑           Amino-telopeptide          "TRAP"

           60                   80                        100
PMGGWLHHQIIPVVSQQTPQNHALQPHHHIPMVPAQQPVVPQQPMMPVPGQHSMTPTQHH

↑                     ↑                   Hydrophobic

          120                  140                         160
QPNLPLPAQQPFQPOSIQPQPHQPLQPHQPLQLQPPMQPLQPLQPLQPQPPVHPIQPLPP

        c  o  r  e              Sequence

          180              198
QPPLPPIFPMQPLPPMIPDLPLEAWPATDKTKREEVD

            -------------↑ Carboxyl-telopeptide
        "LRAP"Sequence
```

FIG. 2. Molecular features of 26 kDa amelogenin, showing: the predicted amino
acid sequence based upon mouse and bovine amelogenin cDNA sequence data, (see
Shimokawa et al 1987, Snead et al 1985); the signal peptide cleavage point; cleavage
of the conserved N-terminal 'TRAP' sequence at amino acid position 45/46 (A.
Fincham, unpublished results); a proposed cleavage point at amino acid residues
62/63 (M. Crenshaw, personal communication); a cleavage point for the C-terminal
telopeptide at residues 184/185 (M. Crenshaw and A. Fincham, unpublished results);
and the 'LRAP' sequence adjacent to the carboxyl telopeptide (Fincham et al 1983).

amelogenins are relatively hydrophobic proteins of low molecular mass (5–30
kDa), and they are enriched in proline, glutamic acid, leucine and histidine.
During enamel matrix formation, amelogenins represent at least 90% of the
total enamel protein. These neutral (pI 6.5–7.5) polypeptides appear to
regulate the size and rates of enamel calcium hydroxyapatite crystal growth.
Several mammalian amelogenin cDNAs have been produced, sequenced and
used to predict the complete primary structure of bovine and mouse amelogenin (Shimokawa et al 1987, Snead et al 1983, 1985). Both amelogenin
cDNA probes hybridize with human genomic DNA (Fig. 2).

Enamelins, in contrast, are relatively hydrophilic proteins of higher
molecular mass (about 68–72 kDa), constitute 1–3% of the total enamel
matrix protein, and are characterized by a high content of glycine, aspartic
acid, serine and glutamic acid (Termine et al 1980). These relatively anionic
molecules appear to be synthesized and secreted before amelogenins, and
may serve to initiate calcium hydroxyapatite crystal formation (D. Deutsch,
personal communication). Ameloblast poly(A)-enriched RNA fractions expressed in a cell-free system show similarties in the physical–chemical properties of the enamel protein translation products in porcine, bovine and
rodent species (M. Zeichner-David, unpublished results).

The biochemical phenotype of the inner enamel epithelia has recently been

TABLE 2   Sequence of enamel protein expression during mouse tooth formation

Days of development (postcoitus)	Enamel proteins molecular mass (kDa)	pI
16	—	—
17	—	—
18	46	5.5
19	72	5.8
	46	5.5
20 (birth)	72	5.8
	46	5.5
	26	6.5–6.7

defined according to the coordinated sequence of enamel proteins synthesized and secreted during ameloblast cytodifferentiation in the embryonic, fetal and postnatal stages of mouse M_1 (Nanci et al 1985, Slavkin et al 1988b). This sequence is summarized in Table 2. According to Kallenbach's description (1971) of inner enamel epithelial 'differentiation zones' I–VI, a 46 kDa enamel protein is first expressed by differentiation zone III–IV at 18 days

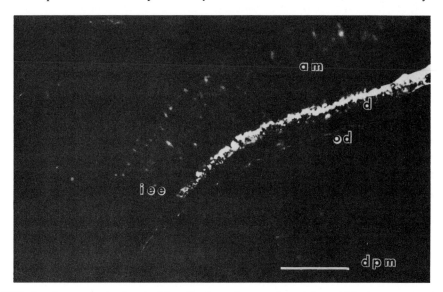

FIG. 3.   Immunocytochemical localization of enamel proteins within inner enamel epithelia (iee, differentiation zones IV-VI) and within ameloblasts (am) during mouse molar tooth development. Anti-mouse amelogenin polyclonal antibodies were used to identify both intracellular and forming enamel extracellular matrix enamel proteins. Dental papilla ectomesenchyme (dpm), odontoblasts (od) and dentine matrix (d) were not stained. Bar line = 50 μm.

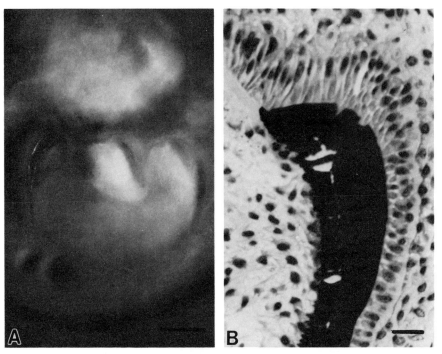

FIG. 4. *In vitro* organ culture model system which is permissive for ECM biomineralization in serumless, chemically-defined medium. (A) E16d cap stage molars cultured
for 21 days. Bar line = 500 μm. (B) Von Kossa histochemical staining indicating
dentine and enamel ECM biomineralization. Bar line = 20 μm.

gestation; the basement membrane is continuous at this stage of tooth development (Fig. 3). At 19 days gestation the 46 kDa enamel protein and
another 72 kDa enamel protein are expressed by differentiation zones III–VI;
the basement membrane is discontinuous in differentiation zone V and completely removed in differentiation zone VI (Kallenbach 1971). At 20 days
(postcoitus), these two relatively anionic (pI 5.5) enamel proteins continue to
be produced, and a third amelogenin polypeptide (26 kDa) is synthesized and
secreted into the enamel matrix by functional secretory ameloblasts (Nanci et
al 1985, Snead et al 1984). Thereafter, increased numbers of amelogenins
ranging from 5–30 kDa, presumably derived from post-translational processing (see Fig. 2), become the major constituent of the enamel ECM (Slavkin
et al 1984, Slavkin et al 1988b, Termine et al 1980). With advanced enamel
mineralization, termed 'enamel maturation', water and protein are removed
from the enamel ECM at the same time as calcium hydroxyapatite crystal
formation is increased: eventually a 99.9% inorganic enamel tooth covering is
formed.

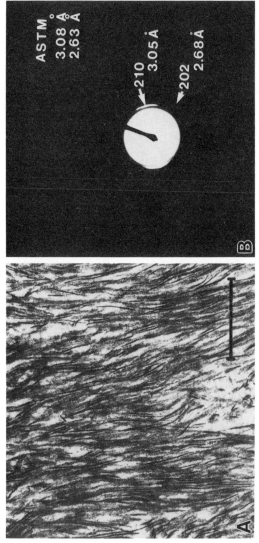

FIG. 5. Enamel-specific ECM-mediated biomineralization during cap stage tooth development in serumless and chemically-defined medium *in vitro*. (A) Enamel crystalite formation. Bar line = 500 nm. (B) Electron diffraction pattern of enamel showing calcium hydroxyapatite.

Intrinsic genetic programme regulates dentine and enamel biomineralization

Over the last 50 years, numerous investigations have demonstrated that isolated mammalian tooth organs can be cultured *in vitro* using chemically-defined medium supplemented with fetal calf sera, horse sera and/or embryonic chick extract. In these *in vitro* conditions, dentine and enamel mineralization were observed. More recently, several researchers reported that late cap stage mouse M_1 would not mineralize in the absence of fetal calf serum (Finkelman & Butler 1985, MacDougall et al 1985b). These results were intrepreted to mean that exogenous serum-derived factors are required to regulate ECM-mediated dentine and enamel biomineralization (Evans et al 1988, Finkelman & Butler 1985, MacDougall et al 1985b).

However, we have recently offered an alternative hypothesis suggesting that enamel and dentine tissue-specific patterns of biomineralization are regulated by intrinsic genetic programmes independent of exogenous humoral factors. To test our hypothesis, we have cultured early cap stage mouse M_1 tooth organs (15/16-days gestation; n=70) as explants in chemically-defined BGJb medium (Dr Sylvia Fitton-Jackson's modification of BGJ medium) for periods up to one month (Fig. 4). The medium was supplemented with 100 µg/ml L-ascorbic acid and 100 units/ml penicillin and streptomycin, adjusted to pH 7.4, and changed every two days. The organ culture was maintained at 37.5 °C with atmospheric conditions of 95% air and 5% CO_2, and optimal humidity. Previous studies used late cap to early bell stages of tooth organs, and cultured explants for periods of 10–14 days. We used a number of strategies and assays for dentine- and enamel-specific patterns of biomineralization including: (i) von Kossa histochemistry to evaluate calcium phosphate salt deposition; (ii) anhydrous fixation and processing of cultured tooth organs; (iii) Ca/P ratio determination by microprobe analysis; and (iv) electron diffraction analysis to identify calcium hydroxyapatite crystal formation. Controls for these studies included 2- and 7-day postnatal M_1 enamel and dentine.

Whereas indications of biomineralization were recorded by 12 days *in vitro* (Evans et al 1988), overt dentine and enamel ECM production with tissue-specific patterns of biomineralization were observed by 21 days using a serumless and chemically-defined medium (Bringas et al 1987, Evans et al 1988). Enamel Ca/P ratio at 21 days *in vitro* was 2.03 (SEM+/− 0.04) as compared with control 7-day postnatal mouse enamel Ca/P values of 2.35 (SEM+/− 0.04). Electron diffraction patterns with a 210° reflection present as a strong meridional pattern of short arc which is indicative of calcium hydroxyapatite crystalites (Fig. 5).

The mechanisms by which embryonic tooth organs cultured in serumless and chemically-defined medium regulate tissue-specific biomineralization are not known. Because serum or serum-derived factors were not employed in

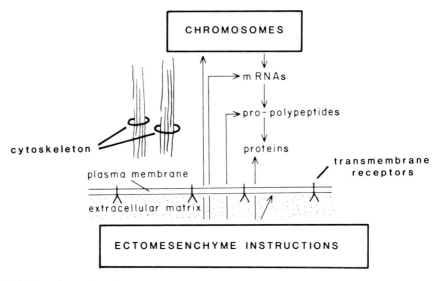

FIG. 6.   General scheme for ectomesenchyme-derived instructions for epithelial cytodifferentiation. Putative signals (e.g. epidermal growth factor, laminin, heparan sulphate proteoglycan, entactin, cytotactin, fibronectin) serve as developmental information providing regional specification for epithelial cytodifferentiation. The transmission enlists cell–cell and/or cell–matrix communication. Ectomesenchyme-derived signals may be received by epithelial integral plasma membrane glycoproteins termed 'transmembrane receptors' (e.g. talin, integrin), which extend from the interior of the epithelial cell, through the plasma membrane, and into the basal lamina and adjacent extracellular matrix. Epithelial reception of the signal(s) effects a series of steps which provide biophysical linkages between the extracellular matrix, transmembrane receptors, the cytoskeleton matrix and the nuclear matrix resulting in differential gene expression.

the culture medium, exogenous hormonal regulation was excluded as a possibility for regulation of this *in vitro* model system. It is not known if proteins dependent for their synthesis on the endocrine system (e.g. a 28 kDa vitamin D-dependent calcium-binding protein (CaBP)) are expressed by odontogenic tissues in organ explants cultured in serumless medium. Serum and/or plasma depletion may invoke *de novo* gene expression of putative regulatory molecules usually found in the serum or plasma. Therefore, we suggest that putative autocrine and/or paracrine factors derived from ectomesenchyme as well as enamel organ epithelia are expressed in a unique temporal and positional sequence providing the required signals for morphogenesis, cytodifferentiation, ECM production, and dentine and enamel biomineralization in this *in vitro* model system (Fig. 6).

Human genetic studies

Amelogenesis imperfecta (AI) and dentinogenesis imperfecta (DGI) are inherited dental disorders which reflect alterations in regulatory and/or structural genes associated with enamel and dentine formation. Each of these genetic disorders affect both primary and secondary dentitions. The recent advances in biochemistry, immunochemistry, developmental biology and recombinant DNA technology applied to human genetic diseases offer a number of approaches towards understanding the molecular genetics of AI and DI. Antibodies are now available which are directed against molecular determinants expressed during normal amelogenesis and dentinogenesis (e.g. cytotactin, N-CAM, L-CAM, integrin, laminin, fibronectin, type VII collagen, vitamin D-binding proteins, calcium-binding glycoproteins, dentine phosphoprotein, enamel proteins). These can be used to screen cDNA expression libraries. Tissue-specific probes which hybridize with human genomic DNA, for example, enable linkage of restriction fragment length polymorphisms (RFLPs) in affected members of a multiple-generation kindred, containing individuals affected with AI or DI. The collaborative opportunities between experimental embryology, molecular biology and human medical genetics may provide for rapid advances in understanding the molecular genetics of inherited dental tissue diseases.

Summary

Cap stage molar tooth organs cultured organotypically express morphogenesis through crown formation, odontoblast and ameloblast cytodifferentiation, dentine and enamel extracellular matrix production, and dentine and enamel tissue-specific patterns of biomineralization in serumless, chemically-defined medium using *in vitro* culture conditions (Bringas et al 1987, Evans et al 1988). We interpret these results to suggest that intrinsic autocrine and paracrine regulatory factors (e.g. fibroblast growth factor, epidermal growth factor, transferrin, insulin-like factor), produced by ectomesenchyme and/or epithelia during *in vitro* morphogenesis, mediate instructive and/or permissive morphogenesis and cytodifferentiation in the absence of serum or exogenous steroid or polypeptide hormones.

Acknowledgements

The authors wish to dedicate this paper to the pioneering efforts of Dame Honor Fell who advanced the methodology of *in vitro* organotypic culture studies and Professor Clifford Grobstein (University of California at San Diego) for his untiring efforts towards understanding epithelial–mesenchymal interactions. Excellent technical assistance by Pablo Bringas Jr, Conny Bessem, Julia Vides and Valentino Santos is gratefully acknowledged. We thank Mr Dwain Lewis for his assistance in the preparation of this manuscript. This work was supported in part by National Institutes of Health research grants DE-02848, DE-06908 and DE-06425. M.L.S. is a recipient of a National Institutes of Health Research Career Development Award.

References

Bissell MJ, Hall HG, Parry G 1982 How does the extracellular matrix direct gene expression? J Theor Biol 99:31–68

Bringas P, Nakamura M, Nakamura E, Evans J, Slavkin HC 1987 Ultrastructural analysis of enamel formation during *in vitro* development using chemically-defined medium. J Scanning Electron Microsc 1:1103–1108

Butler WT, Brown MT, DiMuzio MT, Cothran WG, Linde A 1983 Multiple forms of rat dentin phosphoprotein. Arch Biochem Biophys 225:178–186

Chiquet-Ehrismann R, Mackie EJ, Pearson CA, Sakakura T 1986 Tenascin: an extracellular matrix protein involved in tissue interactions during fetal development and oncogenesis. Cell 47:131–139

DeVries IG, Quartier E, Boute P, Wisse E, Coomans D 1987 Immunocytochemical localization of osteocalcin in developing rat teeth. J Dent Res 66:784–790

Edelman GM 1985 Cell adhesion and the molecular processes of morphogenesis. Annu Rev Biochem 54:135–169

Evans J, Bringas P, Nakamura M, Nakamura E, Santos V, Slavkin HC 1988 Metabolic expression of intrinsic developmental programs for dentin and enamel biomineralization in serumless, chemically-defined, organotypic culture. Calcif Tissue Int 42:220–231

Fincham AG, Belcourt AB, Termine JD, Butler WT, Cothran WC 1983 Amelogenins. Sequence homologies in enamel-matrix proteins from three mammalian species. Biochem J 211:149–154

Finkelman RD, Butler WT 1985 Appearance of dentin gamma-carboxyglutamic acid containing factors in developing rat molars *in vitro*. J Dent Res 64:1008–1015

Kallenbach E 1971 Electron microscopy of the differentiating rat incisor ameloblast. J Ultrastruct Res 35:508–531

Kollar EJ 1983 Epithelial–mesenchymal interaction in the mammalian integument: tooth development as a model for instructive induction. In: Sawyer RH, Fallon JF (eds) Epithelial-mesenchymal interactions in development. Praeger, New York, p 27–50

Lumsden AGS 1987 The neural crest contribution to tooth development in the mammalian embryo. In: Maderson PFA (ed) Developmental and evolutionary biology of the neural crest. John Wiley, New York, p 3–43

MacDougall M, Zeichner-David M, Slavkin HC 1985a Production and characterization of antibodies against murine dentin phosphoprotein. Biochem J 232:493–500

MacDougall M, Zeichner-David M, Bringas P, Slavkin HC 1985b Dentin phosphoprotein expression during *in vitro* mouse tooth organ culture. In: Butler WT (ed) The chemistry and biology of mineralized tissues. EBSCO Media, Birmingham, Alabama, p 177–181

Nanci A, Bendayan M, Slavkin HC 1985 Enamel protein biosynthesis and secretion in mouse incisor secretory ameloblasts as revealed by high resolution immunocytochemistry. J Histochem Cytochem 33:1153–1160

Pierschbacher MD, Dedhar S, Ruoslahti E, Argraves S, Suzuki S 1988 An adhesion variant of the MG-63 osteosarcoma cell line displays an osteoblast-like phenotype. In: Cell and molecular biology of vertebrate hard tissues. Wiley, Chichester (Ciba Found Symp 136) p 131–141

Ruch JV 1985 Epithelial–mesenchymal interactions in formation of mineralized tissues. In: Butler WT (ed) The chemistry and biology of mineralized tissues. EBSCO Media, Birmingham, Alabama, p 54–61

Shimokawa H, Sobel ME, Sasaki M, Termine JD, Young MF 1987 Heterogeneity of amelogenin mRNA in the bovine tooth germ. J Biol Chem 262:4042–4047

Slavkin HC, Snead ML, Zeichner-David M, Bringas P, Greenberg GL 1984 Amelogenin gene expression during epithelial–mesenchymal interactions. In: Trelstad RL (ed) The role of extracellular matrix in development. Alan R. Liss, New York, p 221–253

Slavkin HC, MacDougall M, Zeichner-David M, Oliver P, Nakamura M, Snead ML 1988a Molecular determinants of cranial neural crest-derived odontogenic ectomesenchyme during dentinogenesis. Am J Med Genet, in press

Slavkin HC, Bessem C, Bringas P, Zeichner-David M, Nanci A, Snead ML 1988b Sequential expression and differential function of multiple enamel proteins during fetal, neonatal and early postnatal stages of mouse molar organogenesis. Differentiation, in press

Snead ML, Zeichner-David M, Chandra T, Robson KJH, Woo SLC, Slavkin HC 1983 Construction and identification of mouse amelogenin cDNA clones. Proc Natl Acad Sci USA 80:7254–7258

Snead ML, Bringas P, Bessem C, Slavkin HC 1984 De novo gene expression detected by amelogenin gene transcript analysis. Dev Biol 104:255–258

Snead ML, Lau EC, Zeichner-David M, Fincham AG, Woo SLC, Slavkin HC 1985 DNA sequence for cloned cDNA for mouse amelogenin reveal the amino acid sequence for enamel-specific protein. Biochem Biophys Res Commun 129:812–818

Termine JD, Belcourt AB, Christner PJ, Conn KM, Nylen MU 1980 Properties of dissociatively extracted fetal tooth matrix proteins. I. Principal molecular species in developing bovine enamel. J Biol Chem 255:9760–9768

Thesleff I, Hurmerinta K 1981 Tissue interactions in tooth development. Differentiation 18:75–88

DISCUSSION

Termine: I would support your idea of autoendocrine regulation for the ameloblast. In our laboratory, we studied the same cell preparation from which we cloned the bovine amelogenin cDNAs. We saw substantial messenger RNA for transforming growth factor β (TGF-β) in those cells, suggesting that there may be more than one factor produced locally that could regulate the growth of enamel.

Ruch: Have you correlated the secretion of enamel proteins through the basal lamina with the terminal differentiation of odontoblasts? We have suggested that some modification of the extracellular matrix controlled by the epithelial cells could lead to redistribution of fibronectin in the matrix. This redistribution could lead to polarization of odontoblasts (Ruch 1985).

Slavkin: Major rearrangements of these anionic extracellular matrix glycoproteins occur two days before mineralization in the fetal mouse molar, prior to and during the polarization of the odontoblasts. I have no idea what coordinates these epithelia-derived proteins.

Hauschka: Partanen et al (1985) studied the effects of growth factors on the morphogenesis and differentiation of embryonic mouse teeth *in vitro*. When epidermal growth factor (EGF) was added to the organ culture system the epithelium was mitogenically stimulated whereas the mesenchyme was not. It is interesting that in normal tooth development the preameloblasts arising from the inner enamel epithelium do not reach their full height and secretory activity until a layer of predentine (produced by the mesenchymal odontoblasts) begins to mineralize underneath them. This mineralizing dentine may act as a barrier, hindering the diffusion of EGF, which is apparently produced by the mesenchyme. Thus protected from mitogenic stimulation, the preameloblasts could finally differentiate into secretory ameloblasts and begin to produce enamel matrix. The characteristic temporal sequence in which the appearance of dentine always preceeds that of enamel could be controlled by EGF in this way.

Slavkin: This is an interesting idea because the cells become postmitotic. The postmitotic inner enamel epithelia could use an autocrine process to up- or down-regulate intracellular metabolic functions. Our data, unfortunately, say nothing about EGF production—we identify prepro EGF transcripts only. We are interested in the non-EGF motifs and their putative functions.

Glimcher: In the earliest crystallization of enamel *in vivo*, are the crystals highly oriented, as suggested by published electron microscopy and X-ray diffraction studies, or are they randomly arranged?

Slavkin: *In vivo*, the first position that indicates the future dentine–enamel junction is very reminiscent of shark tooth enameloid development. The crystals are not oriented: they are random. It is a hypomineralized region in the mouse tooth which, one or two μm thickness later, becomes the oriented enamel crystalline material. That first material is very strange. Perhaps the amelogenins which are expressed at this stage, after the initial mineralization, have some influence on the rates and pattern of mammalian enamel crystal growth. In the absence of amelogenins in the shark one observes a so-called aprismatic enameloid. Only anionic glycoproteins seem to represent enameloid in the shark, and these anionic proteins are also initially secreted in the mouse molar model.

Ruch: It is clear that the final volume of a specific tooth is mainly a function of the number of postmitotic odontoblasts and postmitotic ameloblasts and of their spatial distribution. We have shown in the mouse mandibular molar that between the initiation of the dental lamina and the terminal differentiation of the first odontoblast, there might be 14 or 15 cell cycles, and one cycle more for the ameloblasts (Ahmad & Ruch 1988). It is apparently never possible to trigger terminal differentiation before these cells undergo this genetically determined number of cycles. I don't understand how growth factors are working in this context. I agree that intradental control mechanisms exist (Olive & Ruch 1982) and it may be that growth factors just intervene to coordinate growth of the teeth, with other structures acting at the level of the cell cycle duration.

Rodan: Dr Slavkin, you are assuming that other components of the proEGF molecule may have a different function, other than acting as growth factors *per se*.

Slavkin: That's right. First, when you grow these teeth *in vitro* in a serumless environment they are miniatures by cell number and overall size compared to what would occur *in vivo*. Second, the word 'growth factor' is based on the original assays of molecules that would induce tritiated thymidine incorporation into DNA—DNA synthesis being a very appropriate assay. However, this should not bias our thinking. These molecules in early embryonic development may be doing many things other than producing a pro-active enhancement of thymidine incorporation into DNA. The growth factors that are expressed by embryonic tooth cells may be modulating cell elongation, may be actually down-regulating cell division, and may have a number of very interesting biological functions that have nothing to do with DNA synthesis.

Ruch: We also compared cell kinetics *in vitro* in two different culture techniques with *in vivo* cell kinetics (Ahmad & Ruch 1988). We observed exactly the same evolution. The number of cell cycles was the same (*in vivo* and *in vitro*) before terminal differentiation occurred; the decrease in mitotic activity in all the different cellular compartments was the same *in vitro* and *in vivo*; the only difference was an increase in the duration of the cell cycle.

Canalis: Your cultures grow in serum-free medium, Dr Slavkin, and therefore it is likely that growth factors are secreted. I doubt that your cultures last 21 days without growth factor secretion.

Have you attempted to do antibody localization studies with EGF antibodies, hoping that they would penetrate the tissue?

Slavkin: No, we have not done that.

Weiner: You described the rather precise timing of the introduction of amelogenins into the model system. You mentioned the interesting observation that in shark and fish enameloid bundles of oriented crystals which are very similar to normal enamel crystals form in the absence of amelogenin. How does this influence your thinking about the functions of the amelogenins in this tooth organ system?

Slavkin: John Termine has elegantly summarized what the possible roles of these hydrophobic molecules might be. They might remove water and allow crystals to grow uniquely in length and width to obtain the distinctive orientation of mammalian enamel crystals. There is definitely a removal of the amelogenin, concomitant with water removal and enhanced crystal growth. Therefore, it is possible that the amelogenins function as a detergent and facilitate crystal growth. The details of the mechanism are not known.

Weiner: Do the amelogenins have specific secondary structures which fit this detergent concept? I specifically raised the question about the sharks because there you have bundles of oriented crystals that are very similar to enamel and yet there are no amelogenins.

Slavkin: We do not have the secondary or tertiary structure data for the major 26kDa amelogenin.

Butler: You have shown that dentine phosphoproteins are expressed in preodontoblasts (which might be called young odontoblasts) at the same stage that predentine is expressed, but before mineralization occurs. You differentiate between expression and secretion because the phosphoprotein is not outside the cell. Perhaps it is expressed *and* secreted at an early stage before mineralization and simply not found extracellularly.

Slavkin: There are a number of models in cell biology of merocrine-type secretory cells where molecules are synthesized and then retained in cells for significant periods of time before they are secreted. I emphasize that point because, somewhat to our surprise, the Gla protein and the dentine phosphoprotein are secreted *after* the mineralization is initiated, suggesting that they act as modifiers of already initiated and propagating mineralization. Therefore, we argue that they are *not* molecules that would initiate mineralization; their secretion is much later in time and position than the onset of mineralization.

Butler: But Gla protein *is* expressed and secreted by young odontoblasts and osteoblasts before mineralization (Bronckers et al 1987) at a stage when predentine and osteoid are first formed, respectively. It even gets into the circulation. The same might be true for the phosphoprotein; that is, because of its affinity for mineral, it may remain in the tissue only after mineralization has occurred, even though it is synthesized *and* secreted before mineralization.

Slavkin: The advantage of looking at tooth development in serumless chemically defined medium is the absence of the background of serum- and plasma-derived molecules. We have studied dentine phosphoprotein (DPP) in a developmental sequence *in vivo* and *in vitro*. Using immunolocalization we observe that DPP is retained inside the cells until mineralization takes place: it is secreted after initial mineralization.

Veis: Professor Butler and I have both shown that at least three phosphoprotein molecules are present in most species of dentine. The antibodies that we use may not have equal reactivity with all three. We know that at least two of the phosphophoryns are cross-reactive but they are very different in their level of reactivity. It might be that the phosphoprotein that is intracellular may not be the one that is secreted into the extracellular matrix after the cells have changed their shape and differentiated into mature odontoblasts. We have looked very carefully at what is in the first layer of mantle dentine, as opposed to what is in the more mature dentine. The monospecific antibody shows a clear difference in the distribution of the phosphophoryn in those two locations. So the mechanisms which you are hinting at, in relation to the role of molecules outside the cell, may be very different, depending upon whether you are in the mantle dentine or reparative dentine rather than mature normal intertubular dentine. You have to be cautious about using a single probe.

Slavkin: The polyclonal antibodies that we are using cross-react unequivocally with all three of the phosphoproteins found in rat, rabbit, mouse and man.

Veis: Do they react with the same affinity?

Slavkin: We use the same concentration of IgG antibodies and the same conditions for each of these mammalian species. The avidity of the antibody for the epitopes seems comparable in each species.

Veis: The point is that you don't know which phosphoprotein you are looking at at any given time.

Slavkin: In the mouse, and in the other species, we detect one primary translation product.

Urist: In your cultures in serum-free medium, the continuation of development suggests that the explants were predifferentiated and, therefore, beyond the morphogenetic stage of development. This raises the question: have you made extracts of the embryonic predifferentiated cell?

Slavkin: No. We have cultured the second branchial arch segments from the mouse, prior to dental lamina formation, and these explants formed mandibular segments with cartilage, bone and tooth organs.

Urist: Then you are assuming that you have already accounted for all the molecules that are associated with development. You claim that there are no specific morphogens in this system. The best evidence for that concept comes from observations on invertebrates (Kay et al 1983), but the evidence for morphogens in vertebrates is significant (Marek & Kubicek 1981, Kapro 1981, Rowe & Fallon 1982). Some convincing evidence could come from work on extracts of the cells and extracellular matrix in a predifferentiated stage of development, not necessarily postdifferentiated.

Slavkin: This illustrates a classical controversy in developmental biology— preformation versus epigenesis. One can look at a region in a developing embryo and suggest that it is predetermined. If you transplant it to some other part of the embryo, will it form according to the original region or accommodate the new region? Another way of thinking about the process is that it requires reciprocal interactions between dissimilar tissues; the expression of the genetic programme requires temporal and spatial cues. Explants placed in a permissive environment will continue to generate epigenetic cues and mediate a particular developmental pathway.

The notion of a morphogen comes from Hans Spemann's (1938) concept of an organizer in amphibian development. For example, presumptive neural plate cells are not predetermined: they require signals from the notochordal cells. A molecule was thought to induce a particular organ or part of an organ. The idea of one morphogen having that degree of instructiveness is too simplistic. Rather, discrete steps in a spatiotemporal sequence mediate progressive development.

Urist: In the reductionist approach to embryonic development, the assumption is that something initiates the process. The up-grading and down-grading and the discovery of the growth factors enables us to account for the details of development, but it doesn't enable us to account for the initiation. I don't see how you can look at this beautiful explant and not admire how much information it contains before you begin to use it.

Slavkin: All of us who investigate developmental biology recognize that, starting with fertilization, a number of events have taken place before the experiment begins. That is so in this second branchial arch model as well as in other models. Our observations are relative to the state of the tissue that we are starting with; we then monitor the subsequent expression of new phenotypes and new forms.

Pierschbacher: You mentioned the early localization of tenascin. Do you have any idea which cells are making it and what it might be doing there?

Slavkin: Tenascin, myotendinous antigen and cytotactin are identical, having disulphide-linked subunits of 240 kDa. Tenascin (or cytotactin) appears to be expressed by cranial neural crest cells during migration into the branchial arches to form dental ectomesenchyme. It remains with the subset of crest cells that are going to be called the ectomesenchyme—cells from the mesencephalic and rhombocephalic neural tube regions that engage with the oral ectoderm. Therefore, we think that after the cranial neural crest cells have already migrated some distance from the neural tube, some kind of microenvironmental cue turns on this substrate adhesion molecule.

Caplan: You raise a fundamental consideration, namely that molecules with specific structural roles in hard tissue may also have other functions in regulating the differentiation and expression of embryonic cells. For example, we have studied bone Gla protein (osteocalcin) which is a chemoattractant for mesenchymal stem cells (P. Lucas, P. Price & A.I. Caplan, unpublished). The γ-carboxyl units are not responsible for the observed chemotaxis, since they can be removed with no loss in chemotactic potency. This protein may have other domains which affect particular cells. Clearly, osteocalcin is a bone calcium-binding protein, but it may have additional bioactivities.

Termine: The system is probably a lot more sophisticated than any of us would like to believe. When we look at sequences for the amelogenins (at both the cDNA and isolated protein level) from the viewpoint of differential RNA splicing, we observe the addition, deletion and switching of exons. So far, we don't know whether any of those changes occur at different stages of enamel development.

Weiner: You described the very first macromolecules that are introduced into the extracellular matrix as anionic glycoproteins, Dr Slavkin. Could you say into which extracellular matrix they are secreted? What is known about these anionic glycoproteins?

Slavkin: The 72 kDa anionic glycoprotein appears to have a sequence within it with epitopes that cross-react with the N-terminus of the amelogenin. The extracellular matrix at that stage in development is the interface between the epithelium and mesenchyme, prior to predentine and dentine production. These anionic proteins are molecules that are being secreted by inner enamel epithelial cells into an interface, an extracellular microenvironment. They are termed 'anionic' because of their relative isoelectric point (pI = 5.5).

References

Ahmad N, Ruch JV 1988 Comparison of growth and cell proliferation kinetics during mouse molar odontogenesis in vivo and in vitro. Cell Tissue Kinet, in press

Bronckers ALJJ, Gay S, Finkelman RD, Butler WT 1987 Developmental appearance of Gla proteins (osteocalcin) and alkaline phosphatase in tooth germs and bones of the rat. Bone Miner 2:361–373

Kapro EA 1981 Transfilter evidence for a zone of polarizing activity participation in limb morphogenesis. J Embryol Exp Morph 65:185–197

Kay RR, Dhokia B, Jermyn KA 1983 Purification of stalk-cell-inducing morphogens from dictyostelium discordeum. Eur J Biochem 136:51–56

Marek M, Kubicek M 1981 Morphogen pattern formation and development in growth. Bull Math Biol 41:359–370

Olive M, Ruch JV 1982 Does the basement membrane control the mitotic activity of the inner dental epithelium of the embryonic first lower molar? Dev Biol 93:301–307

Partanen AM, Ekblom P, Thesleff I 1985 Epidermal growth factor inhibits morphogenesis and cell differentiation in culture mouse embryonic teeth. Dev Biol 111:84–94

Rowe DA, Fallon JF 1982 Normal anterior pattern formation after barrier placement in the chick leg. J Embryol Exp Morph 69:1–6

Ruch JV 1985 Odontoblast differentiation and the formation of the odontoblast layer. J Dent Res 64:489–498

Spemann H 1938 Embryonic development and induction. Yale University Press, New Haven

Stromal stem cells: marrow-derived osteogenic precursors

Maureen Owen* and A.J. Friedenstein[†]

*MRC Bone Research Laboratory, Nuffield Department of Orthopaedic Surgery, University of Oxford, Oxford, UK and [†] Immunomorphological Laboratory, Gamaleya Institute for Microbiology and Epidemiology, Moscow, USSR

Abstract. Evidence is discussed for the hypothesis that there are stromal stem cells present in the soft connective tissues associated with marrow and bone surfaces that are able to give rise to a number of different cell lines including the osteogenic line. Fibroblastic colonies, each derived from a single colony-forming unit fibroblastic (CFU-F), are formed when marrow cells are cultured *in vitro*. *In vivo* assays of CFU-F have demonstrated that some CFU-F have a high ability for self renewal and multipotentiality whereas some have more limited potential. *In vitro* studies also support the hypothesis and have shown that CFU-F are a heterogeneous population of stem and progenitor cells and that their differentiation *in vitro* can be modified at the colony level. Factors added to the medium can activate osteogenesis in a range of multipotential and more committed precursors. Different stromal cell lines can be promoted under different culture conditions. The number and hierarchy of cell lines belonging to the stromal fibroblastic system are not yet fully elucidated and more specific markers for the different lines are required before a better understanding can be achieved.

1988 Cell and molecular biology of vertebrate hard tissues. Wiley, Chichester (Ciba Foundation Symposium 136) p 42–60

There is evidence that osteogenic cells belong to the stromal fibroblastic system associated with marrow and bone surfaces. This system is part of the stromal microenvironment which has an important role in haemopoiesis and which also includes other cell types, for example macrophages and endothelial cells (Dexter 1982). In the present paper, however, the term stromal will be restricted to cells of the stromal fibroblastic system which are thought to be histogenetically distinct from haemopoietic and endothelial cells in the adult under normal conditions (Le Douarin 1979, Wilson 1983, Simmons et al 1987).

We have proposed a hypothesis for differentiation in the stromal system, that is analogous to that in the haemopoietic system, where stromal stem cells give rise to committed progenitors for different cell lines but their number and hierarchy is not completely known (Owen 1985, 1987). They include fibroblastic, reticular, adipocytic, osteogenic and possibly other lines. Data supporting the existence of these cell lines include the demonstration that they can be established *in vitro* from marrow and soft connective tissues of bone surfaces. Osteogenic and adipocytic lines have been derived from new-born mouse calvaria and fibroblastic, osteogenic and adipocytic lines have been derived from mouse marrow (Kodama et al 1982, Sudo et al 1983, Zipori et al 1985, Benayahu et al 1987). In this paper, the evidence supporting the stromal system hypothesis and its relationship to osteogenesis will be considered, and some recent experiments on the system will be described.

Studies on single cell suspensions of marrow cells: osteogenic precursors in marrow

The demonstration that suspensions of single cells derived from marrow form a fibrous osteogenic tissue when incubated within diffusion chambers implanted *in vivo* was an important advance (Friedenstein 1976). This showed that the capacity for differentiation was not dependent on the structural relationship of the cells *in situ* and that precursor cells determined in an osteogenic direction were present in marrow stroma. Cells of the host do not penetrate the chambers and neither vascularization nor evidence of bone resorption was seen within the chambers (Ashton et al 1980). Consequently the diffusion chamber system is suitable for studying differentiation of the stromal cells *per se*.

When rabbit marrow cells were cultured within diffusion chambers implanted *in vivo*, a fibrous, bone and cartilage tissue was formed in 20 days. This tissue was morphologically identical to its skeletal counterparts according to both light and electron microscopic criteria (Ashton et al 1980, Bab et al 1984b). Using histomorphometric and biochemical techniques, analysis of cell kinetics in the chambers showed that the number of haemopoietic cells decreased to a negligible level during the first twenty days, whereas there was an increase in the total stromal cell population by more than six orders of magnitude between three and twenty days (Bab et al 1986). At three days an average of only fifteen stromal fibroblastic cells was identified within the chamber. Therefore, it was concluded that the mixture of bone, cartilage and fibrous tissues formed within the chambers was generated by a small number of precursor cells with high capacity for proliferation and differentiation, i.e. cells with stem cell characteristics.

In vivo culture of marrow cells within diffusion chambers, therefore, provides a good system for investigating tissue development from early stromal precursors present in the adult. The sequential expression of collagen types I, II, III and IV, laminin and fibronectin during the development of osteogenic tissue from rat marrow cells was found to be similar to that seen during embryonic bone and cartilage development, suggesting that the formation of tissues within chambers by precursor cells derived from adult marrow resembles the normal developmental osteogenic process (Mardon et al 1987). The tissue formed prior to osteogenesis by the stem and progenitor cells was a soft fibrous connective tissue which was strongly reactive for collagen type III and weakly for type I. Then, at sites of developing osteogenic tissue within this fibrous anlage, loss of expression of type III collagen concomitant with an increase in expression of either collagen type I or II only, appeared to signal development of the osteoblastic and chondroblastic phenotype respectively.

Measurement of the alkaline phosphatase activity and the accumulation of mineral within chambers has been developed as an assay for the osteogenic potential of cell populations (Bab et al 1984a). Results with this method demonstrated that osteogenic precursors are found throughout marrow but are more highly concentrated near bone surfaces (Ashton et al 1984).

Bone and cartilage precursors

In diffusion chambers cartilage was generally found towards the centre of the chamber whereas bone was formed near to the millipore filter. This led to the suggestion that the osteogenic precursor population in marrow may form either bone or cartilage depending on the nutritional environment (Ashton et al 1980, Bab et al 1984b). Support for this has been provided in a recent study which demonstrated that bone only developed in chambers where the gap between the filters was narrow (0.1 mm), whereas both cartilage and bone were almost always present in chambers where the gap between opposing filters was wide (2 mm) (Friedenstein et al 1987).

Studies on stromal CFU-F: evidence for stromal stem cells

When suspensions of dispersed marrow cells, obtained by flushing the cells from the marrow cavity and endosteal bone surfaces, are cultured *in vitro* with fetal calf serum, fibroblastic colonies are formed each of which is derived from a single cell, i.e. fibroblastic colony-forming cell (FCFC) or colony-forming unit fibroblastic (CFU-F) (Friedenstein et al 1970). The clonal origin of the colonies has been confirmed using thymidine labelling, time-lapse photography and chromosome markers (Friedenstein 1976). A definitive answer to the question of whether stem cells are present in the stroma of bone and marrow must come from an analysis based on the self-replicating ability and capacity for differentiation of individual CFU-F. This has been investigated by assay of the potential for differentiation of single colonies in open transplant and diffusion chambers *in vivo* (Friedenstein 1980, Friedenstein et al 1987).

Single fibroblastic colonies cultured for 14 to 16 days from mouse marrow cells were transplanted under the renal capsule (Friedenstein 1980). In this open system about 15% of colonies produced several lines — bone, adipose and marrow reticular tissue — with establishment of haemopoiesis by host cells forming a heterotopic bone and marrow organ. About 15% formed bone tissue without associated marrow and the rest formed a soft fibrous tissue or nothing. This experiment provided evidence for the existence among the CFU-F population of multipotential stromal stem cells able to form the haemopoietic microenvironment and for the presence of precursors with more limited potential.

Using rabbit marrow cells a high ability of CFU-F to self-renew, a necessary characteristic of stem cells, was demonstrated (Friedenstein et al 1987). Fibroblasts were grown to confluence in culture from a number of CFU-F (FCFC) and after two to eighteen passages *in vitro* these were assayed in diffusion chambers. The inoculum of harvested fibroblasts able to form osteogenic tissue in the chamber was defined as one osteogenic unit. It was found that the number of osteogenic units which could be harvested from culture after passaging ranged up to about one thousand times the number of initiating CFU-F, thus demonstrating the high self-replicating ability of CFU-F in culture. It was also clear from these experiments that CFU-F retain their capacity for osteogenic differentiation after extensive culture and passaging *in vitro*. The ability of fibroblasts grown from single colonies *in vitro* to form osteogenic tissue in diffusion chambers was also confirmed in these experiments. When fibroblasts harvested from a single colony after two or three passages were incubated within diffusion chambers *in vivo* about 40% of colonies formed osteogenic tissues within the chambers.

Differentiation of CFU-F in vitro

Application of *in vitro* clonal assay methods to early haemopoietic precursors has resulted in major advances in our understanding of differentiation in the haemopoietic system (Metcalf 1984). Similar methods have scarcely been applied to the stromal system. We have initiated studies to investigate whether CFU-F are able to differentiate at the colony level in culture, with a view to developing *in vitro* clonal methods for studying lineage in the marrow stromal system (Owen et al 1987).

We have used young rabbit marrow cells and the expression of alkaline phosphatase activity as a marker for osteogenic differentiation, since the enzyme is an early indicator of developing osteogenic tissue when rabbit marrow cells are cultured in diffusion chambers *in vivo*. The fibroblastic colonies formed when a suspension of rabbit marrow cells was cultured *in vitro* varied widely in size, morphology and level of expression of the enzyme. Alkaline phosphatase activity appeared to originate in the centre of the colony where the cells first reach confluence and where cell growth is arrested. Labelling with tritiated thymidine showed growing cells located

FIGS. 1 to 4. Fibroblastic colonies formed when a suspension of single rabbit marrow cells (10^7 cells/25 cm^2 flask) is cultured in BGJ$_b$ medium with 10% fetal calf serum, and the flasks fixed and stained for alkaline phosphatase (AP) activity (black) at 16 days. Full details are given in Owen et al (1987).

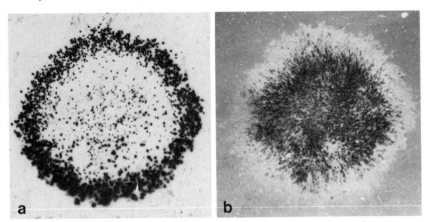

FIG. 1. Colony in culture treated with hydrocortisone from Day 7. [^{3}H]Thymidine (1 μc/ml) was added to the medium 1½ hours before fixation. (a) Contact autoradiograph using Hyperfilm-^{3}H Amersham, exposure time 15 days, (b) stained for AP activity, photographed with reflected light. Note the area of labelled cells in a ring at the edge of the colony around the central area that is positive for AP activity. ($\times$ 9)

mainly at the edge of the colony (Fig. 1). However, this was also the case for colonies which did not express the enzyme so that growth arrest alone is not the trigger for alkaline phosphatase expression. The level of enzyme expression varied: some colonies were entirely negative, whereas in others a large proportion of the cells were positive for the enzyme (Fig. 2a and b).

Hydrocortisone (HC) increases alkaline phosphatase activity in a variety of osteogenic systems (Rodan & Rodan 1984). When added to the culture medium in this system its main effect was to stimulate clonal expression of the enzyme in a wide range of colonies expressing the enzyme at different levels (Owen et al 1987). Examples of colonies where HC was present in the culture medium are shown in Figs 3 and 4. Addition of epidermal growth factor (EGF) increased average colony size and reduced expression of the enzyme to negligible levels (Fig. 2c). Continuous presence of EGF was necessary for this effect and it was suggested that the presence of the factor keeps the cells in cycle, preventing them from moving towards differentiation.

In these experiments the colonies formed are heterogeneous, which is consistent with their derivation from a population of stem and progenitor cells at different stages of a tissue developmental system (Owen et al 1987). A suggested, but not yet proven, hierarchy is that colonies expressing a high level of alkaline phosphatase activity are initiated by CFU-F approaching committed progenitor status for the osteogenic line, and that colonies with

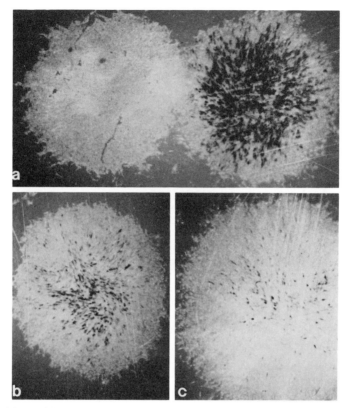

FIG. 2.   (a) and (b), colonies in control culture. (a) Adjacent colonies, colony on left completely negative for the enzyme, colony on right with large proportion of cells positive for the enzyme, and (b) colony with small proportion of cells positive for the enzyme. (c) Colony in culture treated with epidermal growth factor from Day 0. This colony is larger than in control culture and a negligible proportion of cells is positive for the enzyme. (× 8) Reproduced by permission from Owen et al (1987).

fewer cells positive for the enzyme arise from CFU-F which are earlier in the lineage. Markers for other cell lines were not available. Nevertheless, it seems likely that the majority of CFU-F in the present cultures are stem and progenitors, most of them with more than one lineage potential. This has several implications. The results with HC suggest that the programme of osteogenesis may be activated in both multipotential and more 'committed' precursors. It is likely that the cells responsive to EGF also consist of a similar range of precursors. The characteristics of different cloned cell lines will vary, depending on the position in the lineage from which the originating cell was derived.

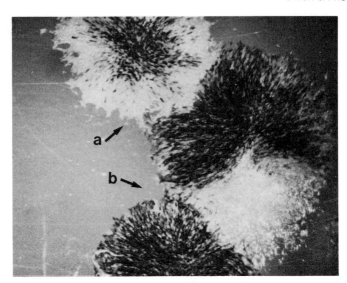

FIG. 3. Four colonies in culture treated with hydrocortisone from Day 0. Colony with enzyme expression mainly in the centre (top left) and two colonies where cells right to the edge of the colony express the enzyme with a negative colony between. (× 9) Areas indicated by a and b are enlarged in Fig. 4.

Discussion

A study of cloned populations from single precursor cells derived from bone surfaces of fetal rat calvariae has demonstrated a wide variation in their characteristics and this is consistent with the present hypothesis (Aubin et al 1982). Differences in proliferative activity, in synthesis of extracellular matrix components and hormonal and biochemical responses by individual clones suggested that their origin is a heterogeneous population of precursors, some of which have multipotentiality (Sodek et al 1985).

Stromal cell lines synthesize haemopoietic colony-stimulating factors and this has stimulated considerable interest in the characterization of these lines from marrow (Zipori et al 1985, Lim et al 1986, Rennick et al 1987, Lanotte et al 1982). Different culture conditions selectively promote the growth of different stromal cell populations. With suspensions of single marrow cells and fetal calf serum, the formation of fibroblastic colonies from precursors designated CFU-F has been confirmed for several species including man, in a number of laboratories. These cells are commonly referred to as marrow fibroblasts (Castro-Malaspina et al 1980). In culture conditions which included 0.9% methylcellulose, leukocyte-conditioned medium and hydrocortisone, human marrow cells gave rise mainly to growth of colonies with a reticular-fibroblastoid morphology from precursors designated CFU-RF (colony forming unit-reticulofibroblastoid) (Lim et al 1986). In this study the

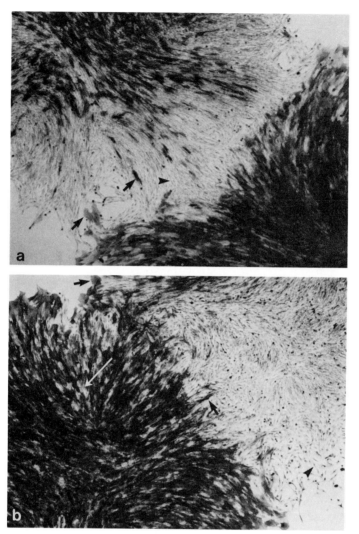

FIG. 4. a and b are high power photographs of areas in Fig. 3. Note that AP positive cells (thick arrows) are generally more polygonal than unstained cells (arrowheads). Some unstained cells (thin white arrow) are also present in colonies which are stained right to the edge. (× 28)

synthesis of a variety of macromolecules by progeny from CFU-F and CFU-RF was compared using immunofluorescent antibody techniques. Both cell populations were highly positive for collagen type I, III and V. However, cells from CFU-RF were, in addition, strongly positive for collagen type IV and laminin. The cells of both populations were negative for factor VIII, an endothelial cell marker.

An important question is: how are cells in culture related to their *in vivo* counterparts? The major cells of the stromal fibroblastic system of marrow and bone surfaces include: osteogenic cells (osteoblasts, bone-lining cells and preosteoblasts) which are found near to bone surfaces and within Haversian canals; fibroblasts and reticular cells which are the main terms applied to cells of the soft connective tissue network throughout marrow and spaces in bone; and adipocytes which are distinguished by their lipid inclusions. Other candidates for this system are the cells of mesenchymal origin which are associated with the walls of small blood vessels. These include pericytes, smooth muscle cells and cells of fibroblastic morphology (Ham 1974). Stromal stem and progenitor cells are not identifiable morphologically *in vivo* but are likely to be found among cells which have an ill-defined spindle-shaped, fibroblastic appearance. These cells are widespread, often in perivascular locations, and in the past have been called undifferentiated mesenchymal cells.

The best characterized cells of the marrow stromal system are osteogenic cells. Both *in vivo* and *in vitro* they are identified by high levels of alkaline phosphatase activity, synthesis of type I collagen (Sandberg & Vuorio 1987), and of the bone-specific protein, osteocalcin (Bronckers et al 1987), and the presence of receptors for parathyroid hormone (Rodan et al, this volume). By comparison there has been little characterization of other stromal cell types. Different cloned populations produce varying combinations of collagen types and other macromolecules, but this is not specific for different phenotypes. Collagen types I and III are found *in situ* widely distributed throughout the reticular-fibroblastic network and connective tissues of blood vessel walls (Bentley et al 1981, 1984). Type IV collagen and laminin are localized in the walls of small blood vessels and associated with basement membrane. It is possible that stromal cells which synthesize these latter components, for example CFU-RF, may be derived from perivascular mesenchymal cells. There is little understanding of the relationship of perivascular cells (i.e. pericytes, smooth muscle cells and fibroblasts of the vessel walls) to other cells of the stromal system. Nor is it known how CFU-F are related to CFU-RF. The situation is complicated by confusion in terminology both *in vivo* and *in vitro*, the term fibroblast, in particular, being very loosely applied, but undue concern about terminology is inappropriate in our present state of knowledge.

Finally, *in vivo* and *in vitro* studies have provided strong circumstantial evidence for the existence of stromal stem cells and a stromal fibroblastic system of which osteogenic cells are a product of one of several cell lines. There is reasonable evidence for the presence in marrow of precursor cells determined for osteogenesis and for adipogenesis (Friedenstein 1976, Patt et al 1982). However, the existence of irreversibly committed progenitors is still hypothetical. For further understanding of differentiation and the hierarchy of cell lineage in this system, the availability of specific markers for the different cell lines will be essential. For rigorous proof of the stem cell hypothesis, marked stem cells with the ability to follow the marker in the stem cell progeny will be necessary. Methods which have been used for haemopoietic stem cells (Abramson et al 1977) including gene transfer techniques (Lemischka et al 1986) are likely to be applied to the problem in the near future and rapid advances in understanding should follow.

Acknowledgements

Grateful acknowledgement is made for the support of this work by the Medical Research Council. The authors thank J. Judge for typing the manuscript and M. Thomas for photography. They are indebted to many collaborators: B.A. Ashton, I. Bab, J. Cave, C.C. Eaglesom, C.R. Howlett, H. Mardon, J.T. Triffitt, M. Williamson, R.K. Chailakhyan, U.V. Gerasimov, N.V. Latsinik and E.A. Luria.

References

Abramson S, Miller RG, Phillips RA 1977 The identification in adult bone marrow of pluripotent and restricted stem cells of the myeloid and lymphoid systems. J Exp Med 145:1567–1579

Ashton BA, Allen TD, Howlett CR, Eaglesom CC, Hattori A, Owen M 1980 Formation of bone and cartilage by marrow stromal cells in diffusion chambers in vivo. Clin Orthop Relat Res 151:294–307

Ashton BA, Eaglesom CC, Bab I, Owen ME 1984 Distribution of fibroblastic colony-forming cells in rabbit bone marrow and assay of their osteogenic potential by an in vivo diffusion chamber method. Calcif Tissue Int 36:83–86

Aubin JE, Heersche JNM, Merrilees MJ, Sodek J 1982 Isolation of bone cell clones with differences in growth, hormone responses, and extracellular matrix production. J Cell Biol 92:452–461

Bab I, Ashton BA, Syftestad GT, Owen ME 1984a Assessment of an in vivo diffusion chamber method as a quantitative assay for osteogenesis. Calcif Tissue Int 36:77–82

Bab I, Howlett CR, Ashton BA, Owen ME 1984b Ultrastructure of bone and cartilage formed in vivo in diffusion chambers. Clin Orthop Relat Res 187:243–254

Bab I, Ashton BA, Gazit D, Marx G, Williamson MC, Owen ME 1986 Kinetics and differentiation of marrow stromal cells in diffusion chambers in vivo. J Cell Sci 84:139–151

Benayahu D, Kletter Y, Zipori D, Wientroub S 1987 Osteoblastic expression of marrow stromal derived cell line: in-vitro and in-vivo. Calcif Tissue Int 41: Suppl. 2, 81

Bentley SA, Alabaster OA, Foidart JM 1981 Collagen heterogeneity in normal human bone marrow. Br J Haematol 48:287–291

Bentley SA, Foidart J-M, Kleinman HK 1984 Connective tissue elements in rat bone marrow: immunofluorescent visualization of the hematopoietic microenvironment. J Histochem Cytochem 32:114–116

Bronckers ALJJ, Gay S, Finkelman RD, Butler WT 1987 Developmental appearance of Gla proteins (osteocalcin) and alkaline phosphatase in tooth germs and bones of the rat. Bone Miner 2:361–373

Castro-Malaspina H, Gay RE, Resnick G et al 1980 Characterisation of human bone marrow fibroblast colony-forming cells (CFU-F) and their progeny. Blood 56:289–301

Dexter TM 1982 Stromal cell associated haemopoiesis. J Cell Physiol Suppl 1:87–94

Friedenstein AJ 1976 Precursor cells of mechanocytes. Int Rev Cytol 47:327–355

Friedenstein AJ 1980 Stromal mechanisms of bone marrow: cloning in vitro and retransplantation in vivo. In: Thienfelder S (ed) Immunobiology of bone marrow transplantation. Springer-Verlag, Berlin, p 19–29

Friedenstein AJ, Chailakhyan RK, Lalykina KS 1970 The development of fibroblastic colonies in monolayer cultures of guinea-pig bone marrow and spleen cells. Cell Tissue Kinet 3:393–402

Friedenstein AJ, Chailakhyan RK, Gerasimov UV 1987 Bone marrow osteogenic stem cells: In vitro cultivation and transplantation in diffusion chambers. Cell Tissue Kinet 20:263–272

Ham AW 1974 Histology, 7th edn. Lippincott, Philadelphia, p 577

Kodama H, Amagai Y, Koyama H, Kasai S 1982 A new pre-adipose cell line derived from newborn mouse calvaria can promote the proliferation of pluripotent hemopoietic stem cells in vitro. J Cell Physiol 112:89–95

Lanotte M, Scott D, Dexter TM, Allen TD 1982 Clonal preadipocyte cell lines with different phenotypes derived from murine marrow stroma: Factors influencing growth and adipogenesis in vitro. J Cell Physiol 111:117–186

Le Douarin NM 1979 Dependence of myeloid and lymphoid organ development on stem-cell seeding: investigations on mechanisms in cell marker analysis. In: Ebert J, Okada T (eds) Mechanisms of cell change. Wiley, New York, p 293–326

Lemischka IR, Raulet DH, Mulligan RC 1986 Developmental potential and dynamic behavior of hematopoietic stem cells. Cell 45:917–927

Lim B, Isaguirre CA, Aye MT et al 1986 Characterisation of reticulofibroblastoid colonies (CFU-RF) derived from bone marrow and long-term marrow culture monolayers. J Cell Physiol 127:45–54

Mardon HJ, Bee J, von der Mark K, Owen ME 1987 Development of osteogenic tissue in diffusion chambers from early precursor cells in bone marrow of adult rats. Cell Tissue Res 250:157–165

Metcalf D 1984 The hemopoietic colony-stimulating factors. Elsevier, Amsterdam.

Owen M 1985 Lineage of osteogenic cells and their relationship to the stromal system. In: Peck WA (ed) Bone and mineral research. Excerpta Medica, Amsterdam, vol 3:1–25

Owen M 1987 The osteogenic potential of marrow. In: Sen A, Thornhill T (eds) Development and diseases of cartilage and bone matrix. Alan R. Liss, New York (UCLA Symposia on Molecular and Cellular Biology New Series) vol 46:247–255

Owen ME, Cave J, Joyner CJ 1987 Clonal analysis in vitro of osteogenic differentiation of marrow CFU-F. J Cell Sci 87:731–738

Patt HM, Maloney MA, Flannery ML 1982 Hematopoietic microenvironment transfer by stromal fibroblasts derived from bone marrow varying in cellularity. Exp Hematol 10:738–742

Rennick D, Yang G, Gemmell L, Lee F 1987 Control of hemopoiesis by a bone marrow stromal cell clone: Lipopolysaccharide — and Interleukin-1-inducible production of colony-stimulating factors. Blood 69:682–691

Rodan GA, Rodan SB 1984 Expression of the osteoblastic phenotype. In: Peck WA (ed) Bone and mineral research. Excerpta Medica, Amsterdam, vol 2:244–286

Rodan GA, Heath JK, Yoan K, Noda M, Rodan SB 1988 Diversity of the osteoblastic phenotype. In: Cell and molecular biology of vertebrate hard tissues. Wiley, Chichester (Ciba Found Symp 136) p 78–91

Sandberg M, Vuorio E 1987 Localization of types I, II, III collagen mRNAs in developing human skeletal tissues by in situ hybridization. J Cell Biol 104:1077–1084

Simmons PJ, Przepiorka D, Thomas ED, Torok-Storb B 1987 Host origin of marrow stromal cells following allogeneic bone marrow transplantation. Nature (Lond) 328:429–432

Sodek J, Bellows CG, Aubin JE, Lineback H, Otsuka K, Yao K-L 1985 Differences in collagen gene expression in subclones and long-term cultures of clonally derived rat bone cell populations. In: Butler WT (ed) The chemistry and biology of mineralized tissues. EBSCO Media, Birmingham, Alabama, p 303–306

Sudo H, Kodama H-A, Amagai Y, Yamamoto S, Kasai S 1983 In vitro differentiation and calcification in a new clonal osteogenic cell line derived from newborn mouse calvaria. J Cell Biol 96:191–198

Wilson D 1983 The origin of the endothelium in the developing marginal vein of the chick wing-bud. Cell Differ 13:63–67

Zipori D, Duksin D, Tamir M, Argaman A, Toledo J, Malik Z 1985 Cultured mouse marrow stromal cell lines. II. Distinct subtypes differing in morphology, collagen types, myelopoietic factors, and leukemic cell growth modulating activities. J Cell Physiol 122:81–90

DISCUSSION

Raisz: Have you looked to see whether indomethacin will in any way replicate the effects of hydrocortisone?

Owen: No, we haven't done many studies yet with this system.

Raisz: There is a tendency in osteoporosis for the marrow to contain more adipocytes than usual. Is it possible that making an adipocyte will steal stem cells? The assumption is that both osteoblasts and the stem cells themselves have receptors for lipoprotein. Catherwood et al (1986) demonstrated receptors for low density lipoprotein (LDL) in osteoblasts, and LDL can induce adipocyte formation in the stem cell population. Has anyone tried to change the lipid environment in rabbits, for example? With a high blood level of LDL in the rabbit, would you get less bone and more adipocytes in the diffusion chamber?

Owen: The effect of changes in lipid environment on the development of tissue in diffusion chambers has not been investigated. You also comment on an increase in marrow adipocytes in osteoporosis and ask whether adipogenesis may have taken place at the expense of osteogenesis. Whether this is a possibility is not known. However, marrow fibroblasts cultured *in vitro* from red and fatty marrow of young and old rabbits retain their phenotypic characteristics when transplanted under the renal capsule (Patt et al 1982). This could suggest that marrow fibroblasts contain populations of precursors determined along particular lineages and that the preponderance of different precursors may change with age.

Testa: The critical thing to know about cell progeny is their clonality. You referred to a paper by Lim et al (1986). They modified the culture conditions and produced cells with different phenotypes to those seen in CFU-F colonies.

Their system is similar to the long-term marrow culture developed in our laboratory in which all those cell populations are seen. Even with the addition of methylcellulose, there is still a fairly high degree of cell mobility. Therefore, the possibility exists that those colonies may not be clonal. I am not yet convinced that reticular-type cells and other cell types have the same clonal origin as the fibroblast cells. We need stringent proof.

Also, we see a lot of species variation. For example, in mouse CFU-F-type colonies large numbers of macrophages, which cannot be of the same clonal origin as fibroblasts, migrate and invade the colony. At later stages of development you see reticular-endothelioid, ill-defined cells growing in association with the fibroblastic cells which may or may not be of the same clonal origin (Xu et al 1983).

Owen: I agree that the question of the origin of different cell lines from a common stem cell progenitor is basic to the stromal system hypothesis. However, studies of the stromal system are at a relatively early stage and, as you are well aware, rigorous proof of clonality is difficult. It was many years before the clonal origin of haemopoietic stem cell progeny was proved, using uniquely marked stem cells (Abramson et al 1977). Professor Friedenstein has provided much good experimental evidence for the clonal origin of the fibroblastic colonies formed from marrow cells.

Comment added in proof by A.J. Friedenstein: The clonality of CFU-F-derived colonies *in vitro* was confirmed by thymidine labelling, time-lapse photography and chromosome analysis in mixed cultures. However, there remained the possibility of colony formation from stromal aggregates that could simulate clonality. This possibly was tested by counting the number of stromal clumps adhering when nylon-filtered bone marrow cell suspensions were introduced into flasks coated with poly-L-lysine. The content of stromal clumps was found to be less than 10^{-5} for mouse and rabbit marrow cells, whereas the mean efficiency of CFU-F colony formation was approximately 0.5 $\times$ 10^{-3}. Consequently, the possibility of colonies formed from stromal aggregates is very low and the presence of a single karyotype within colonies in mixed cultures proves that CFU-F-derived colonies are indeed cell clones (Friedenstein 1976, Friedenstein et al 1987, Latsinik et al 1987).

Rodan: Using the 3T3 calvaria-derived cells that he developed (Kodama et al 1982, Sudo et al 1983), Dr H. Kodama obtained three clonal cell lines: one was adipogenic; the second was osteogenic; and the third was myogenic.

Fleisch: Because the bone formed in diffusion chambers is not mature, it's sometimes difficult to know whether it is bone or ectopic calcification. How do you differentiate between these two possibilities?

Owen: One can distinguish them electron microscopically. By ultrastructural criteria osteogenic tissues formed in the diffusion chambers were identical to those found in the skeleton. They were a mixture of lamellar and a more primitive woven-type bone and cartilage. In areas of bone and cartilage, thick well-banded collagen and thin collagen fibrils typical of these tissues were found and genuine matrix vesicles were present in both areas.

Fleisch: Do you sometimes find it difficult to decide? Do you use markers such as osteocalcin?

Owen: We have seen dystrophic mineralization in diffusion chambers occasionally, but this can usually be distinguished at the light microscope level and with certainty at the electron microscope level. Also, in these chambers there is often a very high content of calcium and an undetectable level of alkaline phosphatase activity, when measured biochemically. We have not yet looked at osteocalcin in diffusion chambers, but we are planning to investigate the stage at which osteocalcin appears in this system.

Martin: You mentioned a possible connection between small vessel endothelial cells and certain of the marrow fibroblast population. Previous studies have attempted to relate the endothelial line to the osteogenic line.

Owen: You are referring to early work where the origin of osteogenic cells was attributed to cells associated with small blood vessels near bone surfaces (Trueta 1963, Mankin 1964). In these studies it was not possible to distinguish endothelial cells from perivascular cells (pericytes, smooth muscle cells and fibroblastic cells) that form a coating around the majority of small blood vessels. These perivascular cells are thought to be of mesenchymal origin (Ham 1974) and are of different origin to endothelial cells in the adult and late embryonic stage, as discussed in our paper.

Martin: A few years ago we found that osteoblasts of various sorts (clonal lines that are malignant, mixed cultures that are not malignant and clonal early osteoblasts) have receptors and responses to atrial natriuretic peptide (Fletcher et al 1986) which one would not expect to act upon osteogenic cells but which does act upon smooth muscle cells and endothelial cells. The receptors are present, and there are large cyclic GMP responses. For that reason we went back to the old literature. There may be something to be learned by studying the nature of the connection between perivascular cells and osteogenesis that you mention.

Owen: That the perivascular cells are undifferentiated mesenchymal cells with potential to give rise to osteogenic and other cell lines is an old idea, but there is little understanding of the nature and function of these cells and this deserves further study.

Hauschka: During the harvesting of marrow cells from bone, one could envisage that osteoblasts might occasionally be stripped from the endosteal surface. What techniques can one use in the clonal assay to distinguish between osteoblast-like colonies, derived from proliferation of such stripped osteoblasts, and true stem cells? Secondly, after harvesting marrow could one study what remains endosteally, which might be thought of as an enriched osteoblastic population? Perhaps one could enzymically strip the cells from the endosteal surface and repeat the CFU-F or CFU-OB assay. Would one see the same kind of clonal growth of these cells?

Owen: We have no means of knowing whether fully differentiated osteoblasts stripped from the endosteal surface revert back to a proliferative stage and form colonies *in vitro*, but that seems unlikely. In the clonal assay, colonies derived from precursors which are advanced towards osteogenic differentiation should express alkaline phosphatase activity and other osteogenic parameters at a higher level than colonies derived from earlier stem cells.

Osteoblast-like populations have been obtained by sequential collagenase digestion of embryonic and neonatal bones (Cohn & Wong 1979). A CFU-F assay has not been done on these cell populations, but I would expect a proportion of the cells to show clonal growth *in vitro*. These cells have characteristics of osteoblasts and make a mineralized tissue in diffusion chambers, although ultrastructural studies have not been made on it (Simmons et al 1982). On the other hand, when cultured *in vitro* for a few weeks, these isolated osteoblast-like populations lose their osteogenic capacity.

Fleisch: It is very interesting that the osteoblast-like population lost its osteogenic capacity. How were the cells cultured? Could there be a way to culture them so they don't lose it?

Owen: Culture conditions may well influence which cells survive. However, loss of osteogenic capacity after a relatively short time in culture by isolated osteoblast-like populations has also been observed by other workers (Moskalewski et al 1983). In contrast, the marrow fibroblast population can be passaged many times and still retain its osteogenic capacity (Friedenstein et al 1987). We speculated that this was consistent with the presence of early stem cells in marrow and of later progenitors of more limited proliferative activity and differentiation potential in isolated osteoblast-like cell populations (Owen & Ashton 1986).

Fleisch: Were the culture conditions the same?

Owen: Probably not; the experiments were done in different laboratories.

Testa: When Dr Friedenstein harvested marrow by trypsinizing the bone surface he obtained four or five times more CFU-F than by washing the marrow out of the cavity. Have you compared the developmental sequence of CFU-F harvested from within the marrow and from the surface of the bone by enzymic treatment? I shall present evidence (Testa et al, this volume) that haemopoietic cell populations are not evenly distributed in the marrow cavity. The CFU-F are not evenly distributed either, but we have no data on the quality of CFU-F population harvested by trypsin treatment of bone. Have you studied that?

Owen: In the experiments to which you refer, marrow tissue, *not* the bone surface, was trypsinized (Friedenstein et al 1982). There is no information, to my knowledge, on CFU-F assay of cell populations harvested by trypsinization of bone. However, in rabbit we found that cells close to the bone surface had a greater number of CFU-F and higher osteogenic potential than cells from the centre of the marrow cavity. These populations were separated mechanically (Ashton et al 1984).

Prockop: We should be careful how we define lineages of cells. You mentioned that colony forming unit-reticulofibroblastoid cells (CFU-RF) make type IV collagen and laminin. We have been surprised to find that, depending on the assays and the age of the donor, ordinary skin fibroblasts also make type IV collagen and laminin. We discovered this because some cell lines from patients with osteogenesis imperfecta make large amounts of type IV collagen (de Wet et al 1983; T. Pihlajaniemi, J. McKeon, S. Gay, R. Gay, W.J. de Wet, J.C. Myers & D.J. Prockop, to be published). We studied cells from young donors as controls and saw that skin fibroblasts from these donors made type IV collagen as well. In fact, with sensitive assays you can find mRNA for type IV collagen in almost any type of fibroblast.

Owen: In the work referred to (Lim et al 1986) the synthesis of a number of collagen types, laminin and fibronectin by skin fibroblasts, endothelial cells and two populations of marrow stromal cells was compared. Different combinations of the macromolecular components were made by the two marrow stromal populations but there was no suggestion that this implied lineage specificity. In that study, skin fibroblasts were negative for type IV collagen and laminin.

Prockop: It's a matter of sensitivity of assay. We need to do quantitative experiments.

Owen: Yes. Those studies were non-quantitative.

Hanaoka: I agree with the importance of the perivascular mesenchymal cells. I am not quite sure whether the precursor of the osteoblast and the precursor of the osteoclast are derived from the same type of perivascular cells, but when I have looked at normal bone marrow tissue in the electron microscope I have always seen very immature cells situated around the endothelial cells.

Williams: We use alkaline phosphatase as a marker for enzyme activity and we assay collagen for the network of the outside of cells. Why do we not have any markers for the cross-linking of the collagen? Why are we not measuring lysine oxidases? Surely lysine oxidase must appear before the tissue is mineralized. These are special copper enzymes. We have examined some teeth from limpets (not from humans) and copper oxidases are expressed before mineralization.

Similarly, if the bone matrix is to be degraded, the zinc collagenases must be introduced in order to produce the balance between cross-linking and collagen filament breaking, making space available for development. These zinc collagenases should be expressed later than the copper proteins. I wonder if we could find a set of marker enzymes for development instead of using an enzyme such as alkaline phosphatase which is simply present to produce phosphate. Has anybody developed such a system with these series of cells?

Rodan: Are you suggesting that we identify tissues by post-translational modifications of common macromolecules?

Williams: Yes. The lysine oxidases must be expressed after the collagen is generated but they must not be expressed too early because their reaction would block off the system. The question is whether they belong to a particular cell type as differentiation progresses.

Prockop: That might be a useful approach. Unfortunately most of these enzymes are extremely hard to assay. The substrates are complex and it's difficult to extract them quantitatively, particularly as the tissue deposits more and more collagen.

Williams: Lysine oxidases could be easy to assay because they catalyse other amine oxidase reactions which produce intense colours. Copper is now easy to locate because we have very sensitive detection systems—electron beams and proton beams.

Prockop: The lysine oxidase involved in collagen cross-linking is a highly specific enzyme and requires an insoluble collagen substrate. Assaying it through its copper content may be possible.

Gazit: We have done preliminary experiments to evaluate the effect of hyperbaric oxygen on the production of bone and mineralized cartilage in diffusion chamber cultures of rabbit marrow cells. The diffusion chambers were incubated intraperitoneally in athymic mice exposed to oxygen-rich hyperbaric conditions for 2 h per day for 20 days. A histomorphometric analysis revealed that in these conditions there was a decrease in the amount of mineralized tissue, compared to the controls. The amount of non-mineralized cartilage and fibrous tissue remained unchanged. Have you any additional data about the role of oxygen during the mineralization process in diffusion chambers?

Owen: No, but those results are very interesting.

Caplan: A cell line called 10t½ was originally derived from 3T3 cells. When these cells are exposed to the DNA methylating agent, 5-azacytidine, adipocytes, chondrocytes, fibrogenic cells and also myogenic cells are detected in clonal assay. Would you be willing to extend your list of cells derived from stromal stem cells to include cells of the myogenic lineage?

Owen: Yes; in particular I would include some of the perivascular cells which have myogenic properties. I was interested to learn recently that a clonal cell population isolated from 21-day-old fetal rat calvaria expressed three distinct phenotypes—multinucleated muscle cells, adipocytes and chondrocytes—when cultured *in vitro* under certain conditions (Grigoriadis et al 1987).

Caplan: The key to dealing with these mesenchymal cells is defining *in vitro* conditions which push these stem cells into particular lineage pathways. We know that mesenchymal stem cells are highly sensitive to their microenvironments. Choosing culture conditions to obtain a particular lineage pathway will be important. Perhaps we haven't yet chosen conditions appropriate for myogenic cells or for some of these other phenotypes.

Owen: There is much work to be done in that area. We find that the colony type promoted *in vitro* in the stromal system is very dependent on culture conditions.

Krane: Dr Caplan (this volume) implied that the osteoblast is oriented towards the basal border of the endothelial cell, suggesting that the osteoblast may make contact with a basement membrane-type material produced by the endothelial cell. One could possibly culture these osteoblasts on basement membrane components such as laminin or type IV collagen, which might influence the proliferation or phenotype of the cells.

Urist: The predominant cell in the intact stroma of bone marrow is said to be the reticular cell. This cell looks like a small lymphocyte, and its marker is its reticular fibres. Has the staining of these clonal cultures with alkaline solutions of reducible silver salts shown reticular fibres?

Owen: The marrow reticular cell has a fibroblastic appearance with long processes extending into the marrow spaces (Weiss 1976) and is mainly a differentiated cell producing an extracellular matrix of reticular fibres. We have not stained the clonal cultures with silver stains.

Urist: If this fibroblast and potentially osteogenic colony comes from reticular cells, does the cell stop making, if it ever does, reticular fibres? There are such fibres in the bone matrix: they can be seen by electron microscopy.

Is it true that if you substitute horse serum for fetal calf serum, the haemopoietic cell lines survive? What component promotes this survival?

Owen: The culture conditions with horse serum, etc. are the long-term bone marrow culture system that Dr Testa will discuss (Testa et al, this volume).

References

Abramson S, Miller RG, Phillips RA 1977 The identification in adult bone marrow of pluripotent and restricted stem cells of the myeloid and lymphoid systems. J Exp Med 145:1567–1579

Ashton BA, Eaglesom CC, Bab I, Owen ME 1984 Distribution of fibroblastic colony-forming cells in rabbit bone marrow and assay of their osteogenic potential by an in vivo diffusion chamber method. Calcif Tissue Int 36:83–86

Caplan AI 1988 Bone development. In: Cell and molecular biology of vertebrate hard tissues. Wiley, Chichester (Ciba Found Symp 136) p 3–21

Catherwood BD, Contreras ST, Elsholtz VC, Lorang MT 1986 Potentiation of the proliferative response to mitogens by lipoprotein (LDL) or exogenous lipid in cultured rat osteoblast (ROB). J Bone Miner Res 1(suppl 1): abstr 151

Cohn DV, Wong GL 1979 Isolated bone cells. In: Simmons D, Kahn A (eds) Skeletal research. Academic Press, New York, p 3–19

de Wet WJ, Pihlajaniemi T, Myers J, Kelly TE, Prockop DJ 1983 Synthesis of a shortened $pro\alpha2(I)$ chain and decreased synthesis of $pro\alpha2(I)$ chains in a patient with osteogenesis imperfecta. J Biol Chem 258:7721–7728

Fletcher AE, Allan EH, Casley DJ, Martin TJ 1986 Atrial natriuretic factor receptors and stimulation of cyclic GMP formation in normal and malignant osteoblasts. FEBS (Fed Eur Biochem Soc) Lett 208:263–268

Friedenstein AJ 1976 Precursor cells of mechanocytes. Int Rev Cytol 47:327–355

Friedenstein AJ, Latsinik NV, Grosheva AG, Gorskaya UF 1982 Marrow microenvironment transfer by heterotopic transplantation of freshly isolated and cultured cells in porous sponges. Exp Hematol 10:217–227

Friedenstein AJ, Chailakhyan RK, Gerasimov UV 1987 Bone marrow osteogenic stem cells: in vitro cultivation and transplantation in diffusion chambers. Cell Tissue Kinet 20:263–272

Grigoriadis AE, Aubin JE, Heersche JNM 1987 Glucocorticoid induces the differentiation of progenitor cells present in a clonally-derived fetal rat calvaria cell population. In: Cohn DV et al (eds) Calcium regulation and bone metabolism: basic and clinical aspects. Excerpta Medica, Amsterdam, vol 9:646

Ham AW 1974 Histology, 7th edn. Lippincott, Philadelphia, p 577

Kodama HA, Amagai Y, Koyama H, Kasai S 1982 A new preadipose cell line derived from newborn mouse calvaria can promote the proliferation of pluripotent hemopoietic stem cells in vitro. J Cell Physiol 112:89–95

Latsinik NV, Grosheva AG, Narovlyanskii AN, Pavlenko RG, Friedenstein AJ 1987 Clonal nature of fibroblast colonies formed by bone marrow stromal cells in culture. Bull Exp Biol Med (Engl Trans Byull Eksp Biol Med) 103:356–358

Lim B, Isaguirre CA, Aye MT et al 1986 Characterization of reticulofibroblastoid colonies (CFU-RF) derived from bone marrow and long-term marrow culture monolayers. J Cell Physiol 127:45–54

Mankin HJ 1964 Osteogenesis in the subchondral bone of rabbits. J Bone Jt Surg 46A:1253–1261

Moskalewski S, Boonekam PM, Scherft JP 1983 Bone-formation by isolated calvarial osteoblasts in syngeneic and allogeneic transplants: light microscope observations. Am J Anat 167:249–263

Owen M, Ashton B 1986 Osteogenic differentiation of skeletal cell populations. In: Yousuf Ali S (ed) Cell mediated calcification and matrix vesicles. Excerpta Medica, Amsterdam, p 279–284

Patt HM, Maloney MA, Flannery ML 1982 Hematopoietic microenvironment transfer by stromal fibroblasts derived from bone marrow varying in cellularity. Exp Hematol 10:738–742

Simmons DJ, Kent GN, Jilka RL, Scott DM, Fallon M, Cohn DV 1982 Formation of bone by isolated, cultured osteoblasts in millipore diffusion chambers. Calcif Tissue Int 34:291–294

Sudo H, Kodama HA, Amagai Y, Yamamoto S, Kasai S 1983 In vitro differentiation and calcification in a new clonal osteogenic cell line derived from newborn mouse calvaria. J Cell Biol 96:191–198

Testa NG, Allen T, Molineux G, Lord BI, Onions D 1988 Haemopoietic growth factors: their relevance in osteoclast formation and function. In: Cell and molecular biology of vertebrate hard tissues. Wiley, Chichester (Ciba Found Symp 136) p 257–274

Trueta J 1963 The role of vessels in osteogenesis. J Bone Jt Surg 45B:402–418

Weiss L 1976 The hemopoietic microenvironment of the bone marrow: an ultrastructural study of the stroma in rats. Anat Rec 186:161–182

Xu CS, Hendry JH, Testa NG, Allen TD 1983 Stromal colonies from mouse bone marrow. Characterization of cell types, optimization of plating efficiency and its effect on radiosensitivity. J Cell Sci 61:453–466

Osteoblastic differentiation

P.J. Nijweide, A. van der Plas and A.A. Olthof

Laboratory of Cell Biology and Histology, Leiden University, Rijnsburgerweg 10, 2333 AA Leiden, The Netherlands

Abstract. The fully differentiated osteoblast may be easily recognized in bone tissue. Its cuboidal shape, its position directly opposed to the bone surface and its capacity to produce calcified bone matrix are characteristic. Three other differentiation stages are also reasonably well defined — the preosteoblast, the osteocyte and the lining cell. These differentiation stages are preceded by an unknown number of precursor, progenitor and stem cell stages. Little is known about the regulation of the transitions between the various osteogenic phenotypes and their reversibility or irreversibility. One of the reasons for this is the lack of adequate tools with which to recognize the various differentiation stages. We have developed a number of monoclonal antibodies (in bone) specifically directed against osteocytes, osteoblasts and as yet unidentified cells in the periosteum. The anti-osteocyte monoclonals were used to recognize osteocytes in bone cell cultures and we obtained purified osteocyte populations for metabolic studies. Osteocytes were shown to have binding sites for parathyroid hormone. The antibodies directed against osteoblasts showed that at present our culture conditions are inadequate to allow osteoblast differentiation *in vitro*.

1988 Cell and molecular biology of vertebrate hard tissues. Wiley, Chichester (Ciba Foundation Symposium 136) p 61–77

Osteoblast lineage

One of the most important cell systems in the formation, modelling and remodelling of bone tissue is the osteogenic cell lineage. Both cartilage and (osteoblastic) bone cells are considered to belong to this lineage. In many ways the two sublines of the osteogenic line are interchangeable. Early progenitor stages of the osteogenic cell line can be induced to differentiate either way (Hall 1972). Recent *in vitro* studies have shown that hypertrophic cartilage cells can transdifferentiate into osteoblasts that produce bone matrix (Thesingh & Scherft 1986). These experiments irrefutably prove the close relationship between cartilage cells and cells of the osteoblastic subline, although they do not yet prove that hypertrophic cartilage cells contribute significantly to the formation of endosteal osteoblasts and bone marrow stromal cells *in vivo*.

The principal progenitor(s) of osteogenic cells migrate(s) from the periosteum into the newly formed bone marrow cavity close behind the osteoclasts responsible for the formation of the marrow cavity. These osteoprogenitors are rather undefined, mesenchymal-like cells with an unremarkable morphology. As yet it is not possible to distinguish them in the periosteum from cells of a more fibroblastic nature. How and through which (discrete or gradual) differentiation stages the osteogenic progenitor gives rise to the formation of osteoblasts is not known, as no tools are available so far for defining those hypothetical differentiation stages (Owen 1985, Nijweide et al 1986).

Preosteoblast, osteoblast, osteocyte, and lining cell

The first recognizable differentiation stage is the preosteoblast (Scott 1967). This is defined by its position in the tissue, close to the osteoblast layer, by its morphology, with similar ultrastructural features to the osteoblast, and by its still present, although probably limited, capacity to proliferate.

The osteoblast plays the pivotal role in bone formation. It is readily recognizable by its morphology, its position directly opposed to the bone surface and its capacity to produce calcified bone matrix. It is often called an end cell, the final non-dividing end stage of the osteoblastic line, which of course it is not. There are many reports that, at least *in vitro*, osteoblasts proliferate readily. Perhaps the presence of calcified matrix has a restraining effect on the proliferative capacity of the osteoblast. Another possibility is that osteoblasts dedifferentiate under certain circumstances (e.g. bone matrix loss) into preosteoblasts.

The real end cell of the osteoblastic differentiation line is the osteocyte. During bone formation a number of osteoblasts become incorporated in the organic bone matrix. Subsequently the bone matrix around the osteocytes is calcified. Especially in the chicken and quail (as compared to mouse and rat) the distance between the osteoid formation and the calcification front is quite large (Nijweide et al 1981). The osteocytes in this osteoid layer share many ultrastructural features with osteoblasts but are sufficiently differentiated to display the osteocyte-specific cOB 7.3 antigen (Nijweide & Mulder 1986). Deeper into the bone, the osteocytes become smaller and lose many of their cytoplasmic organelles. The young, immature osteocytes have been called osteocytic osteoblasts (Nijweide et al 1981) or osteoid osteocytes (Palumbo 1986).

Finally, the fourth recognizable phenotype is the lining cell. Part of the bone surface, especially in the adult animal, is lined not with osteoblasts but with very flattened cells possessing few organelles. They do not take part in bone formation but can be activated to differentiate into active osteoblasts (Miller & Bowman 1981). It is not yet clear whether these lining cells originate directly from the preosteoblast or represent an osteoblast differentiation step different from the osteoblast–osteocyte differentiation. In the male quail they have been shown to incorporate [³H]thymidine on oestrogen stimulation *in vivo*, demonstrating that they pass through a proliferative stage before becoming active bone-forming osteoblasts (Bowman & Miller 1986).

Although these four osteoblastic differentiation stages can be recognized in the tissue, controlled manipulation of the differentiation is difficult in organ culture studies. Therefore cell isolation and culture procedures were designed to study the hormonal regulation of the proliferation, differentiation and function of osteoblastic cells.

Isolation of osteoblast-like cells and differentiation in vitro

Since the report by Peck et al (1964), cultures of osteoblast-like cells have been widely used. Schemes for the isolation of rat, mouse, chicken and human bone cells have been proposed, improved and again changed. Also, (tumour) cell lines that have several characteristics of the osteoblast have been developed (Rodan & Rodan 1984). In his paper (Rodan et al, this volume) Dr Rodan discusses various systems which show many variations in morphology, functional properties and hormonal sensitivities.

One of the central questions in the study of osteoblastic differentiation is: does the osteoblastic cell lineage have a number of discrete differentiation stages, such as preosteoblast, osteoblast and lining cell, perhaps even including the chondroblast and hypertrophic chondrocyte, or are these apparently different phenotypic expressions simply modulations of the same differentiation stage brought about by changing extracellular circumstances? In the latter case, the variation in the results of different groups working with isolated osteoblastic cells would be a reflection of the differences in isolation procedures and culture conditions.

One problem in solving this question is that not enough tools are available with which to assess the differentiation stage of each individual, preferably living, cell in culture. Most of the available methods are used to study populations of living cells or dead, fixed cells, and are not suitable for single living cells. We therefore undertook the production of monoclonal antibodies directed against cells of the osteogenic type.

Monoclonal antibodies directed against osteogenic cells

This is a preliminary report on the production of monoclonal antibodies using osteogenic cell populations from the chick and the rat as immunogens. Chicken osteoblast-like cells (chicken OB) and periosteal fibroblasts (PF) were isolated and cultured as reported earlier (Nijweide et al 1981). The periostea on both sides of calvariae of fetal 18-day-old chickens were removed and the periosteal cells (PF) were liberated from their collagen matrix by collagenase treatment. The cells still present on the calvaria (OB) were also isolated with collagenase. Both cell populations were cultured in αMEM containing 10% fresh chicken serum.

Rat osteoblast-like cells (rOB) were isolated as described by Boonekamp et al (1984) and modified by M.P.M. Herrmann-Erlee & J.M. van der Meer (unpublished work 1987). Calvariae of 18 day-old fetal rats were dissected and sequentially treated with 4 mM EDTA in phosphate-buffered salt solution (twice, 15 min) and collagenase (2 mg/ml) dissolved in the same buffer solution (once for 10 min, twice for 30 min). The two 30 min cell isolates were combined and the other fractions discarded. The cells were cultured in αMEM containing 2% fetal calf serum (FCS).

The *in vivo* immunization and monoclonal antibody production procedure (Köhler & Milstein 1975) has been described (Nijweide et al 1985, Nijweide & Mulder 1986). Recently we have used an *in vitro* immunization method in which mouse spleen cells are added directly to cell cultures according to the method of Ossendorp et al (1986) modified by C.G. Groot (unpublished work 1987).

After a large number of immunizations, we have now prepared a small number of promising monoclonal antibodies. All antibodies were characterized on frozen sections of bone and other tissues using rabbit anti-mouse immunoglobulins IgG and IgM, conjugated to fluorescein isothiocyanate (FITC).

Using either freshly isolated (cOB 23.2) or cultured 4–7 days (cOB 7.3, cOB 11.4) chicken osteoblast-like cells and freshly isolated periosteal cells (cOB 37.4.4, cOB 37.5.3, cOB 37.11.4, cOB 38.5.3) as antigens, we obtained the following monoclonal antibodies:

cOB 7.3. A monoclonal that stains osteocytes in bone tissue (Nijweide & Mulder 1986). Although the monoclonal appears to be a specific anti-osteocyte antibody in bone (Fig. 1a, b), some reaction was also found in other tissues, notably kidney and proventriculus. Rat tissues are negative. The antigen is present on the outside of the living osteocyte, as could be shown by staining unfixed OB cultures (Fig. 1c, d);

cOB 37.4.4 and cOB 37.11.4. Both these monoclonals closely resemble OB 7.3. We do not yet know whether the antigens differ from that of OB 7.3;

cOB 11.4. This stains cartilage and a narrow layer of fibroblast-like cells in the periosteum of long bones (Fig. 2a, b). The antigen is the only one we possess so far that is resistant (to some extent) to fixation procedures using

FIG. 1.   A,B: Frozen sections of trabecular bone of the metaphysis of 18-day-old fetal chicken phalange obtained with cOB 7.3 followed by rabbit anti-mouse IgG-FITC. Osteocytes including cytoplasmic extrusions in the canaliculi are stained. The periosteum and osteoblast zone in the upper left corner (A) are negative. Magnification: A, × 350; B, × 700. C,D: 4-day-old culture of chicken OB cells. C, phase contrast; D, cOB 7.3 and rabbit anti-mouse IgG-FITC. Note the presence of fluorescent and non-fluorescent cells. Magnification × 350.

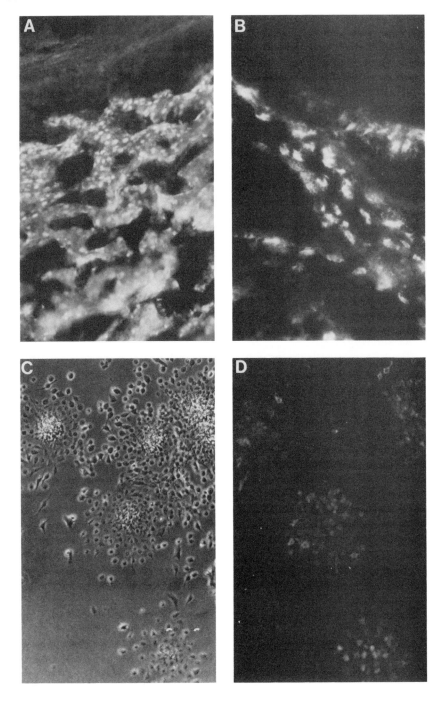

96% alcohol, 4% formaldehyde and 2.5% glutaraldehyde. It is probably an extracellular, matrix antigen. Cross-reactivity occurs in the small intestine, spleen and proventriculus. The antibody stains similar structures in rat long bone;

cOB 23.2. This monoclonal is specific for the osteoblast layer in chicken and quail bone (Fig. 2c). The antibody does not cross-react with rat bone or with chicken organs other than bone (small intestine, proventriculus, spleen, liver, muscle, kidney and pancreas);

cOB 37.5.3. and cOB 38.3.1. These stain the cell layer in the periosteum of long bones also stained by OB 11.4, but are negative for cartilage. The perichondrium and parts of the connective tissue are also stained (Fig. 3a, b). The monoclonals are not specific for bone but also stain structures in organs such as the small intestine and proventriculus. Rat bone tissue is negative.

Recently we started to use rat osteoblast-like cells as immunogen to produce monoclonal antibodies. The rat OB were used either directly after isolation (rOB 42.2) or after 4–7 days of culture (rOB 46.4.1). The third monoclonal was obtained by *in vitro* immunization (rOB 49.1).

rOB 42.2 stains specifically the osteoblast layer in fetal rat bones (Fig. 3c).

rOB 46.4.1 stains a narrow band of material lining the bone collar and the trabeculae in fetal rat bone. The osteocyte lacunae are also positive for this monoclonal (Fig. 3d);

rOB 49.1 reacts with the osteoblast layer in fetal rat bone.

Cross-reactivity tests either with other rat tissues or bone tissue of other species have not yet been performed.

Some of the monoclonals were tested on non-confluent or near-confluent cultures of OB cells of the matching species. The antibody cOB 11.4 stained very fine threads of extracellular matrix between the cells, which indicates that this antibody stains a matrix antigen. OB cell cultures from the rat were negative for rOB 42.2, rOB 46.4.1 and rOB 49.1.

A cOB 23.2-positive reaction could be seen in chicken OB cultures. However, on closer examination groups of osteocytes present in the culture (see later) stained weakly with the monoclonal. All other cells were negative. Freshly isolated OB cells were used to generate cOB 23.2 and rOB 42.2. The fact that isolated and cultured chicken and rat osteoblasts do not react with these antibodies strengthens the possibility that osteoblasts may lose some of their specific markers during culture. Monoclonal rOB 49.1 was raised using cultured rat OB cells. In this case the antibody was raised *in vitro*, so rOB 49.1 may be directed against an OB cell product which is released into the medium in cell culture conditions but is retained in the osteoblast layer *in vivo*.

We have no explanation, as yet, for the lack of reaction of rOB 46.4.1 with its antigen, cultured rat OB. Clearly, there are processes involved in the antibody induction and antibody–antigen reaction that we do not fully understand. It is possible that the amount of antigenic marker per unit of cell

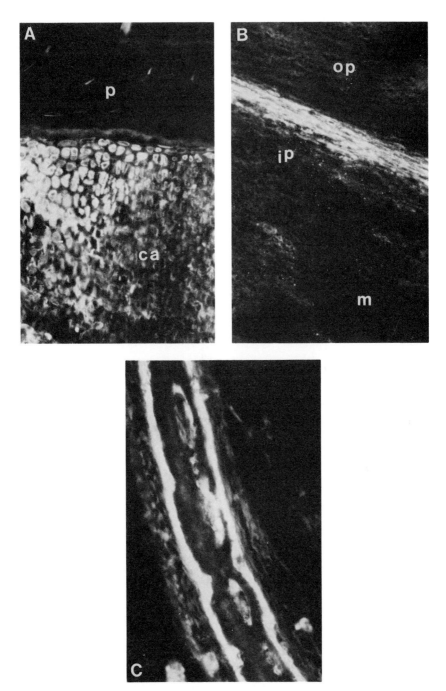

FIG. 2.   A,B,C: Frozen sections of phalange (A,B) or calvaria (C) of 18-day-old fetal chicken stained with cOB 11.4 (A,B) or cOB 23.2 (C) followed by FITC-conjugated second antibody. ca, cartilage; p, periosteum; op, outer periosteum; ip, inner periosteum; m, marrow cavity. Magnification × 350.

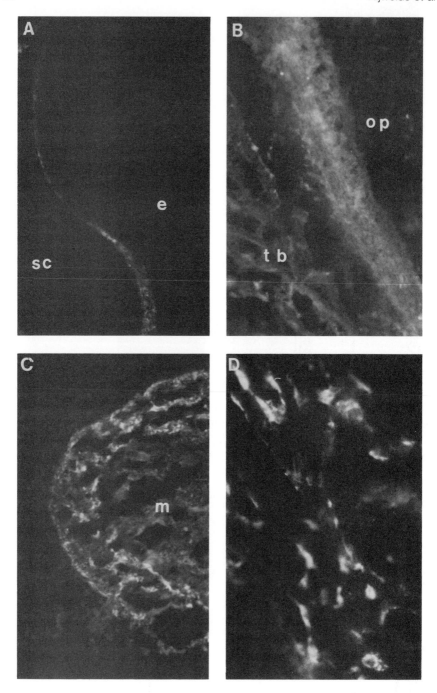

surface in the thinly spread out cells in culture is too low to be visualized but high enough to stimulate antibody formation.

Alternatively, antigens present on specific cells may be masked or otherwise unreactive when the cells are in culture. Perhaps they become unmasked or reactive when the cultured cells are detached from the culture dishes by a short collagenase treatment and are used to immunize mice.

The osteocyte in culture

Monoclonal cOB 7.3 does react with a number of cells in chicken OB cultures (Fig. 1c, d). When we applied the antibody and obtained a positive reaction we realized, for the first time, that chicken osteoblast-like cultures contain a reasonable number of osteocytes. The osteocytes retain their specific morphology in culture. Although their surrounding matrix is removed by the collagenase treatment during the isolation procedure, the osteocytes retain in culture the many cytoplasmic extrusions through which they retain contact with each other (compare Fig. 1a, b with Fig. 1c, d).

During the isolation procedure osteocytes tend to stick together in clumps. This enabled us to purify osteocytes by sieving the isolated cell populations, thereby separating the single cells from the osteocyte clusters. Osteocytes do not appear to proliferate in culture (Nijweide & Mulder 1986) and may therefore be kept under culture conditions which do not promote cell proliferation of contaminating non-osteocytes (low concentrations of serum).

Using partially purified osteocyte populations we have recently been able to study parathyroid hormone (PTH) binding to osteocytes. The cells present in the cultures can be divided into three categories according to their cell shape: firstly, osteocytes, spider-like cells with many cytoplasmic extrusions; secondly, more or less square, osteoblast-like cells; thirdly, elongated, fibroblast-like cells. Cultures containing the three cell types were incubated with HPLC-purified [^{8}Nle, ^{18}Nle, ^{34}Tyr] synthetic bovine PTH(1–34) in the presence or absence of an excess of cold PTH(1–34).

The results shown in Table 1 demonstrate that osteocytes have specific binding sites for PTH, although fewer per cell than the osteoblast-like cells. The elongated, fibroblast-like cells, which probably represent an immature population of progenitor-like cells, because of their high proliferative activity, appear not to be able to bind PTH specifically.

FIG. 3.   A,B: Frozen sections of phalange of 18-day-old fetal chicken stained with cOB 37.5.3. The perichondrium around the epiphyseal head and a layer of cells within the periosteum are positive. sc, synovial cavity; e, epiphysis; tb, trabecular bone; op, outer periosteum. Magnification: A, × 175; B, × 700. C,D: Frozen cross sections of metatarsus of 18-day-old rat embryo stained with rOB 42.2 (C) and rOB 46.4.1 (D) followed by a FITC conjugated second antibody; rOB 42.2 stains the surfaces of the trabeculae, rOB 46.4.1 occasionally stains the bone surface and stains the osteocyte lacunae. m; marrow cavity. Magnification: C, × 175; D, × 700.

Discussion and conclusions

Unlike the haemopoietic lineage, little is known, as yet, about the osteogenic lineage. One of the complicating factors in this field is the apparent instability and reversibility of the various phenotypes of the osteogenic lineage *in vivo* and *in vitro*. Secondly, although many researchers, including ourselves (Nijweide et al 1981, 1982), have tried to isolate populations of more or less homogeneous composition, the populations are still a mixture of different cell types (e.g. Burger et al 1986) or differentiation stages (e.g. this study). Thirdly, we do not have enough reliable methods to establish the differentiation stage of an individual cell. The use of monoclonal antibodies directed against osteogenic cells, preferably against markers on the cell surface, is one of the methods of choice in the near future.

We have reported here on the production of a number of monoclonal antibodies that recognize specific cell types in chicken and rat bone, notably the osteoblast and the osteocyte. The negative staining results in osteoblast-like cell cultures of chicken and rat origin that we obtained with three osteoblast-specific antibodies (cOB 23.2, rOB 42.2, rOB 49.1) may illustrate the instability and variability of the osteoblastic phenotype *in vitro*. The results also show that the culture conditions we have been using are probably incapable of maintaining the osteoblastic phenotype or stimulating differentiation towards that phenotype *in vitro*. We must now study variables such as cell density, the culture medium, and the presence of hormones or cytokines, to find conditions that promote osteoblast differentiation. The

TABLE 1 I^{125} [^{8}Nle, ^{18}Nle, ^{34}Tyr] bTPH (1–34) binding to chicken OB cells, studied with autoradiography

Cell type	Cold PTH (1–34) (M)	Number of grains per cell or corresponding cell-free area		
		Cells (± SE)	Background (±SE)	n
Osteocytes	—	15.5 ± 7.2	5.7 ± 3.3	33
	2.10^{-7}	3.8 ± 2.6	5.4 ± 2.2	
Osteoblasts	—	32.2 ± 12.7	6.5 ± 2.8	32
	2.10^{-7}	5.3 ± 2.4	6.8 ± 2.6	
Fibroblast-like	—	8.2 ± 2.8	6.1 ± 3.2	14
	2.10^{-7}	3.5 ± 2.5	3.7 ± 2.3	

Cells present in four OB cell cultures were divided into three categories according to their cell shape: osteocytes, spider-like cells with many cytoplasmic extrusions; osteoblasts, more or less square cells; and fibroblast-like cells, elongated, spindle-shaped cells. The number of grains on each cell and on a cell-free area close to that cell (background) was counted. n, number of cells or cell-free areas used for counting.

osteocyte, on the other hand, retains its typical morphology and its specific cell surface marker (the cOB 7.3 antigen) in culture.

The monoclonal antibodies we have obtained promise to be of great help in the study of the exact conditions in which a specific osteogenic phenotype can be stabilized. Then the metabolic properties and hormonal regulatory mechanisms of defined differentiation stages can be assessed. For the osteocyte, this strategy is already operational, as we are able to isolate osteocytes in reasonable purity. The osteocytes have been shown to have parathyroid hormone-binding sites, although fewer per cell than osteoblasts. This is in contrast with the results of the study of Silve et al (1982). In their study, poor penetration of the iodinated hormone may have prevented an interaction with osteocytes. Another possible explanation is that they have considered the osteoid–osteocytes to be osteoblasts. The difference in PTH-binding between the osteoblast and osteoid–osteocyte in our study (Table 1) suggests a decrease in PTH binding concomitant with osteoblast–osteocyte differentiation. Thus calcified matrix osteocytes may have lost their receptors.

References

Boonekamp PM, Hekkelman JW, Hamilton JW, Cohn DV, Jilka RL 1984 Effect of culture on the hormone responsiveness of bone cells isolated by an improved sequential digestion procedure. Proc K Ned Akad Wet Ser B 87:371–381

Bowman BM, Miller SC 1986 The proliferation and differentiation of the bone lining cell in estrogen-induced osteogenesis. Bone (NY) 7:351–357

Burger EH, Boonekamp PM, Nijweide PJ 1986 Osteoblast and osteoclast precursors in primary cultures of calvarial bone cells. Anat Rec 214:32–40

Hall BK 1972 Differentiation of cartilage and bone from common germinal cells. Calcif Tissue Res 8:276–286

Köhler G, Milstein G 1975 Continuous cultures of fused cells secreting antibody of predefined specificity. Nature (Lond) 256:495–497

Miller SC, Bowman BM 1981 Medullary bone osteogenesis following estrogen administration to mature male japanese quail. Dev Biol 87:52–63

Nijweide PJ, Mulder RJP 1986 Identification of osteocytes in osteoblast-like cell cultures using a monoclonal antibody specifically directed against osteocytes. Histochemistry 84:342–347

Nijweide PJ, van der Plas A, Scherft JP 1981 Biochemical and histological studies on various bone cell preparations. Calcif Tissue Int 33:529–540

Nijweide PJ, van Iperen-ven Gent AS, Kawilarang-de Haas EWM, van der Plas A, Wassenaar AM 1982 Bone formation and calcification by isolated osteoblast-like cells. J Cell Biol 93:318–323

Nijweide PJ, Vrijheid-Lammers T, Mulder RJP, Blok J 1985 Cell surface antigens on osteoclasts and related cells in the quail studied with monoclonal antibodies. Histochemistry 83:315–324

Nijweide PJ, Burger EH, Feyen JHM 1986 Cells of bone: proliferation, differentiation and hormonal regulation. Physiol Rev 66:855–885

Ossendorp FA, de Boer M, Al BJM, Hilgers J, Bruning PF, Tager JM 1986 Production of murine monoclonal antibodies against thyroglobulin using an in vitro immunization procedure in serum-free medium. J Immunol Methods 91:257–264

Owen M 1985 Lineage of osteogenic cells and their relationship to the stromal system. In: Peck WA (ed) Bone and mineral research. Elsevier Science Publishers, Amsterdam, vol 3:1–23

Palumbo C 1986 A three dimensional ultrastructural study of osteoid-osteocytes in the tibia of chick embryos. Cell Tissue Res 246:125–131

Peck WA, Birge SJ, Fedak SA 1964 Bone cells: biochemical and biological studies after enzymatic isolation. Science (Wash DC) 146:1476–1477

Rodan GA, Rodan SB 1984 Expression of the osteoblastic phenotype. In: Peck WA (ed) Bone and mineral research. Elsevier Science Publishers, Amsterdam, vol 2: 244–285

Rodan GA, Heath JK, Yoon K, Noda M, Rodan SB 1988 Diversity of the osteoblastic phenotype. In: Cell and molecular biology of vertebrate hard tissues. Wiley, Chichester (Ciba Found Symp 136) p 78–91

Scott BL 1967 Thymidine ^{3}H electron microscope radio autography of osteogenic cells in the fetal rat. J Cell Biol 35:115–126

Silve CM, Hradek GT, Jones AL, Arnaud CD 1982 Parathyroid hormone receptor in intact embryonic chicken bone: characterization and cellular localization. J Cell Biol 94:379–386

Thesingh CW, Scherft JP 1986 Bone matrix formation by transformed chondrocytes in organ cultures of stripped embryonic metatarsalia. In: Yousouf Ali S (ed) Cell mediated calcification and matrix vesicles. Elsevier Science Publishers, Amsterdam, p 309–314

DISCUSSION

Osdoby: You observe differences in monoclonal antibody staining between frozen sections and cultured cells. When you stain cultured cells do you permeabilize the cells? With frozen sections you are, in a sense, looking at the entire cell spectrum of proteins, whereas in the cultures, if you are not permeabilizing the cells, all you are attacking is the cell surface and you may not be seeing proteins or antigens that are internal.

Nijweide: We either use the cells when they are still alive, because we would like to recognize living cells and use monoclonals to separate cells, or we dry the cultures before we stain them, and that permeabilizes them. We know they are permeabilized because when we use antibodies against the cytoskeleton, it stains beautifully. Therefore, that is not the reason for the differences we observe.

Caplan: We have been investigating some of these questions. Is it possible that osteocytes differentiate directly from mesenchymal precursor cells without necessarily going through the osteoblast state?

Nijweide: I am not aware of any data to show that. It is an interesting question. However, from the histology of bone, the pathway from osteoblasts to osteocytes seems more reasonable than the one you suggest. When we isolate and culture cells from the periosteum, we never get any calcification and the cells do not become positive for the antibody against osteocytes. When osteoblasts are isolated from chick calvariae and cultured, the cells can calcify matrix *in vitro* and osteocytes are formed.

Caplan: The data are clear that periosteum from young animals has chondrogenic potential which, under appropriate conditions, gives rise to an expressed chondrocyte, implying the presence of stem cells. However, your antibody staining indicates that there is a committed chondrogenic cell or cell layer in the periosteum which might have the same properties as a stem cell. Your antibody cOB 11.4, which stains both cartilage and the special layer, may indicate that the stacked cell layer is not a uniform population of stem cells, but may itself have chondrogenic potential.

Nijweide: We are almost sure that this antibody stains a matrix element. This means that these stacked cells, if they are stacked cells, must make the same type of matrix as the chondrocytes.

Fleisch: We have similar problems with antibodies. Dr Wetterwald found that some of her monoclonal antibodies react in the whole calvaria but not on cultured bone cells. The cause of this discrepancy is unclear. Dr Caplan has suggested that the osteocytes would be predetermined. It is a fascinating idea. Is it right that only 10% of the osteoblasts present will become osteocytes? Could these be predetermined? If so, osteoblasts on a bone surface would be heterogeneous.

Nijweide: There are many theories about that: it may be a random affair, or a special type of osteoblast may turn into an osteocyte. Nobody has solved that question yet.

Butler: How do you distinguish morphologically between a preosteoblast and an osteoblast? What is your definition of a preosteoblast?

Nijweide: Preosteoblasts were described by Scott (1967). Until that time it was thought that the osteoclast and osteoblast had the same progenitor cell. Using electron microscopy, Dr Scott saw two different types of proliferating cells, a and b cells. One was osteoblast-like and the other osteoclast-like. The preosteoblast is described as an osteoblastic cell that has some of the features of the adult osteoblast but is still able to proliferate, whereas the osteoblast is thought not to proliferate anymore.

Raisz: A possible reason why certain monoclonal antibodies react to osteocytes but do not react to their presumed parent, the osteoblast, is that in the osteoblast those antigenic sites are covered. What does the production of plasminogen activator and collagenase by activated osteoblasts achieve? Is something being uncovered? Have you used crude enzyme, purified collagenase, or parathyroid hormone treatment to study whether osteoblasts become reactive to the antibody?

Nijweide: No, we haven't.

Raisz: In cell cultures that would be a generalized explanation. In an organ culture it would explain why the osteoblast, which might be expected to have the same surface antigens as the osteocytes, does not show them.

Nijweide: In cultures the morphology of the osteocyte is quite different from that of the osteoblast. Why shouldn't they have different antigens?

Raisz: With the osteocyte you are losing functions rather than gaining them.

Nijweide: Do you know the function of the osteocyte?

Raisz: No.

Nijweide: Neither do I!

Martin: In your controls for the immunohistochemical studies, for example with cOB 7.3 that stains osteocytes, can you absorb the antibodies out? If so, do you need to use a homogenate of cells or can you absorb them out by incubating them with intact cells growing on the culture dish surface?

Nijweide: We have never tried to remove antibody from the ascites by absorption.

Baron: What is the evidence that any of your antigens are surface antigens?

Nijweide: We stain both living and dead cultures. For example, the anti-osteocyte antibody stains living cultures. We do the staining in the cold.

Baron: I am thinking here of your screening procedure on frozen sections. Perhaps in many cases you are staining an extracellular protein?

Nijweide: Yes; we are more or less sure, for example, that antibody cOB 11.4 reacts with an extracellular protein.

Krane: Could the matrix component stained by antibody cOB 11.4 be type V collagen or type XI collagen? Type XI is a cartilage protein and type V is a non-cartilage protein. They are very similar in that the 1α and 2α chains of the type XI collagen resemble type V collagen and they might share epitopes (Morris & Bächinger 1987). Tom Linsenmayer and his colleagues showed that you can only detect staining with type V antibodies if you remove type I collagen using purified collagenase which doesn't affect type V (Fitch et al 1984, Linsenmayer 1987). One problem with monoclonals is that you may be seeing common epitopes on different proteins.

Nijweide: We have not tested this.

Baron: Could your osteoblast-specific antibody simply be specific for alkaline phosphatase?

Nijweide: There's a lot of alkaline phosphatase activity on the osteoblasts, especially directly after isolation, yet they do not stain with the antibody raised against osteoblast-like cells. But we would like to have an antibody against alkaline phosphatase and to investigate whether that antibody competes with our osteoblast-specific antibodies.

Osdoby: Is there any relationship between what is stained by your antibodies in bone tissue and what happens in other tissues?

Nijweide: This is difficult to determine. We have to rely on frozen sections because almost all the antigens with which the monoclonals react cannot withstand any kind of fixation apart from acetone in the cold. The quality of these sections is often poor, which is why it is often difficult to say whether the antibody reacts with an intracellular or extracellular molecule. We know that several monoclonal antibodies stain structures in different tissues. Perhaps specialists in each of the tissues can say what is stained, but I haven't looked into this.

Caplan: It is very difficult to get osteogenic-specific monoclonal antibodies. It depends on the screening procedure. We attempted to obtain monoclonals to membranes associated with osteogenic cells. We found that these monoclonals are unusually unstable whereas monoclonals in the same fusion to cartilage matrix were stable.

Canalis: I am concerned that we refer to the antibodies as cell surface antibodies when there is still insufficient evidence that these are antibodies to cell surface antigens with any of the bone cells.

Nijweide: The antigen for antibody cOB7.3 is on the surface because living cells in culture stained with this antibody. I do not know for sure about the other antibodies because, in most cases, isolated, cultured cells don't stain with them.

Butler: One way to find out more about the differentiation process might be by detecting known components, such as structural or matrix proteins. For example, the 44K bone phosphoprotein (osteopontin) appears to be a marker for early stages of osteoblast differentiation, presumably preosteoblasts. One might be able to use polyclonal antibodies against these other matrix components.

Nijweide: It was very difficult to get antibodies specific for bone cells. However, we did not have the same problem as Dr Caplan. We don't find that antibodies against osteogenic cells are especially unstable. We are now trying to study the biochemistry of the antigens, to discover what they are. Having polyclonal antibodies against different (matrix) proteins would be helpful because you can combine them with monoclonal antibodies.

Termine: A perplexing problem with monoclonal antibody technology is that it is not always possible to identify antigens by the Western blot technique. It has often happened that useful monoclonal antibodies have been produced whose antigen can never be identified at the molecular level.

Urist: Dr Nijweide, you removed periosteal cells and then cultured cartilage models, and you made this beautiful demonstration of the cells becoming more like osteoblasts. I wondered if what was produced might be called 'chondroid' and whether this preparation was equivalent to what was observed in the deer's antler by Wislocki et al (1947) and in other tissues in a large body of literature reviewed by Beresford (1981).

Nijweide: When I gave this example I wanted to stress the reversibility of the osteogenic cell types. The cells in the hypertrophic area are like osteoblasts; they have a similar morphology and they stain basophilically with haematoxylin. They form a matrix which appears to be very much like osteoid under an electron microscope. In these experiments the matrix didn't calcify, as far as I know.

Urist: Other people have made electron micrographs of the epiphyseal plate, and showed, less convincingly, that not all the cartilage cells in the calcified cartilage were dying, but that some survive. You have demonstrated cell survival of an entire cartilage model.

Nijweide: This is Dr Thesingh's work (Thesingh & Scherft 1986). She has shown that *in vitro* these cells have the potential to stay alive, become osteoblast-like and form bone matrix. We don't know whether *in vivo* the liberated chondrocytes play a role in osteogenesis or in the formation of stromal cells.

Urist: To answer this you could digest the matrix of these cells with collagenase and trypsin, continue cultivation in tissue culture, and then transfer the cultures to a diffusion chamber. If the cells are preosteoblasts, you should find everything that Dr Owen observes with bone marrow in diffusion chambers (Owen & Friedenstein, this volume), *in vivo*. Their progeny should develop into osteoblasts and osteocytes and secrete calcifiable bone matrix. The full potential for bone development is difficult to support under *in vitro* conditions. Something is missing even in the best culture media, particularly when the system is required first to initiate differentiation and subsequently to provide conditions to sustain bone tissue development. Postdifferentiated bone cells including vascular cell remnants will survive and even develop in 'super-supplemented' nutrient culture media. Predifferentiated cells develop only as far as cartilage even in the best media under avascular conditions *in vitro*. This is not to depreciate claims for *de novo* bone production from marrow stroma colony-forming cells (CFC-0) which are differentiated with respect to osteogenetic potential and therefore can develop into a bone-like tissue. One point to emphasize is that predifferentiated cells with a requirement for bone morphogenetic protein (BMP) produce cartilage *in vitro* and cartilage and bone *in vivo*. Another point is that postdifferentiated CFC-0 (like embryonic bone cells) have a lesser requirement for survival, with some growth without BMP *in vitro*.

Hanaoka: In dedifferentiated chondrosarcoma, chondrocytes dedifferentiate into less differentiated cells first, and then differentiate into osteoblasts which form bone. However, in the normal limb bud the cartilage calcifies but does not ossify, as Dr Urist pointed out.

References

Beresford WA 1981 Chondroid, secondary cartilage, and metaplasia. Urban & Schwarzenberg, Baltimore/Munich

Fitch JM, Gross J, Mayne R, Johnson-Wintane B, Linsenmayer TF 1984 Organization of collagen type-1 and type-V in the embryonic chicken cornea: monoclonal-antibody studies. Proc Natl Acad Sci USA 81:2791–2795

Linsenmayer TF 1987 Adv Meat Res 4:39–54

Morris NP, Bächinger HP 1987 Type XI collagen is a heterotrimer with the composition (1α, 2α, 3α) retaining non-triple-helical domains. J Biol Chem 262:11345–11350

Owen M, Friedenstein AJ 1988 Stromal stem cells: marrow-derived osteogenic precursors. In: Cell and molecular biology of vertebrate hard tissues. Wiley, Chichester (Ciba Found Symp 136) p 42–60

Scott BL 1967 Thymidine-^{3}H electron microscope radioautography of osteogenic cells in the fetal rat. J Cell Biol 35:115–126

Thesingh CW, Scherft JP 1986 Bone matrix formation by transformed chondrocytes in organ cultures of stripped embryonic metatarsalia. In: Yousuf Ali S (ed) Cell mediated calcification and matrix vesicles. Excerpta Medica, Amsterdam, p 309–314

Wislocki GB, Weatherford HL, Singer M 1947 Osteogenesis of antlers investigated by historical and histochemical methods. Anat Rec 99:265–295

Diversity of the osteoblastic phenotype

Gideon A. Rodan, Joan K. Heath, Kyonggeun Yoon, Masaki Noda and Sevgi B. Rodan

Department of Bone Biology and Osteoporosis Research, Merck Sharp and Dohme Research Laboratories, West Point, Pennsylvania 19486, USA

Abstract. Studies of bone cells in culture have raised two salient questions: are the findings representative of the *in vivo* situation and can the conflicting data from different cell models be reconciled? Review of the literature indicates that all osteoblastic cells, defined by their origin or by their ability to produce mineralized matrix, have a few common properties: production of type I collagen; increased alkaline phosphatase activity; and parathyroid hormone-stimulated adenylate cyclase. Other features, such as osteocalcin and prostaglandin E production and the response to prostaglandin E, are selectively expressed by certain cell types. Pilot studies on mRNA levels of 'bone proteins' in developing calvaria suggest that such differences may reflect stages in osteoblastic differentiation. Immortalization of calvaria-derived cells using a SV40 large T antigen vector, which may freeze the cells in their particular state of differentiation (as proposed for leukaemia cells), yields phenotypes consistent with that hypothesis. Immortal cell lines may thus help to characterize osteoblastic differentiation. The diversity of osteoblast responses in culture to hormones and growth factors could be due to these phenotype differences but could also represent a subspecialization of differentiated cells. In addition, in the organism regulatory agents act in concert on a heterogeneous interactive cell population. Nonetheless cell cultures can be useful in screening for and predicting *in vivo* responses, as was shown by the $1,25\text{-}(OH)_2D_3$ stimulation of osteocalcin, and for studying the molecular mechanisms of regulatory effects. Cell lines are also convenient for the production of specific proteins and cDNA libraries, and for the expression of specific genes.

1988 Cell and molecular biology of vertebrate hard tissues. Wiley, Chichester (Ciba Foundation Symposium 136) p 78–91

Much research is carried out on isolated cells and cell lines. For the last 20 years, studies of bone-derived cells (Peck et al 1964, Peck et al 1977, Wong & Cohn 1974) have generated substantial information but have raised several pertinent questions. How can one reconcile contradictory findings obtained when the same variables are manipulated in different but analogous systems? Which, if any, of the findings can be extrapolated to the *in vivo* situation? If such extrapolation is impossible, what can one learn from studies of cells in culture? The answer to these questions, in a nutshell, is that the complexity of biological systems requires a constant dialogue between whole organism studies and reductionist approaches focused on cells and molecules. For osteoblasts, for example, *in situ* histochemical studies showed increased alkaline phosphatase activity in cells which, from their histological location and developmental changes, were considered to be bone-forming cells. When these cells were isolated from bone tissue and further separated to produce more homogeneous populations (Wong & Cohn 1975) it was found that the alkaline phosphatase-rich cells produced only type I collagen, which was consistent with the predominant presence of type I collagen in bone. In addition, it was unexpectedly found in culture that the alkaline phosphatase-rich cells responded to parathyroid hormone (PTH) by an increase in cyclic AMP accumulation (reviewed in Rodan & Rodan 1984). This prompted various questions and hypotheses regarding the mechanism by which PTH stimulates bone resorption and could help us investigate the stimulation of bone formation by PTH (Tam et al 1982).

The triad of properties mentioned above — increased alkaline phosphatase activity, synthesis of type I collagen and PTH-stimulated adenylate cyclase — was found to be present in bone-derived cells from many species, including man and chick, and in clonal cell lines derived from osteosarcomas (Rodan & Rodan 1984). The demonstration of these properties *in situ* and the fact that the isolated cells were capable of making mineralized matrix *in vitro* or *in vivo*, in tumours or diffusion chambers (Shteyer et al 1986), helped to establish those features as characteristic of the osteoblastic phenotype.

Cell culture studies and their relevance to bone function in vivo

The relative ease of conducting cell culture experiments and the availability of cell lines provide the opportunity for detailed studies and testing of hypotheses arising from observations made in the organism. The validity of the *in vitro* findings requires subsequent testing *in vivo*. Osteocalcin or bone γ-carboxyglutamic acid-containing protein (BGP) provides a typical example of the *in vivo/in vitro* dialogue. Osteocalcin was discovered in bone extracts as a major non-collagenous matrix protein. Its synthesis and the precursor peptide were studied in ROS 17/2 cells (a line of osteoblastic cells), where it was also observed that its synthesis is stimulated by 1,25-dihydroxyvitamin D_3 (Price & Baukol 1980). This mode of regulation was confirmed *in vivo* and led to the development of clinical assays for estimating bone turnover rates in patients (Gundberg et al 1983). Subsequently, a cDNA for osteocalcin was obtained from the ROS 17/2 cell line and studies of osteocalcin gene expression are in progress (Pan & Price 1985).

Cell culture is particularly useful for mechanistic studies because many variables can be experimentally controlled. As an example, consider the effects of transforming growth factor β (TGF-β) on ROS 17/2.8 cells. TGF-β was isolated from bone (Seyedin et al 1985) during the purification of a chondrogenic activity and was shown by Robey et al (1987) to be produced by osteoblastic cells. Centrella et al (1987) showed that it has biphasic effects on the growth of calvarial osteoblasts and that it promotes collagen synthesis in calvaria (Centrella et al 1986). TGF-β was also reported to stimulate alkaline phosphatase in ROS 17/2.8 cells (Pfeilshifter et al 1987). Taken together, these findings suggest that TGF-β may promote bone formation. To gain additional insight into its mode of action we examined further the effect of TGF-β on alkaline phosphatase in ROS 17/2.8 cells. We found that TGF-β increased the abundance of alkaline phosphatase mRNA in a dose-dependent manner by a process which was actinomycin D sensitive, and without affecting alkaline phosphatase mRNA half-life. This suggested transcriptional control. The TGF-β effect on mRNA was not dependent on *de novo* protein synthesis. In addition, TGF-β increased the abundance of procollagen mRNA and osteonectin mRNA in these cells, but decreased the abundance of fibronectin mRNA, contrary to its effect on fibroblasts. These findings are consistent with the increased expression of osteoblastic features and should be confirmed *in vivo*. The findings also suggest specific patterns of gene regulation, which can be verified *in vivo* and investigated in detail in cell cultures.

Like osteocalcin, osteopontin (bone sialoprotein I or bone phosphoprotein) was first purified from bone matrix and was subsequently shown to be synthesized by ROS 17/2 cells (Prince et al 1987). Its cDNA was cloned from a library prepared from these cells (Oldberg et al 1986). Using an osteopontin cDNA isolated in our laboratory on the basis of the published sequence, we found that its mRNA was increased several fold by $1,25\text{-}(OH)_2D_3$ treatment, which is consistent with the observations by Prince & Butler (1987) at the protein level. The production of osteopontin mRNA was suppressed by glucocorticoid treatment, suggesting hormonal responses similar to those of osteocalcin. Osteopontin mRNA was very abundant in rat kidney (the other major organ which handles calcium and phosphorus). We also observed that during the embryonic development of calvaria the relative abundance of osteocalcin and osteopontin mRNAs lags behind that of collagen, alkaline phosphatase and osteonectin, which raises interesting questions regarding coordinate gene regulation. Other *in vivo* questions related to the *in vitro* observations include: is the depression of osteopontin and osteocalcin synthesis by glucocorticoids in any way related to or reflective of the catabolic effect of these hormones on bone? Do low (physiological) concentrations of $1,25\text{-}(OH)_2D_3$ have a positive effect on bone formation? Do osteopontin levels in the circulation correlate with levels of oesteocalcin and could they serve as another index of bone formation?

These examples have illustrated the use of cell cultures in the development of hypotheses for the role of specific factors in the organism and in the pursuit of questions at the mechanistic level, including studies of gene regulation. Cell cultures can also serve as a convenient source of cell products such as specific bone proteins and of cDNA libraries.

Diversity of osteoblast responses in culture

Let us now consider interpretation of the differences observed among various osteoblastic models *in vitro*. As mentioned, all so-called osteoblast-like cells synthesize type I collagen, have increased alkaline phosphatase activity and have PTH-stimulated adenylate cyclase. The first two properties are probably essential for the production of the bone-specific mineralized matrix, and the third property makes bone responsive to the needs of calcium homeostasis. But several other features were found to differ among osteoblast-like cells. Differences were noticed primarily in prostaglandin production; osteocalcin production; response to prostaglandin E, epidermal growth factor (EGF), TGF-β and interleukin 1 (IL-1); abundance of $1,25-(OH)_2D_3$ receptors as a function of cell density; effects of PTH and $1,25-(OH)_2D_3$ on alkaline phosphatase levels; and ability to produce mineralized matrix in diffusion chambers *in vivo* (Rodan & Rodan 1984). Other differences will probably emerge as studies continue. This diversity was attributed to species differences, to different states of differentiation of the cells under investigation, or to subspecialization of the differentiated cells. These possibilities are not mutually exclusive. If the differences among cell lines are not 'tissue culture artifacts' they could offer meaningful insights into finer aspects of the function of this tissue and its regulation.

Maturation-related changes associated with the appearance of new receptors and new cellular products have been observed in all tissues. Attempts to integrate the available information about osteoblast-like cells obtained from several species and from transformed cell lines suggest the following. Less differentiated cells (such as chick limb buds, early digestion calvaria cells and late passage UMR cells) may respond to both calcitonin and PTH by increased cyclic AMP accumulation. These less differentiated cells produce type III collagen along with type I collagen (early digestion calvaria cells and ROS 25/1 cells), respond to PGE by increased cyclic AMP production and make little or no osteocalcin. Retinoic acid may promote the expression of phenotypic features, such as alkaline phosphatase, in early differentiated cells. Responsiveness to certain growth factors, such as EGF, may also be more pronounced at earlier stages of differentiation. During the most active phase of bone matrix production osteoblastic cells synthesize exclusively type I collagen, and the non-collagenous proteins present in the matrix (osteonectin, osteopontin, proteoglycan, etc.) and possess the highest levels of alkaline phosphatase. Human trabecular bone-derived cells and ROS 17/2.8 osteosarcoma cells seem to be of that type. Relatively little is known about the biochemical properties of lining cells and osteocytes. The MC3T3 cells appear to exhibit a broader range of differentiation changes in culture than rat or human bone-derived cells and could be studied at the various stages. Most studies on MC3T3 cells reported so far were conducted at the mature osteoblastic stage.

A recent article on the phenotypic diversity observed in leukaemia cell lines suggested that the different phenotypes may represent normal stages in lymphocyte differentiation. The transformed cells presumably retained the phenotypic features expressed when transformation occurred. This hypothesis was supported by the identification of cells carrying the leukaemia-related antigens in normal lymphocyte populations (Greaves 1986). To test this assumption in bone we tried to transform calvaria-derived cells and examine the phenotypes of the immortalized cells.

Experiments using transformed calvaria-derived cells

Calvaria cells were separated by sequential digestion and early and late digestion cells were exposed to an immortalizing vector which contained the SV40 large T antigen cDNA and the *neo* gene for selection. The cells which survived in culture media containing 400 μg/ml antibiotic G418 were stably transfected, as demonstrated immunocytochemically with antibodies to the large T. These cells were cloned and screened for alkaline phosphatase production. An alkaline phosphatase-rich clone (RCT3) was further characterized and found to possess the following stable features, which have been maintained for 30 passages in culture so far. At confluence the alkaline phosphatase levels reached 0.5 to 1.0 μmol/mg protein per min. After replating, alkaline phosphatase levels decreased to about 0.1 μmol/mg protein per min and gradually increased with cell density. This pattern is characteristic of several osteoblastic cell lines. In ROS 17/2.8 cells we observed that alkaline phosphatase mRNA levels peak during the phase of active cell growth. The RCT3 cells synthesized collagen which was virtually all type I. They also responded to PTH by an increase in cyclic AMP accumulation. In diffusion chambers, implanted into nude mice for eight weeks, the cells produced substantial amounts of mineralized matrix. RCT3 cells also produced tumours when injected subcutaneously into nude mice. In culture the cells responded to PGE_2 by an increased accumulation of cyclic AMP, but not to calcitonin. The cells produced PGE which is probably mostly E_2. They showed no detectable osteocalcin mRNA in either the presence or absence of 1,25-$(OH)_2D_3$.

These findings show that calvaria cells transformed with the large T antigen continue to express differentiated features. They also suggest that the cells continue to express the genes which were active at the time of transformation and thus provide, in an immortal clonal line, the amplified image of a particular cell type. It is interesting in this context that RCT3 cells, although similar to ROS 17/2.8 cells, differ from them in the PGE_2 stimulation of adenylate cyclase and lack of osteocalcin production.

These properties have been described in embryonic rat calvaria, suggesting that they are genuine characteristics rather than transformation- or immortalization-dependent changes. These cells may represent an earlier stage of osteoblastic differentiation. The functional significance of cellular differences in prostaglandin production and response remain to be established. They may pertain to the specialization of groups of cells for intercellular communication. The study in characterized clonal cell lines of unique features, which can be identified *in vivo*, could reveal important details of cellular changes during differentiation and of the function of cells within the tissue.

The role of osteoblasts in the function of bone tissue

All tissues contain functionally specialized cells — osteoclasts and osteoblasts in bone. However, finer subdivisions among bone cells, probably in the osteoblastic lineage, are totally conceivable. Tissue functions which have to be accounted for, and are not likely to be carried out by the matrix-building mason-osteoblast, include: the detection of changes in mechanical strain and translation of that signal into bone resorption and bone formation messages; detection of rises in PTH and $1,25$-$(OH)_2D_3$ levels and translation of these signals into bone resorption messages; detection of rises in interleukin 1 levels and translation of that signal into a bone resorption message; and detection of reduced oestrogen levels and translation of that signal into increased bone turnover. Some of the features of osteoblast-like cells in culture suggest that they participate in these processes. These features include the presence of PTH, $1,25$-$(OH)_2D_3$, interleukin 1 receptors (Bird & Saklatvala 1986) and oestrogen receptors (Eriksen et al 1987), and the stimulation of prostaglandin synthesis by mechanical perturbations (Somjen et al 1980, Yeh & Rodan 1984). For some of these features a demonstration of these properties *in situ* and how they integrate into the regulatory pathways of the organism is still necessary.

In summary, the information about bone-derived cells in culture is slowly reaching the stage at which some clarity can be perceived in the midst of apparently confusing data. At the same time, useful tools for single cell biochemistry (such as *in situ* hybridization and electron microscopic immunocytochemistry) are becoming increasingly available for dissecting the function of bone tissue at the cellular level.

Acknowledgements

We wish to thank Dr Miles Brown for the large T antigen immortalization vector and Ms Dianne McDonald for preparation of this manuscript.

References

Bird TA, Saklatvala J 1986 Identification of a common class of high affinity receptors for both types of porcine interleukin-1 on connective tissue cells. Nature (Lond) 324:263–266

Centrella M, Massague J, Canalis E 1986 Human platelet-derived transforming growth factor-β stimulates parameters of bone growth in fetal rat calvariae. Endocrinology 119:2306–2312

Centrella M, McCarthy TL, Canalis E 1987 Transforming growth factor β is a bifunctional regulator of replication and collagen synthesis in osteoblast-enriched cell cultures from fetal rat bone. J Biol Chem 262:2869–2874

Ericksen EF, Berg NJ, Graham ML, Mann KG, Spelsberg TC, Riggs BL 1987 Evidence of estrogen receptors in human bone cells. J Bone Miner Res 2:238 (abstr)

Greaves MF 1986 Differentiation-linked leukemogenesis in lymphocytes. Science (Wash DC) 234:697–704

Gundberg CM, Cole DEC, Lian JB, Reade TM, Gallop PM 1983 Serum osteocalcin in the treatment of inherited rickets with 1,25-dihydroxyvitamin D_3. J Clin Endocrinal Metab 56:1063–1067

Oldberg A, Franzén A, Heinegård D 1986 Cloning and sequence analysis of rat bone sialoprotein (osteopontin) cDNA reveals an Arg-Gly-Asp cell-binding sequence. Proc Natl Acad Sci USA 83:8819–8823

Pan LC, Price PA 1985 The propeptide of rat bone γ-carboxyglutamic acid protein shares homology with other vitamin K-dependent protein precursors. Proc Natl Acad Sci USA 82:6109–6113

Peck WA, Birge SJ, Fedak SA 1964 Bone cells: biochemical and biological studies after enzymatic isolation. Science (Wash DC) 146:1476–1477

Peck WA, Burks JK, Wilkins J, Rodan SB, Rodan GA 1977 Evidence for preferential effects of parathyroid hormone, calcitonin and adenosine on bone and periosteum. Endocrinology 100:1357–1364

Pfeilschifter J, D'Souza SM, Mundy GR 1987 Effects of transforming growth factor-β on osteoblastic osteosarcoma cells. Endocrinology 121:212–218

Price PA, Baukol SA 1980 1,25-Dihydroxyvitamin D_3 increases synthesis of the vitamin K-dependent bone protein by osteosarcoma cells. J Biol Chem 255:11660–11663

Prince CW, Butler WT 1987 1,25-Dihydroxyvitamin D_3 regulates the biosynthesis of osteopontin, a bone-derived cell attachment protein, in clonal osteoblast-like osteosarcoma cells. Collagen Relat Res 7:305–313

Prince CW, Oosawa T, Butler WT et al 1987 Isolation, characterization, and biosynthesis of a phosphorylated glycoprotein from rat bone. J Biol Chem 262:2900–2907

Robey PG, Young MF, Flanders KC et al 1987 Osteoblasts synthesize and respond to transforming growth factor type β (TGF-β) in vitro. J Cell Biol 105:457–463

Rodan GA, Rodan SB 1984 Expression of the osteoblastic phenotype. In: Peck WA (ed) Bone and mineral research. Elsevier, Amsterdam, vol 2:244–285

Seyedin SM, Thomas TC, Thompson AY, Rosen DM, Piez KA 1985 Purification and characterization of two cartilage inducing factors from bovine demineralized bone. Proc Natl Acad Sci USA 82:2267–2271

Shteyer A, Gazit D, Passi-Even L et al 1986 Formation of calcifying matrix by osteosarcoma cells in diffusion chambers in vivo. Calcif Tissue Int 39:49–54

Somjen D, Binderman I, Bergen E, Harell A 1980 Bone modeling induced by physical stress is prostaglandin mediated. Biochim Biophys Acta 627:91–100

Tam CS, Heersche JNM, Murray TM, Parsons JA 1982 Parathyroid hormone stimulates the bone apposition rate independently of its resorptive action: differential effects of intermittent and continuous administration. Endocrinology 110:506–512

Wong GL, Cohn DV 1974 Separation of parathyroid hormone and calcitonin-sensitive cells from non-responsive bone cells. Nature (Lond) 252:713–715

Wong GL, Cohn DV 1975 Target cells in bone for parathormone and calcitonin are

different: enrichment for each cell type by sequential digestion of mouse calvaria and selective adhesion to polymeric surfaces. Proc Natl Acad Sci USA 72:3167–3171
Yeh CK, Rodan GA 1984 Tensile forces enhance prostaglandin E synthesis in osteoblastic cells grown on collagen ribbons. Calcif Tissue Int 36:S67–S71

DISCUSSION

Fleisch: We have some results which do not fit yours, Dr Rodan. When Dr Hofstetter immortalized various rat bone cell clones, some were like yours, similar to the original clones, but others changed dramatically. For example, one cell completely lost its alkaline phosphatase activity, despite the fact that it still produced osteocalcin. Our caveat would be that you can have unaltered cells but you can also have 'doctored' cells. One should always have the initial clone to compare the new cells with. This is possible if a homogeneous cell population is used. With an inhomogeneous population, one does not know what one has after immortalization and whether it is representative of what occurs in the non-immortalized cells.

Rodan: Point well taken! But we found it remarkable that some immortalized clones possessed the full spectrum of properties examined.

Canalis: I was surprised by the lack of osteocalcin mRNA in RCT3 cells. With J. Lian and T. McCarty, I looked at osteocalcin levels in osteoblast-rich cells from rat calvaria. There was a small amount of osteocalcin, about 8.0 pg per µg of protein. Did you test the effects of vitamin D on RCT3 cells to see if you could stimulate osteocalcin to a more detectable level?

Rodan: So far, the osteocalcin-negative data are based on hybridization with a probe for osteocalcin mRNA, which is not as sensitive as radioimmunoassay or immunocytochemistry.

Canalis: Is it conceivable that your RCT3 cells do make osteocalcin?

Rodan: It is possible. We measured osteocalcin hybridizable mRNA during the development of calvaria, at 15, 18, 21 days and then at six weeks (Yoon et al 1987). Osteocalcin mRNA was virtually undetectable at 15 days. It appeared at 18 days, was higher at 21 days and could be seen after birth. These RCT3 cells were derived from 19-day-old embryos.

Canalis: We detected osteocalcin at 22 days.

Rodan: Price et al (1981) did radioimmunoassay of rat bones. They found that osteocalcin occurs later in the development of these tissues.

Raisz: Between 19 and 22 days there is a large change in differentiation and osteocalcin production by calvariae. That might explain part of the difference.

Hauschka: *In vitro* the rat osteoblast-like cells obtained by collagenase digestion do make osteocalcin at low levels (Chen et al 1986, Hauschka et al, this volume). The sensitivity of the radioimmunoassay is important. Admittedly the ROS 17/2.8 cells produce tremendous amounts of osteocalcin—hundreds of times more than the primary cultures of normal osteoblast-like cells. Therefore, while there appears to be a large relative difference between the RCT3 cells and ROS 17/2.8 cells, I would challenge the absence of osteocalcin in your RCT3 cells.

Rodan: We have characterized a human osteosarcoma cell (Rodan et al 1987) and looked for osteocalcin with hybridization and with various antibodies, but we were unable to detect it. Those cells make mineral in diffusion chambers. Therefore, perhaps the presence of osteocalcin is not as strong a marker for osteoblastic cells as alkaline phosphatase or type I collagen.

Veis: How specific is alkaline phosphatase to bone cells?

Rodan: The same alkaline phosphatase is found in all tissues (except human intestine and placenta): it is the product of a single gene. The only tissue-specific modifications are post-translational, primarily glycosylation. However, in all tissues examined alkaline phosphatase levels are 10% or less of those in osteoblastic cells. Alkaline phosphatase levels of 1 µmol per mg protein per minute are seen only in hypertrophic cartilage cells, bone cells and kidney cells.

Slavkin: I'm curious about bone as a tissue as opposed to bone as isolated cells grown in monolayer. Have you tried to reconstruct three-dimensional cultures to find out if the cells are confluent and are coupled through gap junctions? Does that change the relative levels of gene products?

Rodan: Tenenbaum & Heersche (1982) have done that. They saw a lower production of matrix and mineralization with periosteal or calvarial cells in monolayer cultures. However, if they folded the periosteum they produced a tissue-like morphology. We haven't studied that in this system.

Slavkin: I am curious what an indefinitely dividing cell looks like when it's surrounded by connective tissue matrix that is mineralizing.

Rodan: That's a very good question. The cytoskeleton is much less organized in the osteosarcoma cells than in the primary cultures. However, even in an immortal cell line not all the cells continuously divide; some retain proliferative capability while others may assume differentiated properties.

Slavkin: Why did you study retinoic acid as a mediator in this system?

Rodan: We were motivated by the fact that certain leukaemia cells and other permanent cell lines show differentiation in response to retinoic acid. Therefore, we used retinoic acid, dimethylsulphoxide, butyrate and glucocorticoids, and stained for alkaline phosphatase to see if an undifferentiated cell could be differentiated by these factors. We have not identified a cell which qualifies as a precursor cell and which can be differentiated by this kind of manipulation. We think we are seeing cells that express the phenotype present at the time of immortalization and that only small modulations of this expression, quantitative rather than qualitative, are produced by the agents that are known to act on bones, such as $1,25-(OH)_2D_3$ and parathyroid hormone.

Caplan: Could we investigate the assumption that you can fix cells at lineage developmental positions *in vivo* by micro-injecting transforming vehicles into embryonic tissues?

Rodan: I don't think that experiment would be straightforward. However, you can get tissue-specific expression in transgenic mice if you transfect the egg cell.

Caplan: What about micro-injecting the transforming vector into the stacked cell layer to fix cells at particular pre-lineage positions?

Rodan: The retroviral transfection is still a statistical phenomenon. Only some of the cells would integrate the vector into their DNA and of those cells only a few may acquire the desired properties.

Weiner: We all feel comfortable about describing the activities of cells as making parathyroid hormone, alkaline phosphatase, or osteocalcin, but the concept of 'making mineral' always worries me. 'Making mineral' refers to many different things. If you take a random bacterial culture and feed it enough phosphate and calcium it will also make mineral. It is very important to describe this process accurately. We should not only look at it by electron microscopy. What sort of crystals are produced and what are their shapes? You have, I believe, studied the ROS 17/2 cells. Do they really make bone, and by what criteria?

Rodan: It is woven bone. We looked at the electron microscopic appearance of the matrix and the mineral at different stages. We compared it visually with the sequence of events in woven bone. They look similar, but it hasn't been done in any other way and that may be a good idea.

Raisz: Gloria Gronowicz has been looking at mineralization in organ culture (Gronowicz et al 1987) and comparing what she sees in the orderly process of this slowly developing organ culture to what happens if β-glycerol phosphate is added. With β-glycerol phosphate at even millimolar levels ectopic calcification occurs, whereas with inorganic phosphate you get mineralization of the bone with an osteoid seam. Adele Boskey (unpublished results) has looked at the X-ray diffraction pattern. The hydroxyapatite in the β-glycerol phosphate-induced ectopic mineralization is much more crystalline than the amorphous material that's in the bone itself. X-ray diffraction of the different parts of the tissue is more useful than electron microscopy because we can see the β-glycerol phosphate-induced ectopic calcification physically separate, so we can analyse it separately.

Rodan: We used alkaline phosphatase cDNA to transfect cells which don't normally contain this enzyme. We obtained cells which produced large amounts of alkaline phosphatase. Those cells precipitated calcium phosphate *in vitro* from β-glycerol phosphate media. However, when the cells were introduced into chambers we saw positive van Kossa stains but they did not resemble mineralized matrix. Therefore, the chamber experiment is a good indication of whether cells can make mineralized matrix.

Termine: You can easily tell that by looking at transmission eléctron micrographs of these cultures. It is easy to detect dystrophic as opposed to normal, tissue-like calcification. You also said that you had found in calvaria development that the mRNAs for type I collagen, alkaline phosphatase and osteonectin precede those for osteopontin and osteocalcin. Have you looked at osteonectin and osteopontin production in RCT3 cells?

Rodan: Yes, both are found.

Martin: I would like to discuss the question of whether the transformed or malignant cell represents a specific point in differentiation. It is partly dependent on the phenotypic stability of the cell that you have transformed. It may be that 30 divisions are not sufficient to see any changes that might take place. I used to think that the malignant cells may represent a specific point in the differentiation pathway, and that ROS 17/2.8 might be late and UMR 106 might be early. However, in some respects the UMR 106 cell is late: it has high alkaline phosphatase activity; it does decrease with retinoic acid (Ng et al 1985). It has a substantial amplitude of PTH-responsive cyclase, and it makes type I collagen (Partridge et al 1983). It does not produce osteocalcin, either constitutively, or with $1,25\text{-}(OH)_2D_3$ and it does, as Paul Price has found, produce matrix Gla protein. The latter are very early or perhaps even chondrocyte features. What concerned me particularly was that very late passage UMR 106 cells developed calcitonin responsiveness, which we were able to subclone to a calcitonin responsive-line (Forrest et al 1985). This again could go back to a chondroblastic or chondrocyte-type phenotype. Therefore these cells have both very early and very late characteristics. It may be the malignant cell is able to select its properties from the whole range. It is inappropriate to deduce from the discovery of a certain property in malignant osteoblasts that it must be a property of the osteoblast. For example, Gray et al (1987) have demonstrated oestrogen responses in UMR 106 cells. Even if the oestrogen receptor is there, that does not convince me that there is an oestrogen receptor in the osteoblast.

Rodan: I considered the same data and thought that UMR106 cells do represent an early phenotype because alkaline phosphatase is one of the earliest manifestations of the transition towards osteoblastic lineage. Calcitonin receptors are seen in the early digestion calvarial cells during sequential digestion. When these cells were put into chambers by Simmons et al (1982) they found cartilage. When they put the late digestion cells into chambers they found bone. It is possible that the UMR 106 cells represent an early stage in differentiation that has the dual capability, which Dr Owen suggested, of becoming osteoblastic and chondroblastic, and that is why they have calcitonin receptors.

Williams: What are the levels of the calcium-binding protein in these cells? Have you any information about ATPases pumping calcium?

Rodan: We have not measured those two constituents in these cells. The reported level of vitamin D-dependent calcium-binding protein in osteoblastic cells is very low (Celio et al 1984). We used hybridization (Northern blots) and radioimmunoassay in rat osteosarcoma cells ROS 17/2.8, and did not detect calcium-binding protein after 1,25-(OH)$_2$vitamin D$_3$ treatment. Calcium ATPases have been described by Shen et al (1983) in rat calvarial cells.

Fleisch: Your cells mineralize *in vitro* in the presence of β-glycerol phosphate, but they don't mineralize in the chambers *in vivo*. This shows that incubating cells with 10mM of glycerol phosphate, an amount of substrate which they will never see *in vivo*, doesn't mean much.

Nijweide: As long as the calcium phosphate precipitation *in vitro* in the presence of β-glycerol phosphate is correlated with matrix formation, genuine mineralization may have occurred. In chicken bone cell cultures, for example, osteoid formation and mineralization only take place when the cells are cultured as a multilayer, not as a monolayer even in the presence of 10mM β-glycerol phosphate.

Glimcher: Dr Louis Gerstenfeld in our laboratory has cultured osteoblasts which synthesize an organic matrix that is similar to *in vivo* bone matrix and that calcifies during organ culture. This matrix calcifies without β-glycerol phosphate, but the rate is much increased by its addition. Therefore, calcification does not occur only because β-glycerol phosphate acts as an extraneous substrate for the release of inorganic orthophosphate ions resulting in simple precipitation.

The cells synthesize type I collagen, osteocalcin and phosphoproteins. The size and shape of the matrix crystals are indistinguishable from those of native *in vivo* bone by electron microscopy, and electron and X-ray diffraction. The crystals mature like those *in vivo* and not like small synthetic apatite crystals *in vitro* or like the crystals formed in organic culture but allowed to mature in the cell culture medium of cells which have been killed. The crystals are located principally within the collagen fibrils, as they are in native bone, and few matrix vesicles, with or without crystals, are observed. The collagen fibrils are orthogonally arranged in layers similar to their architecture in native bone. The relative proportion of mineral, collagen and non-collagenous proteins are similar to that reported in embryonic bone *in vivo*. The reducible and non-reducible cross-links of collagen are like those in embryonic bone. The phosphoproteins contain both Ser(P) and Thr(P), typical only of the phosphoproteins of bone and cementum, not of other mineralized tissue. Finally, peptide mapping showed that the phosphoproteins synthesized in cell culture were identical to the major phosphoproteins extracted from bone tissue. These results suggest that the cultured cells are phenotypically osteoblasts producing characteristic bone organic matrix and that the deposition of Ca-P crystals within the collagen fibrils occurs by the same physical/chemical mechanisms as it does *in vivo* and does not represent, as has been suggested, a simple artefactual precipitation

Rodan: I think that the consensus is that if the process of mineralization occurs in the way in which it should, β-glycerol phosphate may accelerate it, but you can find β-glycerol phosphate-promoted precipitation of calcium phosphate which is not similar to biological mineralization.

Nijweide: The transformed rat cells proliferate in culture. Is that normal for osteoblasts?

Rodan: Young (1962) did thymidine labelling *in vivo* and saw labelling of nuclei in the layer which you would histologically identify as osteoblasts.

Nijweide: Yes, but in that case cells may have taken up the thymidine in the preosteoblast stage and then divided and differentiated.

Rodan: On the basis of the intensity of label and the labelling time it appeared that it was the osteoblast cells that had incorporated the label.

Nijweide: What was the labelling period?

Rodan: Half an hour.

Urist: Have you ever seen anything like normal bone outside the chamber? Heiple et al (1968) implanted the Dunn mouse osteosarcoma in diffusion chambers and demonstrated bone formation inside the chamber and transmembrane normal bone outside. The Ridgeway mouse osteosarcoma, a tumour which has lost the capacity to form bone, does not induce transmembrane bone formation. Both human and Dunn osteosarcomas, which retain the capacity to produce bone and secrete bone morphogenetic protein into the chamber fluid, also induce transmembrane normal bone formation on the outside (Hanamura & Urist 1978, Bauer & Urist 1981). We have not seen anything in the literature on rat osteosarcoma-inducing bone formation.

Rodan: We were aware of that study. We always look histologically both inside and outside, but we have not seen it outside.

References

Bauer FCH, Urist MR 1981 Human osteosarcoma derived soluble bone morphogenetic protein. Clin Orthop 154:291–295

Celio MR, Norman AW, Heizmann CW 1984 Vitamin-D-dependent calcium-binding protein and parvalbumin occur in bones and teeth. Calcif Tissue Int 36:129–130

Chen TL, Hauschka PV, Cabrales S, Feldman D 1986 The effects of 1,25(OH)$_2$ vitamin D$_3$ and dexamethasone on rat osteoblast-like primary cell cultures: receptor occupancy and functional expression patterns for three different bioresponses. Endocrinology 118:250–259

Forrest SM, Ng KW, Findlay DM et al 1985 Characterization of an osteoblast-like clonal cell-line which responds to both parathyroid hormone and calcitonin. Calcif Tissue Int 37:51–56

Gray KM, Garrison L, Gray TK 1987 Endothelial cell-conditioned medium influences the growth and differentiation of the UMR106 cell line, an osteoblastic model. Clin Res 35:A23

Gronowicz G, Woodiel F, Raisz LG 1987 Mineralization in cultured fetal rat calvaria. J Bone Miner Res 2(suppl. 1):abstr 428

Hanamura H, Urist MR 1978 Osteogenesis and chondrogenesis in transplants of Dunn and Ridgeway osteosarcoma cell cultures. Am J Pathol 91:277–298

Hauschka PV, Chen TL, Mavrakos AE 1988 Polypeptide growth factors in bone matrix. In: Cell and molecular biology of vertebrate hard tissues. Wiley, Chichester (Ciba Found Symp 136) p 207–225

Heiple KG, Herndon CH, Chase SW, Wattlesworth A 1968 Osteogenic induction by osteosarcoma and normal bone in mice. J Bone Jt Surg 50:311–325

Ng KW, Livesey SA, Collier F, Gummer PR, Martin TJ 1985 Effect of retinoids on the growth, ultrastructure and cytoskeletal structures of malignant rat osteoblasts. Cancer Res 45:5106–5113

Partridge NC, Alcorn D, Michelangeli VP, Ryan G, Martin TJ 1983 Morphological and biochemical characterization of four clonal osteogenic sarcoma lines of rat origin. Cancer Res 43:4388–4394

Price PA, Lothringer JW, Baukol SA, Reddi AH 1981 Developmental appearance of the vitamin K-dependent protein of bone during calcification. J Biol Chem 256:3781–3784

Rodan SB, Imai Y, Thiede M et al 1987 Characterization of a human osteosarcoma cell line (Saos-2) with osteoblastic properties. Cancer Res 47:4961–4966

Shen V, Kohler G, Peck WA 1983 A high affinity calmodulin-responsive ($Ca^{2+}+Mg^{2+}$) ATPase in isolated bone cells. Biochim Biophys Acta 727:230–238

Simmons DJ, Kent GN, Jilka RJL, Scott DM, Fallon M, Cohn DV 1982 Formation of bone by isolated cultured osteoblasts in millipore diffusion chambers. Calcif Tissue Int 34:291–294

Tenenbaum H, Heersche JNM 1982 Differentiation of osteoblasts and formation of mineralized bone in vitro. Calcif Tissue Int 117:76–79

Yoon K, Buenaga R, Rodan GA 1987 Tissue specificity and developmental expression of rat osteopontin. Biochem Biophys Res Commun 148:1129–1136

Young RW 1962 Cell proliferation and specialization during endochondral osteogenesis in young rats. J Cell Biol 14:357–370

The regulation of osteoclastic development and function

T.J. Chambers

Department of Histopathology, St George's Hospital Medical School, Cranmer Terrace, London SW17 0RE, UK

Abstract. Cells of the osteoblastic lineage exert a dominant influence on osteoclastic bone resorption. They form a communicating network of osteocytes, surface osteocytes and osteoblasts that seems well placed to monitor the structure and performance of bone and to judge where bone formation or resorption is appropriate. Osteoblasts produce prostaglandins (PGs) which strongly inhibit osteoclastic resorption. None of the agents that stimulate resorption in intact bone, such as parathyroid hormone (PTH), interleukin 1 (IL-1), 1,25-$(OH)_2$vitamin D_3 (1,25-$(OH)_2D_3$) or tumour necrosis factors, affects isolated osteoclasts, but all induce osteoblastic cells to produce osteoclastic resorption stimulatory activity (ORSA) that acts directly on osteoclasts. Osteoblasts seem to initiate resorption as well as stimulating or inhibiting it. Contact with bone mineral appears to be necessary: osteoclasts resorb mineralized but not unmineralized bone. All bone surfaces are lined by unmineralized organic material. Osteoblastic cells secrete neutral proteases, including collagenase, in response to hormonal stimulators of bone resorption. Incubation of osteoblasts, in the presence of PTH, on such surfaces or preincubation of the bone with collagenase predisposes bone to osteoclastic resorption. Agents that stimulate resorption in organ cultures seem to share these osteoblast-mediated mechanisms for induction and stimulation of resorption but 1,25-$(OH)_2D_3$ stimulates it through an additional mechanism. We have found that osteoclasts can be induced from haemopoietic tissue (including haemopoietic spleen cells) in the presence of 1,25-$(OH)_2D_3$ — PTH and IL-1 have no effect in this system. Because osteoclasts lack receptors for 1,25-$(OH)_2D_3$ these results suggest either that osteoclast precursors lose 1,25-$(OH)_2D_3$ receptors during differentiation, or that a 1,25-$(OH)_2D_3$-responsive accessory cell in bone marrow induces osteoclastic differentiation in the presence of 1,25-$(OH)_2D_3$.

1988 Cell and molecular biology of vertebrate hard tissues. Wiley, Chichester (Ciba Foundation Symposium 136) p 92–107

Osteoclastic resorption of bone can be regulated on at least three distinct levels (see Fig. 1): induction of differentiation; initiation of bone resorption; and modulation of the rate of bone resorption. Although osteoblastic cells do not seem to be essential for osteoclastic differentiation, they appear to be the major agents responsible for mediation of the effects of environmental stimuli on bone resorption by osteoclasts already formed.

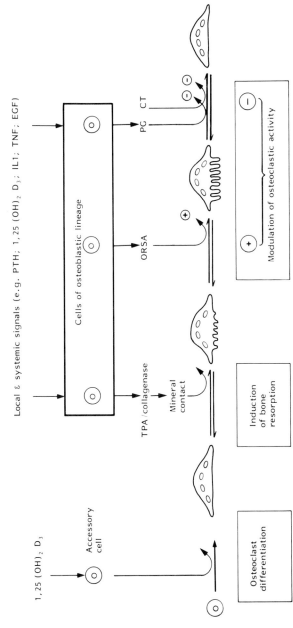

FIG. 1. Diagrammatic representation of the mechanism of regulation of osteoclastic differentiation and function. Among systemic regulators, only calcitonin (CT) acts directly on osteoclasts. All the agents so far tested that stimulate bone resorption in organ culture act primarily on cells of the osteoblastic lineage. These cells are able to initiate and subsequently modulate osteoclastic resorption. PTH, parathyroid hormone; IL1 interleukin 1; TNF, tumour necrosis factor; EGF, epidermal growth factor; PG, prostaglandin; TPA, tissue plasminogen activator.

Modulation of osteoclastic bone resorption

If osteoclasts are disaggregated mechanically from neonatal mammalian long bones and sedimented onto slices of cortical bone, excavation of the bone surface by osteoclasts commences within a few hours, and by 18 hours each osteoclast has generally produced 1–3 excavations. The extent of bone resorption can be quantified by inspecting the whole surface of each bone slice in a scanning electron microscope and measuring the number and surface areas of the resorption cavities. This forms the basis for a bioassay by which the effects of calcium-regulating and other hormones on 'spontaneous' osteoclastic resorption can be assessed (Chambers et al 1985a).

Calcitonin (CT) inhibits bone resorption through a direct effect on osteoclasts. These cells possess CT receptors (Warshawsky et al 1980, Nicholson et al 1986) and CT inhibits disaggregated osteoclasts from excavating bone slices at concentrations (half-maximal, 2 pg/ml for human CT) within the physiological range (Zaidi et al 1987). This, together with the lack of convincing evidence that other cell types in bone have CT receptors or responses, suggests that inhibition of osteoclastic bone resorption is the sole function of this hormone in bone.

Parathyroid hormone (PTH) and $1,25\text{-}(OH)_2D_3$ are without influence on the 'spontaneous' resorption of cortical bone slices by isolated osteoclasts (Chambers et al 1985a). This suggests that the populations of cells that we use are deficient in some cell type that is present in intact bone, and mediates hormone responsiveness. This deficiency can be restored by incubating osteoclasts with either osteoblastic cell lines or primary cultures of osteoblastic cells. Such co-cultures respond to concentrations of PTH as low as 10^{-4} IU/ml, with increases in bone resorption to two to four-fold that seen without the hormone (McSheehy & Chambers 1986a).

The enhanced bone resorption in osteoblast–osteoclast co-cultures can be accounted for by the production within the supernatant, in the presence of PTH, of osteoclast resorption stimulating activity (ORSA) that directly stimulates osteoclasts: addition to osteoclasts of supernatants from osteoblastic cells previously incubated with PTH or $1,25\text{-}(OH)_2D_3$ increases osteoclastic resorption to an extent similar to that seen in co-cultures (McSheehy & Chambers 1986b, 1987).

The major impediment to the identification of ORSA is the lack, at present, of an assay that is sufficiently efficient or reliable. The current assay takes several days to assess whether a single sample is stimulatory. In our experience, light microscopical assessment (Dempster et al 1987) is less accurate than scanning electron microscopy and more prone to observer error. Moreover, the scanning electron microscopy resorption assay is so sensitive to changes in culture conditions that the confidence with which changes in osteoclast activity observed on incubating osteoclasts in supernatant extracts can be interpreted as due to ORSA is reduced. For example, a fall in pH from 7.4 to 6.8 can increase resorption 14-fold (Arnett & Dempster 1986). We have found that alcohol at a concentration of 1:10 000 doubles bone resorption (unpublished observations). Our strategy for the identification of ORSA is to develop improved assays (this will probably require the development of osteoclastic cell lines), and to screen agents known to stimulate bone resorption in organ culture for a direct effect on osteoclasts. ORSA does not seem to be a prostaglandin (PG): despite their undoubted ability to stimulate resorption in intact bone (after an initial inhibition) (Conaway et al 1986), PGs are strong direct inhibitors of osteoclasts (Chambers et al 1985a) and indomethacin does not impair ORSA production by osteoblasts (McSheehy & Chambers 1986b). Leukotrienes B_4, C_4 and D_4 do not influence resorption by isolated osteoclasts (unpublished). Interleukin 1 (IL-1), tumour necrosis factor and lymphotoxin, all potential local messengers for the stimulation of osteoclasts by osteoblasts, are all without direct effect, but stimulate resorption in co-cultures (Thomson et al 1986) and induce the production of ORSA by osteoblastic cells (Thomson et al 1987). Although these agents do not represent the mechanism through which osteoblasts stimulate osteoclasts, the sensitivity of co-cultures to their presence is sufficient to suggest a physiological role. This might be either as agents of systemic catabolism, as part of the systemic response to injury (and the agents might then incidentally cause non-adaptive, pathological bone resorption, if they are produced inappropriately or locally by neoplastic or inflammatory tissues), or as messengers between bone cells (for example, as transmitters of the physiological responses of osteocytes to the bone surface).

Initiation of osteoclastic bone resorption

Cells of the osteoblastic lineage also seem to be required to initiate osteoclastic bone resorption. Native bone surfaces have a layer of unmineralized organic material between bone lining cells and bone mineral (Raina 1972, Fornasier 1980, Vanderwiel 1980). In the bone slices used in our experiments described above, mineral is artificially exposed, by the process of cutting, on the bone surface and the substrate is 'spontaneously' resorbed by osteoclasts. We have found that osteoclasts were not induced to resorptive activity by incubation on native, uncut bone surfaces (the calvariae of adult rats) unless the unmineralized surface layer was first removed by mammalian or bacterial collagenase (Chambers et al 1985b). Since osteoclasts are capable of destroying both organic and mineral components of bone (Chambers et al 1984), this result indicates that contact with bone mineral, but not unmineralized organic material, induces osteoclasts to resorptive activity. We also found that osteoblasts were able to remove the unmineralized organic layer *in vitro* to expose (but seemed incapable of resorbing) subjacent mineral. PTH accelerates this process and bone so modified has an increased susceptibility to osteoclastic resorption. This susceptibility is abrogated by demineralization (Chambers & Fuller 1985).

These observations suggest that the unmineralized organic layer functions to prevent osteoclastic contact with bone mineral, and that osteoblastic cells (probably usually 'surface osteocytes') may locally produce collagenase that acts to provide osteoclastic access to resorption-stimulating bone mineral. The ability of anorganic bone to stimulate resorption suggests that mineral alone, and not the products of matrix degradation, is the stimulus for bone-resorbing activity by osteoclasts. Osteoblastic cells are known to produce collagenase and tissue plasminogen activator, and production of both is increased by PTH and other agents that stimulate osteoclastic bone resorption (Hamilton et al 1984, Heath et al 1984). We believe that, rather than enabling osteoblastic cells to act as alternative bone-resorbing cells, protease secretion may be the final common pathway by which osteoblastic cells initiate osteoclastic bone resorption. Once resorption is under way, cells of the same lineage continue to regulate osteoclastic resorption by producing PG and ORSA (see Fig. 1).

Induction of osteoclastic differentiation

A third mechanism by which osteoclastic resorption may be influenced is through the regulation of osteoclastic differentiation. Parabiosis experiments, quail–chick chimeras, and bone marrow and spleen cell transplantation experiments have established that osteoclasts take origin from haemopoietic tissue (see Marks 1983). Mononuclear phagocytes were initially favoured as the haemopoietic lineage most likely to give rise to osteoclasts, but cells of the mononuclear phagocyte system seem increasingly unlike osteoclasts: they are unable to resorb bone; they lack calcitonin responsiveness; and they show major antigenic differences to osteoclasts (see Chambers 1985).

We have recently attempted to induce osteoclastic differentiation from haemopoietic tissues. We identified osteoclasts using two markers that seem specific for osteoclasts and are absent from mononuclear and multinuclear macrophages: an osteoclast-specific monoclonal antibody (Horton et al 1985), and the ability to excavate the surface of cortical bone slices. We incubated rabbit bone marrow cells or cells from neonatal rabbit (haemopoietic) spleen for up to four weeks on plastic coverslips and bone slices in the presence of PTH, IL-1, 1,25-$(OH)_2D_3$, or combinations of these hormones. Excavations and antibody-positive cells were seen only in cultures of bone marrow or spleen cells incubated with 1,25-$(OH)_2D_3$: neither PTH nor IL-1 was able to induce osteoclastic differentiation (Fuller & Chambers 1987). Osteoclasts spontaneously resorb such bone slices within hours without hormone addition. Our results show both that osteoclasts were absent from the bone marrow and spleen cell suspensions used to initiate the cultures, and

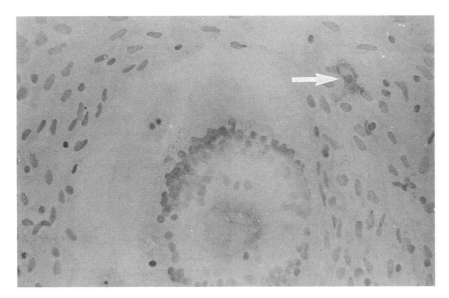

FIG. 2(a). Immunoperoxidase staining of bone marrow cells after 21-day incubation with 1,25-$(OH)_2D_3$, using osteoclast-specific monoclonal antibody (mAb) 23C6 (Horton et al 1985). A multinucleate giant cell occupies most of the centre of the photograph, with a typical circular array of nuclei. This multinucleate cell is mAb-negative. A mononuclear cell (arrow) is mAb-positive. ($\times$ 130)

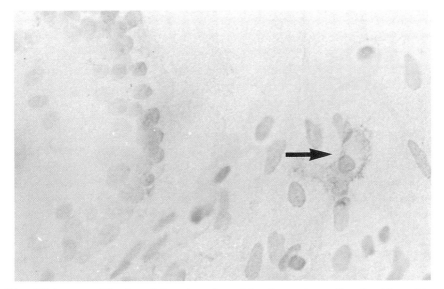

(b). Higher magnification of the edge of a multinucleate giant cell and mAb-positive mononuclear cell (arrowed) outlined by (cell surface) peroxidase product. ($\times$ 260)

also that $1,25\text{-}(OH)_2D_3$ seems to be an absolute requirement for osteoclastic differentiation.

Multinucleate giant cells also developed in the cultures, especially with $1,25\text{-}(OH)_2D_3$, but these seemed unlikely to be the resorptive cell type because we saw no bone resorption in cultures (containing giant cells) incubated without the hormone. The giant cells also failed to bind osteoclast-specific monoclonal antibody (Fig. 2) (rather, the giant cells bind macrophage-specific antibodies, which fail to bind osteoclasts (Horton et al 1986). Surprisingly, antibody-positive cells formed a sparse population of distinctive mononuclear cells and cells of low multinuclearity, morphologically quite unlike the multinucleate giant cells and the much larger numbers of (mononuclear) macrophages in these cultures. It seems likely that $1,25\text{-}(OH)_2D_3$ is exerting distinct effects on two different cell populations in these cultures: it may induce osteoclastic differentiation from an unknown cell type, while coincidentally exerting its known ability to induce fusion of mononuclear phagocytes.

If multinuclearity is used as a criterion for osteoclasts (Roodman et al 1985, Pharoah & Heersche 1985), it will not merely include cells that are not osteoclasts; it will exclude cells that are. Multinuclearity was originally the means of identifying osteoclasts in bone, but is probably not essential to the phenotype: mononuclear cells may have all the attributes of osteoclasts, including CT receptors and responsiveness, osteoclast-specific monoclonal antibody positivity, and the capacity to excavate bone. It seems likely that the osteoclast is a bone-resorptive cell that is initially mononuclear, and may remain so, fully functional, but generally becomes multinucleate, possibly for improved performance or regulation.

The effects of $1,25\text{-}(OH)_2D_3$, PTH and IL-1 on osteoclastic differentiation are distinctively different from their effects as modulators of bone resorption by mature osteoclasts. Whereas the hormones all stimulate resorption by mature osteoclasts through osteoblastic ORSA production, the induction of the osteoclastic phenotype seemed to be specific for and dependent upon $1,25\text{-}(OH)_2D_3$ — the other hormones were completely ineffective. Mature osteoclasts lack $1,25\text{-}(OH)_2D_3$ receptors, and it may be that osteoclastic precursors possess receptors but lose them during maturation. An alternative, more likely, possibility is that there is an accessory cell, among the many that respond to $1,25\text{-}(OH)_2D_3$, that is present not only in bone but also in spleen (and is therefore unlikely to be osteoblastic) that mediates osteoclastic induction.

Acknowledgements

The work described in this paper was supported by the Medical Research Council and the Wellcome Trust.

References

Arnett TR, Dempster DW 1986 Effect of pH on bone resorption by rat osteoclasts *in vitro*. Endocrinology 119:119–124

Chambers TJ 1985 The pathobiology of the osteoclast. J Clin Pathol (Lond) 38:241–252

Chambers TJ, Fuller K 1985 Bone cells predispose endosteal surfaces to resorption by exposure of bone mineral to osteoclastic contact. J Cell Sci 76:155–165

Chambers TJ, Revell PA, Fuller K, Athanasou NA 1984 Resorption of bone by isolated rabbit osteoclasts. J Cell Sci 66:383–399

Chambers TJ, McSheehy PMJ, Thomson BM, Fuller K 1985a The effect of calcium-regulating hormones and prostaglandins on bone resorption by osteoclasts disaggregated from neonatal rabbit bones. Endocrinology 116:234–239

Chambers TJ, Darby JA, Fuller K 1985b Mammalian collagenase predisposes bone surfaces to osteoclastic resorption. Cell Tissue Res 241:671–675

Conaway HH, Diez LF, Raisz LG 1986 Effects of prostaglandin and prostaglandin E1 (PGE1) on bone resorption in the presence and absence of parathyroid hormone. Calcif Tissue Int 38:130–134

Dempster DW, Murrills RJ, Horbert W, Arnett TR 1987 Chicken calcitonin is a potent inhibitor of rat but not chick osteoclasts. J Bone Miner Res 2(Suppl 1): abstr 156

Fornasier VL 1980 Transmission electron microscopy studies of osteoid maturation. Metab Bone Dis & Relat Res 25:103–108

Fuller K, Chambers TJ 1987 Generation of osteoclasts in cultures of rabbit bone marrow and spleen cells. J Cell Physiol 132:441–452

Hamilton JA, Lingelbach SR, Partridge NC, Martin TJ 1984 Stimulation of plasminogen activator in osteoblast-like cells by bone-resorbing hormones. Biochem Biophys Res Commun 122:230–236

Heath JK, Atkinson SJ, Meikle MC, Reynolds JJ 1984 Mouse osteoblasts synthesise collagenase in response to bone resorbing agents. Biochim Biophys Acta 802:151–154

Horton MA, Lewis D, McNulty K, Pringle JAS, Chambers TJ 1985 Monoclonal antibodies to osteoclastomas (giant cell bone tumors): definition of osteoclast-specific cellular antigen. Cancer Res 45:5663–5669

Horton MA, Rimmer EF, Chambers TJ 1986 Giant cell formation in rabbit long-term bone marrow cultures: immunological and functional studies. J Bone Miner Res 1:5–14

McSheehy PMJ, Chambers TJ 1986a Osteoblastic cells mediate osteoclastic responsiveness to PTH. Endocrinology 118:824–828

McSheehy PMJ, Chambers TJ 1986b Osteoblast-like cells in the presence of parathyroid hormone release soluble factor that stimulates osteoclastic bone resorption. Endocrinology 119:1654–1659

McSheehy PMJ, Chambers TJ 1987 1,25-Dihydroxyvitamin D_3 stimulates rat osteoblastic cells to release a soluble factor that increases osteoclastic bone resorption. J Clin Invest 80:425–429

Marks SC 1983 The origin of osteoclasts. J Oral Pathol 12:226–256

Nicholson GC, Moseley JM, Sexton PM, Mendelsohn FAO, Martin TJ 1986 Abundant calcitonin receptors in isolated rat osteoclasts. J Clin Invest 78:355–360

Pharoah MJ, Heersche JM 1985 1,25-dihydroxyvitamin D_3 causes an increase in the number of osteoclastic cells in cat bone marrow cultures. Calcif Tissue Int 37:276–281

Raina V 1972 Normal osteoid tissue. J Clin Pathol 25:229–232

Roodman GD, Ibbotson KJ, MacDonald BR, Kuehl TJ, Mundy GR 1985 1,25 dihydroxyvitamin D_3 causes formation of multinucleated cells with several osteoclastic characteristics in cultures of primate marrow. Proc Natl Acad Sci USA 82:8213–8217

Thomson BM, Saklatvala J, Chambers TJ 1986 Osteoblasts mediate interleukin 1 responsiveness of bone resorption by rat osteoclasts. J Exp Med 104:104–112

Thomson BM, Mundy GR, Chambers TJ 1987 Tumor necrosis factors α and β induce osteoblastic cells to stimulate osteoclastic bone resorption. J Immunol 138:775–779

Vanderwiel CJ 1980 An ultrastructural study of the components which make up the resting surface of bone. Metab Bone Dis & Relat Res 25:109–116

Warshawsky H, Goltzman D, Rouleau MF, Bergeron JJM 1980 Direct *in-vivo* demonstration by autoradiography of specific binding sites for calcitonin in skeletal and renal tissues of the rat. J Cell Biol 85:682–694

Zaidi M, Fuller K, Bevis PJR, GainesDas RE, Chambers TJ, MacIntyre I 1987 Calcitonin gene-related peptide inhibits osteoclastic bone resorption — a comparative study. Calcif Tissue Int 40:149–154

DISCUSSION

Raisz: PGE_1 was clearly more potent an inhibitor than PGE_2, and you said that PGI_2 was more potent than either. Does anyone know if the osteoclast adenylate cyclase activity parallels that structure–activity relationship?

Martin: The osteoclast in the rat is more sensitive to PGI_2 than to PGE_2. In that respect it resembles the platelet and it would be predicted from Professor Chambers' results that PGI_2 might be the locally important prostanoid in osteoclast control.

Raisz: In the presence of parathyroid hormone (PTH) and of osteoblasts or other cells, does PGE_2 still inhibit osteoclasts? We could only see that initial inhibition of resorption in organ culture in the absence of PTH treatment (Conaway et al 1986).

Chambers: We haven't done that experiment.

Glowacki: Which non-osteoblastic cell types (either with or without PTH receptors) have been negative in the PTH co-culture experiments?

Chambers: We haven't looked at any other cell types apart from calvarial cells and UMR cells in that system. We have looked at fibroblasts, but not using PTH.

Krane: In your study of fibroblasts in co-culture, how many different types of fibroblasts did you test?

Chambers: We used fibroblasts obtained by collagenase digestion of neonatal rat skin. We have not yet made a serious attempt to test the influence of non-osteoblastic cells on osteoclastic functions.

Testa: It is known that both mononuclear and multinucleated cells can resorb bone. You show positivity for some multinucleated cells with an osteoclast-specific antibody. Those may be the early multinucleated cells that are just being generated in the culture. I hope you are not saying that those multinucleated cells are not osteoclasts. You said that bone resorption is not related to the degree of multinuclearity. Is that right?

Chambers: I don't think that multinucleation is any guide to the presence of osteoclasts in long-term bone marrow cultures. Macrophages can produce multinucleate cells, and I don't believe that macrophages have a close relationship to osteoclasts.

Rodan: What proportion of macrophage polykaryons and osteoclasts are there in the bone marrow culture?

Chambers: It varies enormously and they are quite independent of each other. Macrophage fusion is very unpredictable: sometimes huge numbers appear in one culture, while in others there are none at all. I think the same applies to osteoclast differentiation: some cultures show many, others few. The number of osteoclasts that develop does not seem related to the number of macrophages present.

Martin: We have been working in collaboration with T. Suda in Tokyo, who has been looking at the generation of osteoclasts in mouse marrow cultures. He has looked for calcitonin receptors by autoradiography as a marker for the developing osteoclast. Roughly 10–15% of the cells that are multinucleated cells in those experiments appear to be osteoclasts, as judged by calcitonin receptors.

Chambers: Those are very interesting data. Would you take it to suggest that macrophages are the precursor for osteoclasts?

Martin: No, I wouldn't interpret it that way at all, either positively or negatively.

Hanaoka: Dr Suda's group recently changed their concept and presumed that Dr Mundy's scheme (Mundy & Roodman 1987) is correct. That is: CFU-GM proliferates into the promonocyte which in turn differentiates into two cell lineages. One is a lineage of monocyte-macrophage which proliferates into the giant cell, and the other is a lineage of the early preosteoclast, which proliferates via the late preosteoclast into the osteoclast. They no longer believe that giant cells formed from the macrophage are osteoclasts.

Nijweide: How much serum did you put in these cultures where you used 10^{-8} M vitamin D? I think this is specific for rabbit; it's not a general phenomenon, because in chick and quail you can stimulate osteoclast formation without any vitamin D, apart from what is already in the serum.

Chambers: We added 10% fetal calf serum.

Canalis: Do you think that the factor that is released by the osteoblast and allows the osteoclast to respond to PTH is collagenase?

Chambers: I don't think so. The osteoclast resorption stimulating activity (ORSA) stimulates bone resorption on our bone slices, which don't need collagenase for osteoclasts to initiate resorption.

Canalis: I am worried by the low molecular weight of the factor. Have you done polyacrylamide gel electrophoresis of the osteoblast medium to see if there are bands below an M_r 1000? It could be that it just stuck to the Amicon membrane.

Chambers: Absolutely!

Glimcher: Could you describe that soft tissue, the osteoid between the osteoblast and the mineralized matrix? How much is there? Do you always see it?

Chambers: Yes, it's about 10 µm when osteoblasts are actively producing bone. When bone formation stops, the mineral front advances towards the bone surface, but a layer of about 0.1 µm between the cells and the mineral is always visible in the electron microscope.

Glimcher: In your model, the osteoblasts remove this layer with collagenase or other proteinases. How does the osteoclast then get by the osteoblast to resorb bone?

Chambers: I don't know. Perhaps the osteoclast normally lives under the osteoblast.

Termine: It is possible to over-generalize the collagenase idea. Collagenase expression is not universal in all osteoblasts in all species. It has a tendency to be very high in fetal osteoblasts, where collagenase expression is perhaps involved in the remodelling of the mesenchyme. There is no mRNA for collagenase in adult human or rat osteoblasts, which would be the ones operative in normal bone turnover. This does not eliminate that mechanism as a possibility in fetal modelling, but I am not sure about adult remodelling.

Chambers: I think this depends on our definition of osteoblasts. I am not sure that one necessarily sees the cells that might produce collagenase in cultures of adult human bone cells. It's likely that only a very small proportion of osteoblastic cells produce collagenase *in vivo* at any particular time, at sites appropriate for resorption.

Termine: One needs to do better studies than have been done so far. The models that have been proposed may be suitable for one kind of cell at one stage of development, but we have no *in situ* hybridization for example, localizing collagenase activity *in vivo* at any point along the osteoblast differentiation pathway, to support them.

Hauschka: When you look at the action of osteoclasts *in vitro* on a slice of bone there is essentially no osteoid but mineralized bone right up to the surface. However, in our calvaria resorption models, the calvaria is covered with osteoblasts on its surface, and yet you appear to be putting osteoclasts on that. Are those osteoblasts first stripped off with collagenase?

Chambers: No, osteoblasts are first removed with water or ammonium hydroxide.

Hauschka: Therefore those osteoclasts applied to a calvaria are apparently attacking a true osteoid surface, as best you can produce it. Can you artificially produce an osteoid-like surface on a bone slice by transient demineralization with acid and generate the same kinds of action?

Chambers: We have done those experiments. We have pretreated bone slices with agents like collagenase that remove the organic material and agents like acid that remove the mineral, or neither. We compared resorption, by osteoclasts, of those three different sorts of substrates. Osteoclasts didn't excavate the surface of bone slices that had been treated with acid to remove the mineral.

Baron: We have repeated your experiments, using a slightly different system, and we arrived at exactly the same conclusions: if we add PGE_2, IL-1, PTH or $1,25-(OH)_2D_3$, we don't see any stimulation of resorption. If we add calcitonin, in chick cultures, we do get inhibition of bone resorption. However, there are many problems in interpreting the results: for instance, I doubt that you can get pure osteoclasts in your bone slices. Also, you have not tried any controls in terms of specificity of the osteoblasts to restore the osteoclastic response.

Many electron microscopists would attest that a thin layer of non-mineralized material is not found on *every* surface of bone. It depends on factors such as how old the surface is, and how the tissues are prepared. Your experiment on the appearance of the bone nodules could be interpreted in two ways. You have used calvariae, covered with osteoid tissue, to test your hypothesis. As you know, osteoclasts do not attach to the osteoid layer. The appearance of the nodules could mean that the soft tissue has been removed by collagenase, as you suggest, or that the osteoid has mineralized *in vitro* up to the surface. You have a lot of interesting data: the observations are true and reproducible. Your interpretations, however, are often too dogmatic.

Chambers: We have never claimed that our osteoclastic preparations are pure. We use an entirely functional approach. We get no response with parathyroid hormone in our osteoclastic preparations, and therefore we add something back and we get a response. We take it that it is the cells that we add back that make the difference.

Those people who have looked carefully at bone surfaces for this unmineralized organic material have all reported its presence. I have not seen any work that offers convincing evidence of its absence. This layer is removed by osteoblasts in culture. We observe the smooth fibrillary surface of osteoid becoming disrupted, and eventually being removed from the surface of the calvaria, exposing nodules of mineral. Osteoid removal correlates with the potential of the substrate to induce osteoclastic resorption on subsequent incubation with osteoclasts.

We haven't used cells other than osteoblasts extensively in our co-cultures, because we have never really seen any need to do so. Cells of the osteoblastic lineage are in bone, close to osteoclasts on bone surfaces, and it is likely that these osteoblastic cells regulate osteoclasts in intact bone. If we found that the same sort of regulation could also be achieved by skin fibroblasts, for example, that wouldn't alter my attitude about how bone is regulated, because skin fibroblasts are not normally in the vicinity. A lot of systems depend on proximity for specificity.

Hanaoka: I would like to present some of our work on the origin of osteoclasts. A previous study on the fate of hypertrophic chondrocytes (Hanaoka 1976) taught me the difficulty of interpreting experiments using markers. A phenomenon in which a few nuclei in the osteoclast are labelled while the rest are unlabelled does not necessarily mean that an initially labelled mononuclear cell is a precursor of the multinucleated osteoclast. This is easily understandable if one remembers that the hypertrophic chondrocyte or osteocyte cannot be a precursor cell of the osteoclast (Hanaoka 1976, 1979). However, many researchers (such as Buring 1975, Fischman & Hay 1962, Göthlin & Ericsson 1973, Jee & Nolan 1963) have interpreted the transfer of markers from nuclei of haemopoietic cells to a few out of many osteoclast nuclei as evidence for the haemopoietic origin of the osteoclast. They did not take into account all the possible origins of the unlabelled nuclei.

If an initially labelled mononuclear cell is proved to transform into either a preosteoclast or a mononucleated osteoclast, or if all nuclei in the osteoclast are labelled by fusion of only initially labelled mononuclear cells, then we would agree with the haemopoietic origin of the osteoclast. All the results of previous experiments with markers should be interpreted to mean that mononuclear cells with markers have fused with the preexisting osteoclast. Zambonin Zallone et al (1984) observed by microcinematography that monocytes do fuse with preexisting osteoclasts. Successful treatment of an osteopetrotic female patient by transfusion of male bone marrow cells (Coccia et al 1980, Sorell et al 1981), and experiments using giant lysosomes in beige mice (Ash et al 1980, Marks & Walker 1981) should be interpreted similarly. In the treatment of a female with male bone marrow cells, the female nuclei are naturally of host origin. Even if nuclei with Y chromosomes in the osteoclast are of bone marrow origin, it does not necessarily mean that the host female nuclei are of haemopoietic origin because bone marrow cells are a mixture of various cells, including osteoclast precursor cells (Burger et al 1987). It is presumed that incompetent preexisting female osteopetrotic osteoclasts are activated by fusion with healthy male mononuclear cells.

We did three experiments (Yabe & Hanaoka 1985): (1) six-day-old quail limb buds were directly implanted into chick chorioallantoic membrane, in a manner similar to the experiment of Kahn & Simmons (1975); (2) viable or devitalized Dunn osteosarcomas were implanted onto chorioallantoic membrane; (3) six-day-old quail limb buds in diffusion chambers were implanted onto chorioallantoic membrane by use of exogenous parathyroid hormone stimulation. In these experiments, osteoblasts and osteoclasts were identified from host (chorioallantoic membrane) and implant (quail limb bud). Our

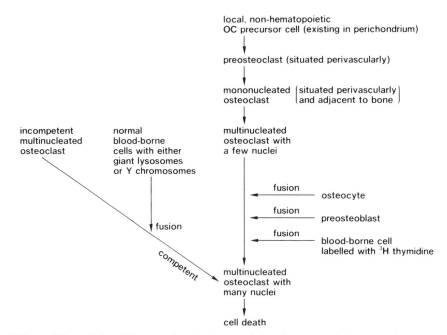

FIG. 1. (*Hanaoka*). The postulated origin and development of the osteoclast.

results proved that the experiments using quail–chick nuclear markers by others (Jotereau & Le Douarin 1978, Kahn & Simmons 1975, Thesingh & Burger 1983) have provided insufficient evidence for the haemopoietic origin of the osteoclast. In addition, we proved that the osteoclast arises without either bone marrow or haemopoiesis. This indicates that the precursor cells of the osteoclast are in the perichondrium. It has been thought that immature haemopoietic cells such as haemoblasts are also present in the perichondrium. Our experiments, and those of Crelin & Koch (1965, 1967), showed no bone marrow formation or haemopoiesis but only the appearance of the osteoclast. This indicates that these two types of precursor cells behave differently from one another, even if they both exist in the perichondrium. Therefore it is more natural to presume that they are different cells belonging to different cell lines.

Our experiments could not rule out the possibility that the osteoclast derives from mature circulating haemopoietic cells. However, recent reports (Burger et al 1987, Mundy & Roodman 1987) are against this possibility. Integration of all this information indicates that there is no definite evidence for the haemopoietic origin of the osteoclast. It seems more logical to presume that the osteoclast is derived from local, non-haemopoietic cells, just like various tumour giant cells, possibly perivascular mesenchymal cells. The osteoclasts degenerate in the way we observed in the brown tumour of hyperparathyroidism (Hanaoka & Bun 1984), and they will perhaps finally die in the same manner as the giant cell in the giant cell tumour of bone (Hanaoka et al 1970). Fig.1 illustrates this concept of the origin of osteoclasts.

References

Ash P, Loutit JF, Townsend KMS 1980 Osteoclasts derived from haematopoietic stem cells. Nature (Lond) 283:669–670

Burger EH et al 1987 Fetal bone conditioned medium stimulates osteoclast precursor cell growth but has no effect on macrophages. In: Cohn DV et al (eds) Calcium regulation and bone metabolism: basic and clinical aspects. Excerpta Medica, Amsterdam vol 9:308–313

Buring K 1975 On the origin of cells in heterotopic bone formation. Clin Orthop 110:293–302

Coccia PF, Krivit W, Cervenka J et al 1980 Successful bone-marrow transplantation for infantile malignant osteopetrosis. N Engl J Med 302:701–708

Conaway HH, Diez LF, Raisz LG 1986 Effects of prostacyclin and prostaglandin E1 on bone resorption in the presence and absence of parathyroid hormone. Calcif Tissue Int 38:130–134

Crelin ES, Koch WE 1965 Development of mouse pubic joint in vitro following initial differentiation in vitro. Anat Rec 153:161–172

Crelin ES, Koch WE 1967 An autoradiographic study of chondrocyte transformation into chondroclasts and osteocytes during bone formation in vitro. Anat Rec 158:473–484

Fischman DA, Hay ED 1962 Origin of osteoclasts from mononuclear leucocytes in regenerating newt limbs. Anat Rec 143:329–337

Göthlin G, Ericsson JLE 1973 On the histogenesis of the cells in fracture callus: electron microscopic autoradiographic observations in parabiotic rats and studies on labelled monocytes. Virchows Arch B Cell Pathol 12:318–329

Hanaoka H 1976 The fate of hypertrophic chondrocytes of the epiphyseal plate: an electron microscopic study. J Bone Jt Surg 58A:226–229

Hanaoka H 1979 The origin of the osteoclast. Clin Orthop 145:252–263

Hanaoka H, Bun H 1984 Ultrastructure of a brown tumor of bone. In: Cohn DV et al (eds) Endocrine control of bone and calcium metabolism. Excerpta Medica, Amsterdam vol 8A:417

Hanaoka H, Friedman B, Mack RP 1970 Ultrastructure and histogenesis of giant cell tumor of bone. Cancer (Phila) 25:1405–1423

Jee WSS, Nolan PD 1963 Origin of osteoclasts from the fusion of phagocytes. Nature (Lond) 200:225–226

Jotereau FU, Le Douarin NM 1978 The developmental relationship between osteocytes and osteoclast: a study using the quail-chick nuclear marker in enchondral ossification. Dev Biol 63:253–265

Kahn AJ, Simmons DJ 1975 Investigation of cell lineage in bone using a chimaera of chick and quail embryonic tissue. Nature (Lond) 258:325–327

Marks SC Jr, Walker DG 1981 The hematogenous origin of osteoclasts: experimental evidence from osteopetrotic (microphthalmic) mice treated with spleen cells from beige mouse donors. Am J Anat 161:1–10

Mundy GR, Roodman D 1987 Osteoclast ontogeny and function. In: Peck WA (ed) Bone and mineral research, Excerpta Medica, Amsterdam vol 5:209–279

Sorell M, Kapoor N, Kirkpatrick D et al 1981 Marrow transplant for juvenile osteopetrosis. Am J Med 70:1280–1287

Thesingh CW, Burger EH 1983 The role of mesenchyme in embryonic long bones as early deposition site for osteoclast progenitor cells. Dev Biol 95:429–438

Yabe H, Hanaoka H 1985 Investigation of the origin of the osteoclast by use of transplantation on chick chorioallantoic membrane. Clin Orthop 187:255–265

Zambonin Zallone A, Teti A, Primavera MV 1984 Monocytes from circulating blood fuse in vitro with purified osteoclasts in primary culture. J Cell Sci 66:335–342

Osteoclast development: the cell surface and the bone environment

Philip Osdoby, Merry Jo Oursler, Teresa Salino-Hugg and Marilyn Krukowski

Department of Cell Biology, Washington University School of Dental Medicine, 4559 Scott Avenue, St. Louis, Missouri 63110, USA

Abstract. Bone development and remodelling processes depend on complex interactions between bone cell precursors, mature bone cells, extracellular matrix molecules, growth factors, the immune system and humoral factors. The exact molecular nature of many of the cell–cell and cell–matrix interactions occurring during bone remodelling remains to be resolved. Cell surface molecules are likely to have important roles in both bone cell differentiation and regulatory processes. However, little is known about changes in the osteoclast cell surface during development and there is only limited information on the cell surface composition of the mature cell phenotype. We describe how one osteoclast-specific monoclonal antibody has been used to identify, characterize and purify a 96 kDa/140 kDa osteoclast membrane protein. The antibody has also been used as a phenotypic marker in studies designed to identify soluble and matrix-related bone factors involved in the terminal stages of osteoclast differentiation. In parallel studies using marrow-derived giant cells and the chick chorioallantoic membrane (CAM), immunohistochemical and enzyme-linked immunoassays (ELISA) have been used to investigate the influence of calvaria, calvaria-conditioned medium, bone matrix, and bone matrix components on osteoclast development. Marrow-derived giant cells express osteoclast-specific cell surface antigens when co-cultured with live calvariae or when exposed to calvaria-conditioned medium. In the richly vascularized and mesenchymal cell-containing CAM, intact bone matrix induces the formation of giant cells that express the osteoclast-specific antigens. In contrast, isolated bone matrix components implanted on the CAM recruit only mononuclear cells which are not recognized by the osteoclast-specific antibody.

1988 Cell and molecular biology of vertebrate hard tissues. Wiley, Chichester (Ciba Foundation Symposium 136) p 108–124

Dynamic and complex processes regulate bone development and remodelling. Interactive endocrine, paracrine and autocrine signals act in concert with genetic programming to orchestrate differentiation and govern the activity of bone-forming osteoblasts and bone-resorbing osteoclasts. Plasma membrane receptors are primarily responsible for sensing such signals and converting them into appropriate cellular responses. In addition, a number of enzyme activities are associated with the plasma membrane and are critical for cell function. Osteoclasts are known to possess high levels of Na$^+$/K$^+$-ATPase (Baron et al 1986) and calcitonin receptors (Warshawsky et al 1980). Beyond this, little is known about osteoclast plasma membrane components.

Recently, a number of laboratories have taken advantage of improved osteoclast isolation procedures and monoclonal antibody technology to tentatively identify both osteoclast-specific cell surface epitopes and epitopes shared by osteoclasts and other cells in the mononuclear phagocyte developmental pathway (Oursler et al 1985, Nijweide et al 1985, Horton et al 1985, Sminia & Digkstra 1986, Athanasou et al 1986). So far, no specific function has been ascribed to any of the antigens recognized by these antibodies. However, the osteoclast-specific antibodies might be used to identify osteoclast-specific molecules and to determine their role, if any, in osteoclast regulatory processes and activity. These antibodies may also serve as cell-specific markers, identifying factors responsible for the terminal stages of osteoclast differentiation. Lastly, a number of the mononuclear cell-reactive antibodies may serve to refine our understanding of the developmental and functional relationships between circulating mononuclear cells, tissue macrophages, and osteoclast precursors.

Although the monocyte has been considered the most likely immediate precursor to osteoclasts, some investigators have suggested that the osteoclast lineage diverges long before the promonocyte stage (Burger et al 1984, Horton et al 1985). This latter hypothesis seems inconsistent with reports that osteoclast precursors are present in the circulation and share antigenic determinants unique to the monocyte-macrophage family. Ferrero et al (1983) reported that adult peripheral blood contains primitive colony-forming unit granulocyte-macrophage progenitor cells (CFU-GM) that differentiate into a more mature CFU-GM which no longer circulates and which then constitutes the majority of the marrow CFU-GM pool. In addition, CFU-GMs exist as subpopulations that differ in size, density, stage of cell cycle, and responsiveness to different stimuli. Therefore, it is quite possible that osteoclast precursors, more primitive than the monocyte, are found in the circulation and within the marrow (see Fig. 1).

Osteoclast precursors may respond to specific humoral signals and factors emanating from bone. Local factors may derive from bone cells, vascular cells, immunocompetent cells associated with bone, or bone matrix components. Cell–cell and cell–matrix interactions generate a wide variety of local soluble, membrane-mediated and matrix-related signals (Canalis 1983, Hauschka et al 1986). The cascade of events and signals that guide osteoclast development through precursor proliferation, chemotaxis, attachment to bone, fusion and differentiation (Fig. 1) is likely to be dependent, in part, on the osteoblast (Kahn et al 1981, Thesingh & Burger 1983). In this context, we have reported that mononuclear-derived giant cells express osteoclast-specific antigens in the presence of live bone (Osdoby et al 1987). In this paper we

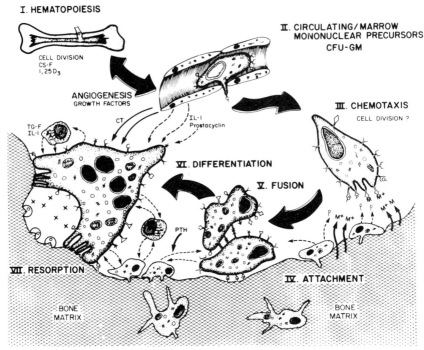

I. HEMATOPOIESIS

II. CIRCULATING/MARROW
MONONUCLEAR PRECURSORS
CFU-GM

CELL DIVISION
CS-F
1,25 D₃

ANGIOGENESIS
GROWTH FACTORS

III. CHEMOTAXIS
CELL DIVISION ?

TG-F
IL-I

CT

IL-I
Prostacyclin

VI. DIFFERENTIATION

V. FUSION

PTH

VII. RESORPTION

IV. ATTACHMENT

BONE
MATRIX

BONE
MATRIX

FIG. 1.   Diagram illustrating the possible sequence in osteoclast differentiation. A simplified view of the complex cell–cell and cell–matrix interactions required for this process is also depicted. CS–F, colony-stimulating factor; CT, calcitonin; IL-l, interleukin l; PTH, parathyroid hormone; TG-F, transforming growth factor. The process is dependent on transduction of general and specific signals via general and specific enzymes and receptors on the plasma membrane.

describe the results of experiments designed to identify the cells and/or matrix factors that may influence the terminal stage of osteoclast cytodifferentiation.

Methods

Cell isolation/culture methods and monoclonal antibody development

All cell culture procedures were carried out as previously described (Osdoby et al 1982, 1987). Chick monocytes, and marrow-derived giant cells, were generated *in vitro* from Ficoll-Hypaque fractions of circulating or marrow mononuclear cells, respectively.

   Monoclonal antibody-producing hybridomas were prepared from mice immunized with chick osteoclasts. Osteoclast antibody specificity was determined by screening monoclonal antibodies against various cell populations, including monocyte and marrow-derived giant cells. The antibody specificity was further analysed on bone tissue sections and compared against other tissues, including liver and whole chicken embryos. Immunohistochemical-immunocytochemical methods (Baron et al 1986), enzyme-linked immunosorbent assay (ELISA), and radioimmunoassay (RIA) were used for screening (Oursler et al 1985, Osdoby et al 1987). One antigen, recognized by a monoclonal antibody designated 121F, has been purified by antibody immunoaffinity protocols and characterized biochemically.

Co-culture and conditioned medium experiments

Chick hatchling calvariae or conditioned medium from cultured calvariae were used to test the effect of bone (osteoblasts) on the expression of the osteoclast 121F antigen in monocyte and marrow mononuclear cell populations. Calvariae were prepared for culture as shown in Table 3. In one series of experiments, monocytes or marrow mononuclear cells were co-cultured with calvariae. In a second series, calvaria-conditioned culture medium (diluted 1:2 with fresh medium) was added to mononuclear cell cultures. In both experiments fibroblast-conditioned medium served as the control and cultures were analysed immunohistochemically and by ELISA for the expression of the 121F antigen (Oursler et al 1987).

Matrix influence: marrow and chorioallantoic membrane experiments

The richly vascularized chick chorioallantoic membrane (CAM) was used as an implant site, *in ovo*, to investigate the role of bone matrix and bone matrix components in the recruitment and differentiation of osteoclasts. Implanting was as described by Krukowski & Kahn (1982). Materials implanted onto the CAM included mineralized bone matrix, demineralized matrix, hydroxyapatite crystals and charged beads. In parallel experiments, purified matrix components, such as osteocalcin, osteonectin, osteopontin and cartilage proteoglycan, were incorporated into the slow-release polymer ELVAX as described by Feinberg & Bebe (1983). The composite material was sterilized by ultraviolet light and either placed onto the CAM or put into culture with marrow mononuclear cells. For CAM studies, frozen sections were prepared and stained for the osteoclast-specific 121F antigen and for an antigen, designated 29C, common to osteoclasts, giant cells and macrophages. The marrow cultures were screened for 121F and 29C expression both histochemically and by ELISA.

Results

Osteoclast antibody localization and specificity

In our original publication describing the osteoclast-specific monoclonal antibody library, 30 monoclonals were identified as osteoclast-specific. One of

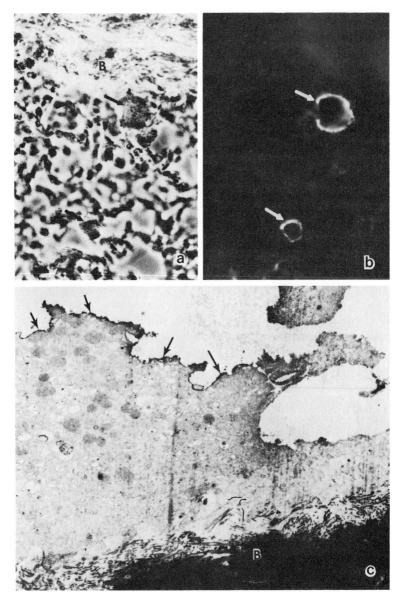

FIG. 2.   Fig. 2a is a phase-contrast micrograph of a frozen section from chick bone (B) revealing marrow and osteoclasts (arrows). Fig. 2b is the same field illuminated by ultraviolet light to visualize the 121F antibody on the osteoclast plasma membrane (× 400). Fig. 2c is a transmission electron micrograph of a bone section prepared for peroxidase localization of the 121F antigen. Peroxidase localization (arrows) is primarily on the osteoclast plasma membrane (× 6750).

TABLE 1   **Comparison of 121F antibody binding to intact and permeabilized cells by ELISA (mean ± SD)**

Cell type	Intact cells	Permeabilized cells
Day 1 monocytes	73 ± 1	50 ± 3
Day 11 monocytes (MNGC)	125 ± 7	66 ± 6
Stage 24 mesenchyme	128 ± 6	58 ± 1
Osteoclasts	636 ± 19	466 ± 5

the antigens recognized by the antibody designated 121F has been further examined by ELISA, RIA and Western blot analysis for cell specificity. We have also investigated whether the 121F epitope is restricted to the osteoclast plasma membrane. Light microscopic localization of the 121F antibody-reactive antigen in either frozen or paraffin sections showed that it is so restricted (Fig. 2a and b). Electron microscopic immunocytochemistry of bone sections supports the plasma membrane localization of the antigen (Fig. 2c). However, given the low level of reaction product and the normal dense inclusions in cell vacuoles, including lysosomes, we cannot discern whether intracellular vacuoles or endosome membranes also possess the antigen. There is evidence against this possibility: ELISA values for the antigen in permeabilized cells are not higher than values determined in unpermeabilized cells (Table 1).

Immunoaffinity purification of the 121F antigen and its analysis by silver staining or autoradiography on one- and two-dimensional gels has identified a single 96 kDa protein with isoelectric points of pH 5.9 and 6.1 (Fig. 3). Monosaccharide hapten inhibition studies, neuraminidase digestion and periodate treatment of osteoclasts does not abolish 121F antibody binding, which suggests that the antibody recognizes a peptide epitope of the molecule

TABLE 2   **Antibody epitope analysis (for carbohydrate composition)**

Antibody	Periodate		Neuraminidase[a]	
	Treatment	Control	Treatment	Control
121F	20 780 ± 785[b]	17 272 ± 911	386 ± 48	393 ± 43
29C	1618 ± 595	5836 ± 263	679 ± 59	722 ± 60

Sixteen sugars used in hapten inhibition studies failed to inhibit 121F antibody binding to osteoclasts.
[a] Neuraminidase values represent separate experiments.
[b] Mean ± SD.

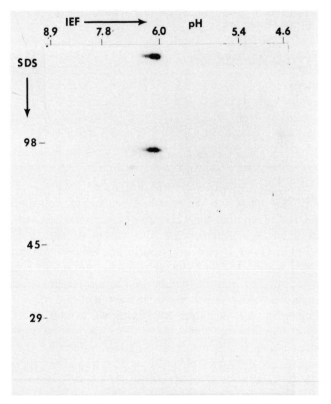

FIG. 3. Purified 121F 96 kDa antigen was chloramine T-iodinated and subjected to two-dimensional polyacrylamide gel electrophoresis and autoradiography. The first isoelectric focusing dimension reveals two closely associated isoelectric points. Both components have apparent molecular masses of 96 kDa.

(Table 2). More recently we have added other protease inhibitors to our osteoclast extraction protocol and have found that the apparent molecular mass of the 121F antigen shifts to 140 kDa on SDS-polyacrylamide gels. This is being further investigated.

Co-culture and conditioned medium experiments

Table 3 summarizes the effects of various calvarial preparation techniques on the expression of the 121F antigen when calvariae are co-cultured with bone marrow cells. The 121F antigen is only expressed in marrow cells that have been exposed to live, periosteum-stripped calvariae. In frozen/thawed preparations of calvariae cultured with marrow, or in calvarial cultures lacking marrow, antigen expression is not detected, suggesting that the stimulation of antigen expression is a conditioned-medium response. It has also been re-

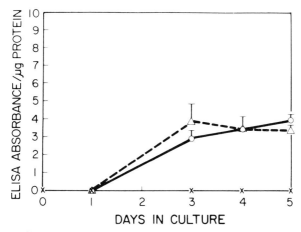

FIG. 4. The effect of PTH on the time course of 121F antigen expression in marrow-derived giant cells exposed to calvaria-conditioned medium. (-△-) regular conditioned medium; (-○-) PTH-stimulated conditioned medium; (X) control medium.

ported that parathyroid hormone (PTH) may augment osteoclastic bone resorption via diffusible osteoblast secretory products. PTH-induced osteoblast signals may also stimulate osteoclast differentiation. To examine these possibilities we cultured calvariae alone, with and without PTH. Conditioned medium taken from these cultures was then added to adherent marrow cells. The results of such experiments confirm that 121F antigen expression is induced by calvaria-conditioned medium. Moreover, PTH did not affect the level or the time course of 121F antigen expression in marrow cultures (Fig. 4).

Immunohistochemical localization of the 121F antibody in marrow cultures exposed to calvaria-conditioned medium verified that the 121F antigen was present on the membranes of a high percentage (70–90%) of the

TABLE 3   Expression of the 121F antigen in marrow–calvaria cultures measured by ELISA

Calvaria preparation	121F (ELISA)
Intact	15 ± 12
Periosteum-stripped	476 ± 20
Collagenase-treated	26 ± 8
Frozen/thawed periosteum-stripped	12 ± 15
Calvaria alone (no marrow cells)	39 ± 46
Marrow cells alone (no calvaria)	64 ± 39

[a] Mean (± SD) values determined on cell attached to plastic.

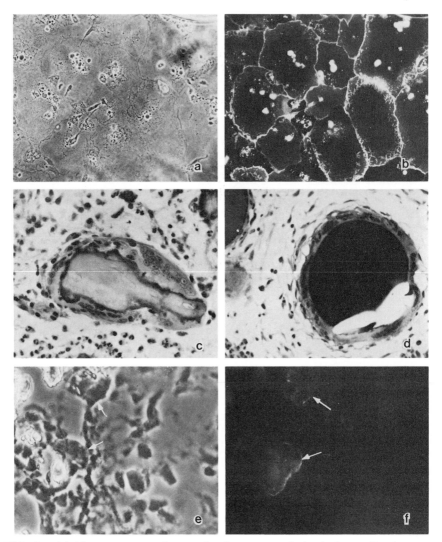

FIG. 5. (a) Phase-contrast micrograph of a marrow-derived giant cell culture exposed to calvaria-conditioned medium (× 195). (b) Fluorescence pattern of 121F antibody localization in the same field as (a) (× 195). (c) Section of chick CAM illustrating the multinucleated giant cells that form in association with mineralized bone particles (× 453). (d) Section through the CAM showing giant cells that form in association with charged beads (× 350). (e) Phase-contrast micrograph of a frozen section prepared from the chick CAM implanted with mineralized bone particles (rat) and stained for the localization of the 121F osteoclast-specific antibody. Osteoclasts are arrowed (× 195). (f) Same field as (e) illustrating 121F antibody localization on the plasma membrane of giant cells associated with bone (× 195).

mononuclear-derived giant cells (Fig. 5a and 5b). Fibroblast-conditioned medium did not induce expression of the 121F antigen; therefore, 121F expression is not caused by depletion of an inhibitor of antigen expression. Parallel experiments using cultures of circulating mononuclear cells demonstrated that under the culture conditions described above, when co-cultured with calvariae or when exposed to calvaria-conditioned medium, monocytes fail to express the 121F antigen.

Matrix influence: marrow cultures and chorioallantoic membrane experiments

The observation that devitalized bone failed to induce osteoclast antigen expression *in vitro* suggests that the bone matrix does not play a role in the terminal stages of osteoclast differentiation, as marked by 121F antigen expression. However, it is quite possible that the *in vitro* conditions were less than optimal to achieve such a matrix effect. Fig. 5c demonstrates that when mineralized bone particles are placed on the CAM, multinucleated giant cells form in association with the bone. Immunohistochemical staining for the 121F antigen reveals that the antigen is present on the cell surface of the bone matrix-associated giant cells (Fig. 5e and 5f). CAM giant cells formed in response to hydroxyapatite crystals or to beads that create a positive counterionic environment (Fig. 5d), express the 29C antigen common to osteoclasts, giant cells and macrophages, but fail to express the 121F osteoclast-specific antigen (results not shown).

Table 4 summarizes the results of experiments designed to determine whether specific matrix components, either alone or in combination, might influence the recruitment of mononuclear cells positive for 29C antigen expression (putative osteoclast precursors). Table 4 also displays the results of the immunohistochemical localization of the 121F osteoclast antigen in the CAM studies, and the results of ELISA analysis of 121F antigen expression in marrow mononuclear cultures maintained in the presence of various preparations of bone matrix and bone matrix components. The observations derived from the CAM are based on an eight-day exposure to test material, whereas the marrow cultures were exposed to test material for 10 days.

Implantation of most of the bone matrix components into the CAM resulted in the appearance of mononuclear cells positive for the 29C antigen. However, certain matrix components appeared to elicit a stronger response than others. The most effective components were osteocalcin, osteonectin, osteopontin and hydroxyapatite. Few, if any, 29C-positive cells were found in association with ELVAX alone. In the CAM studies, large numbers of giant cells formed only in response to mineralized matrix, hydroxyapatite crystals, and beads creating a positive counter-ionic charge. A few giant cells were observed in response to osteopontin.

Marrow cells cultured with various preparations of bone particles or with matrix components all formed giant cells which failed to express the 121F

TABLE 4 Matrix influences on osteoclast development

Matrix component	CAM studies			Marrow cultures	
	Histology	29C expression	121F expression	Histology	121F expression
1. Mineralized matrix	+ GC	+++	+++	GC	---
2. Demineralized matrix	– GC	++	---	GC	---
3. Hydroxyapatite	+ GC	+++	---	GC	---
4. Type I collagen	– GC	++	---	GC	---
5. Pro(α) type I collagen	– GC	++	---	GC	---
6. Osteocalcin	– GC	+++	---	GC	---
7. Osteopontin	+ GC	++	---	GC	---
8. Osteonectin	– GC	++	---	GC	---
9. Mix of 6–8	– GC	++	---	GC	---
10. Bone-derived growth factor	– GC	+	---	GC	---
11. Cartilage proteoglycan	– GC	+	---	GC	---
12. Hyaluronic acid	– GC	+	---	GC	---
13. Cartilage-inducing factor	– GC	+	---	GC	---
14. Laminin	– GC	+	---	GC	---
15. Fibronectin	– GC	+	---	GC	---
16. ELVAX alone	– GC	–	---	GC	---

GC, Giant cell. CAM values based on immunohistochemistry; marrow culture values based on ELISA.

antigen; these findings are in line with the results of the calvaria co-culture studies.

Discussion

The development of osteoclast-specific monoclonal antibodies is the first approach in our attempts to define the osteoclast cell surface and to understand the signals that influence osteoclast activity and development. At this point in our studies we are using the term 'osteoclast-specific' with caution. It is clear that more extensive analysis will be necessary to ensure that the epitopes recognized by these antibodies are specific. One caveat, relevant to all antibody studies which claim a phenotypic-specific antigen, is that the antigen may be present on other cells at levels below the sensitivity of the assays utilized. Similarly, although we may have identified a unique surface osteoclast epitope, we do not yet claim that this epitope is found on a protein unique to osteoclasts. Amino acid sequence data and the development of cDNA probes should help to resolve this issue. Nonetheless, we have been able to use one of the osteoclast monoclonal antibodies to identify and purify a 96 kDa/140 kDa osteoclast membrane protein which, so far, has not been detected on other cells of the mononuclear phagocyte lineage pathway. In preliminary experiments, not presented here, we have used the purified antigen to identify an 80 kDa serum factor. Exposure of osteoclasts to this factor elevates osteoclast cyclic AMP levels (Oursler & Osdoby 1986). Further characterization of the 80 kDa ligand and the osteoclast response elicited from the 'receptor'–ligand interaction is in progress.

The marrow mononuclear cell/calvaria/calvaria-conditioned medium studies reported here demonstrate that soluble factors found in the bone environment can promote the terminal stages of osteoclast differentiation *in vitro*. These studies can be interpreted to indicate that the signals responsible for this induction are produced by the cells themselves, and/or that the cells are actively releasing one or more component(s) from the extracellular matrix. The failure of PTH to augment antigen expression indicates that any action of PTH in promoting osteoclast differentiation may occur earlier in the sequence. The inability of mineralized bone matrix or its components to influence antigen expression under these *in vitro* conditions supports earlier *in vitro* studies (Ko & Bernard 1981, Burger et al 1984), in which it was shown that live bone was required for osteoclast differentiation.

In vitro differentiation studies must always be interpreted with caution. The differentiation process probably depends on a sequential cascade of events involving spatiotemporal relationships that may be difficult or even impossible to reproduce in culture. The calvaria-conditioned medium is a complex mixture of molecules that contains a group of factors which together may promote a sequence of events leading to 121F antigen expression in marrow-derived giant cells. Therefore, the failure of bone matrix to promote marrow-derived giant cell expression of the 121F antigen *in vitro* does not necessarily exclude a role for matrix constituents in the terminal stages or earlier events required for osteoclast differentiation.

Bone extracellular matrix is mineralized and type I collagen is a major component. Additional matrix proteins have been identified; some are products of osteoblasts, whereas others derive from circulating proteins trapped in the matrix. A number of growth factors and differentiation molecules are also found in the matrix. Many of these molecules have yet to be assigned a function. The chick CAM studies reported here demonstrate that bone matrix and many of its components serve as chemotactic attractants for circulating mononuclear cells. Once these cells leave the bloodstream they develop the 29C antigen and many ultimately attach to bone particles and hydroxyapatite crystals. The fact that giant cells form in response to mineralized rat and chick bone, hydroxyapatite crystals, and beads that generate a positive counterionic environment suggests that the charge of the particles may influence the recruitment or fusion of mononuclear cells. Because mineralized matrix-induced giant cells express the 121F osteoclast-specific antigen, whereas hydroxyapatite or charge-induced giant cells fail to do so, components of the bone matrix (perhaps in association with bone mineral) may participate in the terminal stages of osteoclast cytodifferentiation. Furthermore, the observation that both chick and rat bone can promote this induction suggests that the response is not a xenograft reaction — nor is it species specific.

The CAM studies differ from the *in vitro* studies in that bone matrix devoid of live osteoblasts induces osteoclast-specific antigen expression. The CAM contains mesenchymal cells and blood vessels. It is thus possible that the bone particles induce ectopic osteoblast expression on the CAM. We have not detected the emergence of any alkaline phosphatase-positive cells or any indication of *de novo* mineralized matrix formation. However, under these conditions, the bone matrix without functional osteoblasts can promote osteoclast differentiation. Mesenchymal cell and vascular cell factors, as well as humoral signals transported via the CAM vasculature (some of which may originate from osteoblasts), may provide general signals which, in the presence of bone matrix, promote the final steps in osteoclast development.

Further dissection of the bone environment and the selective re-association of components for testing could lead to a better understanding of those factors in bone that play a role in osteoclast precursor recruitment, fusion and differentiation. This, coupled with monoclonal antibody technology to probe the osteoclast surface and with cultured mononuclear precursors to assess behaviour in a variety of bone-stimulated environments, could produce a clearer picture of the osteoclast at its earliest inception, in its developing stages, and finally as the mature resorbing cell in skeletal tissue.

Acknowledgements

The authors are grateful to Ms Earlene Schmitt, Ms Lue Vurn McPeters, Ms Anita Summerfield, Mr Michael Wilkemeyer and Mr Fred Anderson for their expert technical assistance and Ms Terri Moulton for preparing the manuscript. We are indebted to the following investigators for contributions of matrix components: Dr Julie Glowacki (mineralized and demineralized rat bone), Dr Peter Hauschka (osteocalcin), Dr Ahnders Franzén (osteopontin), Drs Larry Fischer and John Termine (Pro-alpha type I collagen and osteonectin), Drs Arnold Caplan and Glenn Syfstestad (hyaluronic acid, cartilage proteoglycan, and cartilage inducing factor), Dr Ernesto Canalis (bone-derived growth factor, alpha-2 microglobulin) and Dr Hinda Kleinman (laminin). This work was supported by NIH grants DE-06891, AR-32927, AR-01474, GM-07067, and DE-05471.

References

Athanasou N, Quinn H, Gatter K, Mason D, McGee J 1986 Osteoclasts contain macrophage and megakaryocyte antigens. J Pathol 150:239–246

Baron R, Neff L, Roy C, Boisuert A, Caplan M 1986 Evidence for a high and specific concentration of (Na^+, K^+) ATPase in the plasma membrane of the osteoclast. Cell 46:311–320

Burger E, Van Der Meer J, Nijweide P 1984 Osteoclast formation from mononuclear phagocytes: role of bone forming cells. J Cell Biol 99:1901–1906

Canalis E 1983 The hormonal and local regulation of bone formation. Endocr Rev 42:62–71

Feinberg R, Bebe D 1983 Hyaluronate in vasculogenesis. Science (Wash DC) 220:1177–1179

Ferrero D, Broxmeyer H, Pagbsardi G et al 1983 Antigenically distinct subpopulations of myeloid progenitor cells (CFU-GM) in human peripheral blood and marrow. Proc Natl Acad Sci USA 80:4114–4118

Hauschka P, Mavrakos A, Iafrati M, Doleman S, Klagsbrun M 1986 Growth factors in bone matrix. J Biol Chem 261:12665–12674

Horton M, Rimmer E, Moore A, Chambers T 1985 On the origin of the osteoclast: the cell surface phenotype of rodent osteoclasts. Calcif Tissue Int 99:46–50

Kahn A, Simmons D, Krukowski M 1981 Osteoclast precursor cells are present in the blood of preossification chick embryos. Dev Biol 95:429–438

Ko J, Bernard G 1981 Osteoclast formation in vitro from bone marrow mononuclear cells in osteoclast-free bone. Am J Anat 161:415–425

Krukowski M, Kahn A 1982 Inductive specificity of mineralized bone matrix in ectopic osteoclast differentiation. Calcif Tissue Int 34:474–479

Nijweide P, Vrijheide-Lammers T, Mulder R, Blok J 1985 Cell surface antigens on osteoclasts and related cells in the quail: studies with monoclonal antibodies. Histochemistry 83:315–324

Osdoby P, Martini M, Caplan A 1982 Isolated osteoclasts and their presumed progenitor cells, the monocytes, in culture. J Exp Zool 224:331–344

Osdoby P, Oursler M, Salino-Hugg T, Wilkemeyer M, Anderson F 1987 The cell surface and osteoclast development. In: Sen S & Thornhill T (eds) Development and diseases of cartilage and bone matrix. Alan R. Liss, New York, p 221–231

Oursler M, Osdoby P 1986 Identification of a serum protein reactive with an osteoclast-specific membrane molecule. J Cell Biol 103:218a

Oursler M, Bell L, Clevinger B, Osdoby P 1985 Identification of osteoclast-specific monoclonal antibodies. J Cell Biol 100:1592–1600

Oursler M, Salino-Hugg T, Wilkemeyer M, Osdoby P 1987 Soluble bone factors induce osteoclast expression in vitro. J Bone Miner Res 2:258

Sminia T, Digkstra A 1986 The origin of osteoclasts: an immunohistochemical study on macrophages and osteoclasts in embryonic rat bone. Calcif Tissue Int 39:263–266

Thesingh L, Burger E 1983 The role of mesenchyme in embryonic long bones as early deposition site for osteoclast progenitor cells. Dev Biol 95:429–438

Warshawsky H, Goltzman D, Rouleau M, Bergeron J 1980 Direct in vitro demonstration by radioautography of specific sites for calcitonin in skeletal and renal tissue of the rat. J Cell Biol 85:682–694

DISCUSSION

Nijweide: When you studied the effects of calvariae stripped of their perios-teum on osteoclast formation in the bone marrow cultures, did all the cells become positive for the 121F antibody?

Osdoby: About 70% of the cells were positive.

Nijweide: I find it hard to believe that all these cells are osteoclasts. We have difficulty in getting such a large number of osteoclasts in culture.

Osdoby: We do not begin with osteoclasts. We begin with marrow mononuc-lear cells that fuse in culture and, in the presence of a bone extract, express osteoclast-specific antigens.

Nijweide: So is the 121F antibody a marker for osteoclasts or not?

Osdoby: We consider it a marker for osteoclasts and we conclude that diffusible components are released from bone that may be responsible for the induction of antigen expression—that is, for the terminal stages of osteoclast differentiation.

Nijweide: You presume that 70% of your cells in this culture are osteoclast precursors?

Osdoby: I would not comment on the number of the precursors present.

Martin: You mentioned that the ligand for the 121F antibody increased cyclic AMP in the multinucleate cells. Was it a substantial effect that could exclude the contaminating mononuclear cells or osteoblasts? Do other factors, such as calcitonin, increase cyclic AMP?

Osdoby: We have examined monocyte giant cells and monocytes that do not have this antigen on the surface, and we do not see an increase in cyclic AMP levels in response to the ligand. That would suggest that this effect is associated with the osteoclasts.

Chick osteoclasts appear to respond to calcitonin but we have not published the data because the experiments were difficult to reproduce. When we isolate osteoclasts we usually expose them to serum. If we label the osteoclasts using lactoperoxidase cell surface iodination, before we expose them to serum, we see one profile of cell surface proteins. If we expose the osteoclasts to serum and then label the surface we see a different profile. A possible interpretation is that there are proteins in serum in the osteoclast isolation procedures or in culture that are masking or down-regulating calcitonin receptors. Why this should occur only in the chicken, I do not know.

Raisz: How do you calculate the abundance of the antigen in the osteoclast as compared to the mononuclear cell? The ELISA (Table 1) showed a positive reaction in the mononuclear cell macrophage family—even though it was one-fifth that of the osteoclast.

Osdoby: In our hands the ELISA has a background of between 50 and 100 units. We don't consider the values to be positive until they rise above the background levels. The values are subtracted for background but are within the error of that background.

Raisz: The standard errors were small. If a large multinucleated osteoclast expresses five times as much antigen as a macrophage, that difference, while very important, isn't conclusive, and the macrophage may be expressing this antigen at a lower level.

Osdoby: I agree that the reality of a specific protein has yet to be determined. There may be relative levels of expression. That there are differences between osteoclasts and monocytes is important, as Dr Baron has shown with Na^+/K^+-ATPase (Baron et al 1986). It may be just as important for other antigens as well, but we cannot say unequivocally that a cell is not expressing something, from these ELISA values.

Raisz: Have you taken your 80 kDa serum factor and seen whether it has any effect on cyclic AMP in macrophages?

Osdoby: Yes, it has no detectable effect.

Pierschbacher: Do osteoclasts bind to bone more tightly than to, say, hydroxyapatite? Did you try taking purified osteopontin mixed with hydroxyapatite?

Osdoby: Those experiments are in progress.

Glowacki: We have some preliminary data from our work with rat implants that might contribute to all the hypotheses about the factors involved in osteoclast differentiation. Fig.1 shows the characteristics of osteoclasts.

Two weeks after implantation of bone particles, recruited multinucleated cells have the following features: they resorb bone matrix; their activity is modulated by calcitonin, heparin, protamine and glucocorticoids; ultrastructurally they have ruffled borders and clear zones; they are multinucleated; they have giant centrospheres, lysosomes and vacuoles; they are positive for tartrate-resistant acid phosphatase (TRAP) activity; they are negative for non-specific esterase activity.

With Steve Goldring we have shown that these bone particle (BP)-elicited multinucleated cells have calcitonin receptors (Goldring et al 1988). By time-lapse phase-contrast, as well as differential interference microcinematography, these cells can be shown to have calcitonin-induced shape changes *in vitro*. In addition, they are reactive to an antibody provided by S. Toverud that was generated against rat bone tartrate-resistant acid phosphatase. None of these features is demonstrated by foreign body giant cells. Therefore, we feel that we have ways of distinguishing between subcutaneous recruitment and the differentiation of osteoclasts generated by bone particles and foreign body giant cells that can be generated in response to particulate polymethylmethacrylate, polyethylene, or hydroxyapatite. We have used various forms of hydroxyapatite in these studies: ashed bone; deorganified bone; and chemically synthesized hydroxyapatite.

Previously, with Jane Lian, we had found a decreased recruitment of these

OSTEOCLAST

- Multinucleated, bone-resorbing cell
- Formed by the fusion of mononuclear precursors
- Lysosomal enzymes
- Tartrate-resistant acid phosphatase
- Calcitonin receptors
- Ruffled borders and clear zones
- Regulated by PTH, $1,25\text{-}(OH)_2D_3$, PGE_2, IL-1

FIG. 1 (Glowacki).   Characteristics of osteoclasts.

osteoclast-like cells in response to bone particles that were produced from warfarin-treated animals. The only abnormal feature of these bone particles was their deficiency in osteocalcin. That led us to the hypothesis that osteocalcin may be one of the bone particle components that contributes to the generation of the osteoclast phenotype.

We have prepared crystalline hydroxyapatite in the presence or absence of bovine serum albumin or rat osteocalcin. Two weeks after implantation of the hydroxyapatite particles, they remain large with straight, sharp edges and are surrounded with multinucleated cells that are TRAP-negative. The experiment with the bovine serum albumin produced very similar results. Those hydroxyapatite particles that were admixed with osteocalcin to a final concentration of 0.1% osteocalcin were partially resorbed (they were much smaller after two weeks) and some of the multinucleated cells were TRAP-positive. Electron micrographs demonstrate the membrane specialization of these cells. In response to the hydroxyapatite particles, one can see clear zones of attachment of the multinucleated cells to the particles, but with the composite hydroxyapatite/osteocalcin substrate, membrane foldings, suggestive of ruffled borders, were seen along the surfaces of these multinucleated cells, particularly adjacent to concavities on the surface of the particle. These results with synthetic substrates support the hypothesis that one of osteocalcin's functions may be in the recruitment and differentiation of the osteoclast.

References

Baron R, Neff L, Roy C, Boisuert A, Caplan M 1986 Evidence for a high and specific concentration of (Na^+,K^+) ATPase in the plasma membrane of the osteoclast. Cell 46:311–320
Goldring SR, Roelke M, Glowacki J 1988 Multinucleated cells elicited in response to implants of devitalized bone particles possess receptors for calcitonin. J Bone Miner Res 3:117–120

General discussion I

Osteoclast activity

Williams: As I see it, osteoclasts land on a bone surface and there they make an acid zone (Fig.1). At the same time the cell surface develops villi or becomes extremely ruffled. There will be enzyme sorting, due to the changing surface tension (curvature) in the membranes. Certain proteins will move into the zones of different curvature. We shall not understand the function of the bone resorption and the attack on bone unless we know which enzymes move round the surface of the cell and into the dissolution zone. If there is an acid zone, I don't think collagenase would work in that zone at all. Again, perhaps osteoclasts eject acid phosphatases into the acid zone to destroy the phosphorylated proteins of the bone.

Hydroxyapatite is also interesting in this context. It is a proton conductor, and if the acidity of the matrix is altered, protons will diffuse into the bone. Metal ions may be released, not into the trap where the acid is, but into the remoter regions at the outside of the cell zone (see Fig. 1). If all this is true, the cell has completely reorganized its membrane on binding to the bone, and it won't be sufficient to look at it with antigens that have been developed against cells in free solution.

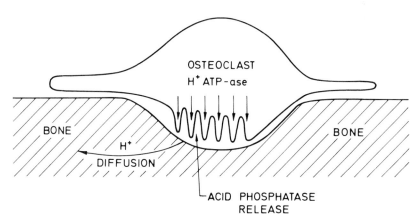

FIG. 1 (Williams). A schematic representation of an osteoclast showing the microvilli, the H^+ ATPase in these villi, the trapped acid aqueous zone into which acid phosphatase is excreted, and the movement of the H^+ in the bone matrix to release calcium at the edges of the cell.

125

Baron: It's clear that there are two membrane domains that are able to segregate membrane proteins. A 100 kDa protein which might be associated with the proton pump system is found in lysosomal membrane. We have also shown a very high concentration of Na^+/K^+-ATPase at the basolateral membrane domain of the osteoclast.

We have demonstrated that there is an acid pH in the bone-resorbing lacuna and that a number of lysosomal enzymes, not only acid phosphatase, are secreted into this extracellular compartment and digest the matrix outside.

Williams: The acid phosphatases are quite different from the alkaline phosphatases and demand a different kind of mineral chemistry in the cell. The alkaline phosphatases are zinc-dependent enzymes, the acid phosphatases are not: they are manganese-and iron-dependent. There might be a special uptake mechanism, via a small cofactor for example, into this cell for manganese. If so, the osteoclast would have some special cellular requirements for mineral elements.

Raisz: Paula Stern (1985) showed that addition of manganese could enhance PTH stimulation of bone resorption.

Osdoby: We are just beginning to explore the polarity of osteoclasts and its importance in their function. Perhaps we can begin to understand what is being kicked into that zone to break down the collagen and other components. There does not appear to be collagenase in osteoclasts. However, if they are producing a relatively acid environment, that may be enough to denature the collagen to make it susceptible to other proteases. Collagenase as such might not be necessary.

Baron: Regarding the pH issue, one has to think in terms of the gradient of proton concentration from the cell membrane, where protons would be pumped out, towards the matrix. Morphologists see demineralization and degradation decreasing with distance from the membrane. So it is very possible that there is collagenase in this matrix which could be activated by some proteolytic cleavage, or any other means, and yet still function at close to neutral pH if it is at some distance from the membrane of the osteoclast's ruffled border.

Krane: Dr Osdoby, does your 121F antibody cross-react with the Na^+/K^+-ATPase?

Osdoby: Doug Fambrough provided some ATPase antibody and we have not been able to see any cross-reactivity to the 121F antigen. When we have the sequence information and the cDNA we shall be able to confirm this.

Raisz: After the osteoclast has apparently finished it's job, there is a mononuclear cell which has been found in the 'reversal' phase. I gather that it is not a typical preosteoblast.

Baron: We are still trying to find out what these cells are. There are a number of possibilities. The bone surface after resorption may be in such a state that it chemoattracts macrophages that would work on it. We have very indirect suggestive evidence for that. Alternatively, there is some evidence for a turnover of nuclei in osteoclasts. In other words, as osteoclasts incorporate precursors they also 'lose' precursors at a similar rate. We have data that could indicate that once the osteoclast finishes its work, it goes into mononuclear units again. The main difficulty is that these reversal sites are relatively rare and difficult to characterize as being in that precise stage of differentiation.

Prockop: The acid at the resorption site will solubilize the collagen, but I don't think it will denature it. The melting temperature of collagen is lowered when you go from neutral pH to acid pH, but only by a few degrees. If you make the solution acidic, you still need a collagenase, particularly if the collagen is cross-linked. The protein at acid pH is very resistant to proteinases.

Williams: Isn't collagenase a typical zinc protease with a bell-shape curve for its pH dependence and not really active at pH 4?

Prockop: That's correct.

Williams: So collagenase won't work, not because it is denatured—you can reverse the activity by returning it to pH 7 in the same way as the other zinc proteases—but because catalysis needs a deprotonation reaction which can't happen at pH 4.

Termine: Steve Teitelbaum has reported an acid collagenase which does not function in the same way as normal mammalian collagenase. It is possible that the osteoclastic acid environment triggers this other protein, which evidently breaks collagen down to 10K fragments without specifically going through a ¾ and ¼ length cleavage.

Krane: This activity appears to be similar to cathepsin B1 described by Burleigh et al (1974). Cathepsin B1 is a protease that degrades collagen at acid pH.

Veis: At that pH and physiological temperature you do not necessarily need a classical collagenase.

Weiner: Is there any indication that osteoclasts have a preference for bone of a specific age? Can they resorb bone that was formed a few minutes before, or do they only resorb bone that has been mineralized to a certain degree, or where the collagen is cross-linked or deformed to some extent?

Hauschka: Both the demineralized bone powder implant system that Julie Glowacki and A.H. Reddi have studied, and a reconstituted osteoinductive implant system which we have characterized, demonstrate that the resorption of very recently made bone tissue (≤ one week old) is quite rapid and robust.

Termine: After bone is laid down, a substantial degradation of the matrix occurs. While this is happening, there is a long-term accretion of mineral. In Dr Glimcher's laboratory the presence of very high density bone mineral particles has been shown. Perhaps these are the sites that the osteoclast seeks when it initiates a resorption cavity.

Hanaoka: Does the preosteoclast have a PTH receptor or not? Osteoclasts are generally thought not to have a PTH receptor.

Osdoby: How do we identify the preosteoclast? From everything that has been presented on osteoclasts, and if you believe that they come from marrow haemopoietic precursors, which you don't, it would be unlikely that they have such receptors.

Chambers: We have recently found that if populations of spleen cells, which do not contain osteoclasts, are incubated with vitamin D_3, osteoclasts (as judged by excavations on bone slices) develop. Therefore, there must be precursors present in such populations that differentiate into osteoclasts. If these cultures are incubated with PTH there is no increase in excavations or osteoclast numbers. This suggests that osteoclast precursors have no functional response to PTH, and, therefore, that preosteoclasts are unlikely to possess receptors for PTH.

Osdoby: Functional response is based on what you assay. Until you explore all the possibilities you can't say, one way or the other.

Chambers: I am not sure what the point of a receptor is without a functional response.

Osdoby: What do you define as the functional response?

Chambers: Bone resorption.

Baron: Earlier, Dr Hanaoka described the concept (p 104) that precursors of the osteoclast fuse and that this fusion might require some local microenvironment which could induce the fusion of different cell types. The lack of evidence to support this idea is a problem, but I think the suggestion should be emphasized. It's possible that many cell types are incorporated into this nuclei pool. The issue of a very specific osteoclast precursor might, therefore, be experimentally confusing.

Caplan: Taking an example from another cell system, the fusion of myogenic cells is a highly specialized enterprise. Myoblasts of different developmental lineages do not cross-fuse. One would hope that the osteoclast is as selective as a myogenic cell and that the spurious fusion of other cell types into it doesn't occur. We can force a lot of fusions with ethylene glycol and other noxious agents, but no one has indicated that such agents exist *in vivo*.

Krane: Part of the evidence that osteoclasts are derived from haemapoietic precursors is obtained in osteopetrosis. Bone marrow from her mother was given to a girl with the disease and there was a clinical response: bone resorption increased (Coccia et al 1980). Biopsy after transplantation showed that some, but not all, of the nuclei in actively resorbing osteoclasts were derived from the donor. The male nuclei were identified by fluorescent-body analysis. Under those circumstances you could propose that the clinical response was not due to the donor cell becoming an osteoclast, but that some other transplanted cell induced normal function in the osteoclast.

The patients were immunosuppressed for the bone marrow transplantation and we know that viruses can produce heterokaryons. These cells might be considered heterokaryons. It is not a bad argument.

Hanaoka: There are various bone tumours which produce tumour giant cells, such as malignant fibrous histocytoma, giant cell tumour of bone, and aneurysmal bone cysts, and rhabdomyosarcomas of soft tissue. Those tumour giant cells are not formed from the haemopoietic cells. Perhaps we have over-emphasized the monocyte–macrophage system regarding the formation of the osteoclast.

Testa: You can have false negatives; for instance when using fluorescence of Y chromosomes. How definite are these results on the origin of the cells? There is experimental evidence using mice which have a similar cytoplasmic marker in the haemopoietic cells to the marker in patients with Chediak-Higashi syndrome. Bone marrow from these mice, transplanted to osteopetrotic mice, cured the osteopetrosis, and the osteoclasts showed the marker giant lysosomes characteristic of the donor haemopoietic cells (Ash et al 1981). The area of cytoplasm occupied by the giant lysosomes was comparable to that in the donor mice, which would support the concept that significant heterologous fusion did not take place, at least once repopulation was completed.

Urist: On the question of the fate of osteoclasts, Dr Baron suggests that fission and extrusion of nuclei may occur one at a time, and that an osteoclast can reduce itself in size. The idea that osteoclasts explode and die arose from photomicrocinematography of bone in tissue culture media in the 1960s. By modern-day standards, these media would be nutritionally deficient and the results would be difficult to extrapolate to *in vivo* conditions. There is one condition where you can be sure that an osteoclast dies: lead poisoning. In bones of children with lead poisoning, giant cells have pyknotic nuclei. Does lead poisoning create a bad environment for osteoclasts, or giant cells, or both? I would like to know whether in lead poisoning the cells are osteoclasts or giant cells.

Baron: To summarize our discussion, it appears that the haemopoietic origin of osteoclasts is fairly well established. The mechanisms of fusion are still an open question. We are beginning to understand the function of the osteoclasts, but their fate is totally unknown. We are making clear progress towards explaining osteoclast regulation. We have identified agents that do not activate the cell alone, which might mean that we need a cocktail of agents or intermediary target cells. The osteoblast or lining cell are good candidates for this regulation, but we lack the final piece of evidence. There are many other candidates; probably more than one cell type is involved in the fine regulation.

References

Ash P, Loutit DF, Townsend KMS 1981 Osteoclasts derive from hematopoietic stem cells according to marker, giant lysosomes of beige mice. Clin Orthop Rel Res 155:249–258

Burleigh MC, Barrett AJ, Lazarus GS 1974 Cathepsin-B1—lysosomal enzyme that degrades native collagen. Biochem J 137:387

Coccia PF, Krivit W, Cervenka J et al 1980 Successful bone-marrow transplantation for infantile malignant osteopetrosis. N Engl J Med 302:701–708

Stern PH 1985 Biphasic effects of manganese on hormone-stimulated bone resorption. Endocrinology 117:2044–2049

An adhesion variant of the MG-63 osteosarcoma cell line displays an osteoblast-like phenotype

M.D. Pierschbacher, S. Dedhar, E. Ruoslahti, S. Argraves and S. Suzuki

Cancer Research Center, La Jolla Cancer Research Foundation, 10901 North Torrey Pines Road, La Jolla, California 92037, USA

Abstract. MG-63 human osteosarcoma cells were selected for attachment and growth in increasing concentrations of a synthetic peptide containing the cell attachment-promoting Arg-Gly-Asp (RGD) sequence derived from the cell-binding region of fibronectin. Cells capable of attachment and growth in 5 mM concentrations of a peptide having the sequence Gly-Arg-Gly-Asp-Ser-Pro overproduce the cell surface receptor for fibronectin. No increase in fibronectin receptor gene copy number was detected by Southern blot analysis. The peptide-resistant MG-63.3A cells look very different from the MG-63 cells and resemble osteocytes. The resistant cells also grow more slowly than MG-63 cells. The enhanced expression of the fibronectin receptor on the resistant cells indicates that cells can regulate the amount of this receptor on their surface in response to environmental factors and that this may affect the phenotypic properties of the cell. MG-63.3A cells differ from MG-63 cells in their ability to form a calcified matrix *in vitro* and in their increased synthesis of type I collagen. The MG-63.3A cells synthesize 50–100-fold less prostaglandin E_2, a mediator of bone resorption, than MG-63 cells. There is an overall down-regulation of chrondroitin sulphate proteoglycans in MG-63.3A cells. These results are consistent with the hypothesis that such proteoglycans interfere with calcium phosphate deposition and with the observation that chondroitin sulphate is increased in a wide variety of neoplasms but is absent or in small amounts in normal tissue. We conclude that MG-63.3A cells represent a more differentiated cell type with osteoblast-like properties.

1988 Cell and molecular biology of vertebrate hard tissues. Wiley, Chichester (Ciba Foundation Symposium 136) p 131–141

The discovery that the structure in the fibronectin molecule that is primarily recognized by cells consists of the three amino acid sequence arginine-glycine-aspartic acid (RGD) (Pierschbacher et al 1981, Pierschbacher et al 1983, Pierschbacher & Ruoslahti 1984a, Pierschbacher & Ruoslahti 1984b, Ruoslahti & Pierschbacher 1986, Yamada & Kennedy 1984) has enabled us to identify the structure at the cell surface that serves as the fibronectin receptor (Pytela et al 1985a). This structure turns out to be related to a protein(s) that has been identified in chick embryos using monoclonal antibodies that disrupt cell adhesion (Knudsen et al 1981, Greve & Gottlieb 1982, Horwitz et al 1984, Brown & Juliano 1985, Giancotti et al 1985). The further findings implicating RGD sequences as the cell attachment determinants of some other extracellular matrix proteins (Ruoslahti & Pierschbacher 1987, Hynes 1987) have led also to the identification of a number of other RGD-directed receptors. Thus, a receptor recognizing the RGD sequence in vitronectin (Pytela et al 1985b) and another binding collagen type I through its RGD sequences have been isolated (Dedhar et al 1987a). Moreover, an RGD-directed receptor present on platelets that binds fibronectin, vitronectin, fibrinogen, and von Willebrand factor has also been purified. This receptor is the IIb/IIIa protein complex (Pytela et al 1986). Obviously the latter receptor does not show the ligand specificity shown by the first two receptors, and some possible explanations for this have been discussed (Ruoslahti & Pierschbacher 1987, Hynes 1987). We feel that the most likely explanation is that some receptors recognize the RGD sequence in only one or a few restricted environments whereas the binding site of the IIb/IIIa protein accepts the RGD sequence in a large variety of conformations. In fact, recent evidence indicating that receptor selectivity can be achieved with short, conformationally restricted, synthetic peptides would seem to confirm this hypothesis (Pierschbacher & Ruoslahti 1987).

The complete primary structure of the fibronectin receptor has been deduced from cDNA, and a number of physical properties of the receptor can be determined from this sequence (Argraves et al 1986, 1987). These are summarized in Fig. 1. The protein exists at the cell surface as a heterodimeric complex (though the larger polypeptide is enzymically processed) having both polypeptide chains inserted into the membrane. Each chain extends 30–40 residues into the cytoplasmic space, and at least one of the cytoplasmic peptides appears to interact with the cytoskeleton (Horwitz et al 1986). The larger of the two polypeptides, that which we call the α subunit, contains a number of regions that are structurally similar to calmodulin and that apparently mediate the binding of calcium to the receptor. The presence of such divalent cations is required for the receptor to bind ligand. The β subunit is somewhat smaller and has a structure compacted by numerous intra-chain disulphide bonds. The cytoplasmic domain of the β subunit contains a potentially phosphorylated tyrosine (Hynes 1987, Hirst et al 1986).

Partial sequences from other RGD-directed receptors as well as from other adhesion receptors, the ligands for which remain unknown, have revealed the existence of a superfamily of cell surface proteins that share a high degree of structural similarity and probably also functional similarity (Ruoslahti & Pierschbacher 1986, Ruoslahti & Pierschbacher 1987, Hynes 1987). The members of this superfamily of proteins that are currently known are listed in Fig. 2, where they are subdivided according to shared β subunits.

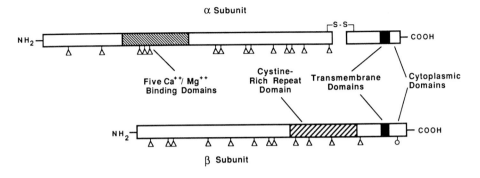

FIG. 1.   Properties of the fibronectin receptor.

Because small synthetic peptides that contain the RGD sequence are quite effective at dislodging most types of cells from the tissue culture substrate as well as in preventing those cells from reattaching (Hayman et al 1985), we asked whether by selecting cells that are capable of attachment and growth in high levels of RGD-containing peptide we could effect a quantitative or qualitative change in the adhesion receptors expressed on the surface of such cells. An osteogenic sarcoma cell line called MG-63 (Billiau et al 1977, Dedhar et al 1987b) was used as the parent cell line because we had isolated a number of adhesion receptors from this cell line and knew that it attached well to both fibronectin and vitronectin. Addition of 1 mg/ml (1.7 mM) of a

FN-R Family

Receptor	Ligand
FN-R (Integrin 1 VLA-5)	FN
VLA- 1	?
VLA- 2	Collagens?
VLA- 3 (CSAT)	FN,LM?
VLA- 4	?

VN-R Family

Receptor	Ligand
VN-R	VN,OP
gpIIb/IIIa	FBG,FN, vWF,VN

Type I Collagen receptor
Discoidin I receptor
Drosophila position specific antigens

Leukocyte Receptor Family

Receptor	Ligand
LFA-1	ICAM-1
Mac-1	C3bi
p150/95	?

FIG. 2.   The RGD receptor superfamily. FN, fibronectin; VN, vitronectin; FBG, fibrinogen; OP, osteopontin; Mac-1, LFA-1, leucocyte adhesion molecules; ICAM-1, intercellular adhesion molecule 1; VLA, very late activation antigens; vWF, von Willebrand factor; CSAT, chicken adhesion receptor complex.

peptide having the sequence Gly-Arg-Gly-Asp-Ser-Pro (GRGDSP) to confluent cultures of these cells resulted in the detachment of all of the cells in the
culture dish and their subsequent death. However, 0.5 mg/ml (0.85 mM) of
the same peptide allowed about 1% of the cells to remain attached and grow
even in the continued presence of peptide. When these cells reached confluence, the concentration of GRGDSP peptide was increased to 1 mg/ml, at
which time less than 1% of them were found to remain attached. This process
was continued at 0.85 mM increments, each time growing up those cells that
remained attached to the substrate. After about five months, the cells were
growing in 5 mM peptide but at a much slower rate than the parent cells. The
morphology of the peptide-resistant cells, however, was strikingly different
from that of the MG-63 cells, even though karyotypic comparison indicated
that the peptide-resistant cells were indeed derived from the parent MG-63
cell line (Dedhar et al 1987b).

When the selected cells (now called MG-63.3A) were examined biochemically, a number of interesting features emerged. The MG-63.3A cells carry
about six times as many fibronectin receptors on their surface as do the parent
MG-63 cells while the number of RGD-directed vitronectin receptors remains
constant. We have found no evidence of amplification of the fibronectin
receptor gene in the selected cells. Rather, an increased amount of messenger
RNA encoding the receptor protein appears to be the explanation for increased receptor number, indicating that cells are capable of regulating the
amount of fibronectin receptor on their surface in response to environmental
factors and that this may in turn affect the phenotypic properties of the cell.
Curiously, the fibronectin receptor number remains elevated on the MG-
63.3A cells and the cells remain resistant to detachment by peptide even when
they are grown for long periods in the absence of added peptide (Dedhar et al
1987b), suggesting that a new steady state may have been reached.

We are currently examining the selected cells biochemically and are finding
evidence that a number of genes other than that for the fibronectin receptor
appear to have been either entirely activated or inactivated. Most strikingly,
it appears that the MG-63.3A cells deposit calcium phosphate in culture in the
absence of exogenous glycerol phosphate (Dedhar et al 1988). The nature of
this deposit is now under investigation. Furthermore, there is an overall
down-regulation of chondroitin sulphate proteoglycans in the MG-63.3A
cells. These data are consistent with the hypothesis that such proteoglycans
interfere with calcium phosphate deposition and with the observation that
chondroitin sulphate is increased in a wide variety of neoplasms but is absent
or occurs in small amounts in normal tissue. Our hypothesis is that an
increased signal from the fibronectin receptor or, more indirectly, from the
altered cell shape has a profound influence on gene expression in these cells.
These cells should prove a valuable model for bone development as well as
for studying the influence of extracellular matrix molecules on gene
expression.

References

Argraves WS, Pytela R, Suzuki S, Millan JL, Pierschbacher MD, Ruoslahti E 1986 cDNA sequences from the α subunit of the fibronectin receptor predict a transmembrane domain and a short cytoplasmic peptide. J Biol Chem 261:12922–12924

Argraves WS, Suzuki S, Arai H, Thompson K, Pierschbacher MD, Ruoslahti E 1987 Amino acid sequence of the human fibronectin receptor. J Cell Biol 105:1183–1190

Billiau A, Edy VG, Heremans H, Van Damme J, Desmyter J, Georgiades JA, DeSomer P 1977 Human interferon: mass production in a newly established cell line, MG-63. Antimicrob Agents Chemother 12:11–15

Brown PJ, Juliano RL 1985 Selective inhibition of fibronectin-mediated cell adhesion by monoclonal antibodies to a cell-surface glycoprotein. Science (Wash DC) 228:1448–1451

Dedhar S, Ruoslahti E, Pierschbacher MD 1987a A cell surface receptor complex for collagen type I recognizes the Arg-Gly-Asp sequence. J Cell Biol 104:585–593

Dedhar S, Argraves WS, Suzuki S, Ruoslahti E, Pierschbacher MD 1987b Human osteosarcoma cells resistant to detachment by an Arg-Gly-Asp-containing peptide overproduce the fibronectin receptor. J Cell Biol 105:1175–1182

Dedhar S, Stallcup WB, Pierschbacher MD 1988 The osteoblast-like differentiated phenotype of a variant of MG-63 osteosarcoma cell line correlated with altered adhesion properties. In preparation

Giancotti FP, Tarone G, Knudsen KA, Damsky CH, Comoglio PM 1985 Cleavage of a 135 kD cell surface glycoprotein correlates with loss of fibroblast adhesion to fibronectin. Exp Cell Res 156:182–190

Greve JM, Gottlieb DI 1982 Monoclonal antibodies which alter the morphology of cultured chick myogenic cells. J Cell Biochem 18:221–230

Hayman EG, Pierschbacher MD, Ruoslahti E 1985 Detachment of cells from culture substrate by soluble fibronectin peptides. J Cell Biol 100:1948–1952

Hirst R, Horwitz A, Buck C, Rohrschneider L 1986 Phosphorylation of the fibronectin receptor complex in cells transformed by oncogenes that encode tyrosine kinases. Proc Natl Acad Sci USA 83:6470–6474

Horwitz AF, Knudsen KA, Damsky CH, Decker C, Buck CA, Neff NT 1984 Adhesion-related integral membrane glycoproteins identified by monoclonal antibodies. In: Kennett RH et al (eds) Monoclonal antibodies and functional cell lines. Plenum Publishing, New York, p 103–118

Horwitz A, Duggan K, Buck C, Beckerle MC, Burridge K 1986 Interaction of plasma-membrane fibronectin receptor with talin — a transmembrane linkage. Nature (Lond) 320:531–533

Hynes RO 1987 Integrins: a family of cell surface receptors. Cell 48:549–554

Knudsen KA, Rao PE, Damsky CH, Buck CA 1981 Membrane glycoproteins involved in cell–substratum adhesion. Proc Natl Acad Sci USA 78:6071–6075

Pierschbacher MD, Ruoslahti E 1984a The cell attachment activity of fibronectin can be duplicated by small synthetic fragments of the molecule. Nature (Lond) 309:30–33

Pierschbacher MD, Ruoslahti E 1984b Variants of the cell recognition site of fibronectin that retain attachment-promoting activity. Proc Natl Acad Sci USA 81:5985–5988

Pierschbacher MD, Ruoslahti E 1987 Influence of stereochemistry of the sequence Arg-Gly-Asp-Xaa on binding specificity in cell adhesion. J Biol Chem 262:17294–17298

Pierschbacher MD, Hayman EG, Ruoslahti E 1981 Location of the cell-attachment site in fibronectin with monoclonal antibodies and proteolytic fragments of the molecule. Cell 26:259–267

Pierschbacher MD, Hayman EG, Ruoslahti E 1983 A synthetic peptide with the cell attachment activity of fibronectin. Proc Natl Acad Sci USA 80:1224–1227

Pytela R, Pierschbacher MD, Ruoslahti E 1985a Identification and isolation of a 140 kilodalton cell surface glycoprotein with properties expected of a fibronectin receptor. Cell 40:191–198

Pytela R, Pierschbacher MD, Ruoslahti E 1985b A 125/115 KD cell surface receptor specific for vitronectin interacts with the Arg-Gly-Asp adhesion sequence derived from fibronectin. Proc Natl Acad Sci USA 82:5766–5770

Pytela R, Pierschbacher MD, Ginsberg MH, Plow EF, Ruoslahti E 1986 Platelet membrane glycoprotein IIb/IIIa is a member of a family of Arg-Gly-Asp-specific adhesion receptors. Science (Wash DC) 231:1559–1562

Ruoslahti E, Pierschbacher MD 1986 Arg-Gly-Asp: a highly versatile cell recognition signal. Cell 44:517–518

Ruoslahti E, Pierschbacher MD 1987 New perspectives in cell adhesion: RGD and integrins. Science (Wash DC) 238:491–497

Yamada KM, Kennedy DW 1984 Dualistic nature of adhesive protein function: fibronectin and its biologically active peptide fragments can autoinhibit fibronectin function. J Cell Biol 99:29–36

DISCUSSION

Slavkin: Have you looked at the possibility of other molecules such as chondroitin sulphates binding to the fibronectin receptor?

Pierschbacher: We haven't studied that. I wouldn't have confidence in such experiments. For example, galactose will inhibit cell adhesion, but I don't believe it has any physiological role in that process.

Slavkin: Does heparin interfere with the binding of the RGD sequence to the 140 kDa receptor site?

Pierschbacher: No, it doesn't.

Rodan: There has been a recent report about another fibronectin receptor (Peters & Mosher 1987).

Pierschbacher: I haven't looked at the interaction of cells with other areas of fibronectin. However, I do not believe that the entire purpose of fibronectin is to hold the RGD sequence of fibronectin in the right conformation. There are probably other interactions of fibronectin with the cell surface.

Caplan: In the selected MG-63.3A cells, even though you don't find gene amplification, is there perhaps some modification in the arrangement or frequency of the calcium-binding domain of the receptors, which gives a more effective calcium concentration centre in the receptor and allows the spectacular alizarin red staining that you see?

Pierschbacher: I haven't actually looked at that in any way that would allow me to answer that question.

Caplan: It would also be interesting to look at the electron microscopic immunolocalization of the receptor.

Testa: The role of proteoglycans may be more than just involvement in calcification. There are some recent reports that haemopoietic growth factors may bind selectively to some types of proteoglycans *in vitro*, and that may well occur *in vivo* also. We think there is a large wave of differentiation at the level of progenitor cells occurring quite close to the bone surface. In that sense I think components of the extracellular matrix are likely to be important.

Pierschbacher: There are a large number of cell types in bone. I would not presume to say which cells are actually responsible for mineralization and which cells are making what extracellular matrix products, or which cells are making which proteoglycans.

Testa: The levels of certain proteoglycans close to the surface of bone may be directing haemopoietic cell differentiation towards one lineage or another by their capacity to bind growth factors selectively.

Pierschbacher: We have done experiments showing that chondroitin sulphate inhibited the attachment of cells to fibronectin. We interpreted that to be a result of masking of the cell attachment-promoting site, because if we took fragments of fibronectin that did not contain the glycosaminoglycan binding site, the chondroitin sulphate proteoglycan had no effect on cell binding to these fragments. It may be that chondroitin sulphate on the cell surface is acting to mask certain things on the cell surface or to mediate or modulate the interaction of cells with that surrounding matrix.

Williams: The RGD sequence has many possible conformations. When it is cyclized there are probably at least ten comformations still available. Has anybody looked at the cyclic compound by nuclear magnetic resonance (NMR) spectroscopy?

Pierschbacher: We have a good NMR structure on the cyclic peptide. We did circular dichroism studies on the linear peptide and found that it seemed to prefer reversed turn type conformations, but I agree that they are totally floppy peptides. That can account for the big loss of affinity towards the receptor of the short, synthetic, linear peptide. I think it is interesting, however, that upon cyclization the peptide gained affinity for the vitronectin receptor.

Williams: After obtaining the new structure, did you compute which conformations had been lost on cyclization?

Pierschbacher: We are modelling the structure that we got from NMR and minimizing the structure of the linear peptides so that we can start looking at that. We haven't got the results yet. The cyclization was achieved with disulphide bonding of penicillamine to the cysteine—a method which, because of the two extra methyl groups close to the ring, does restrict the isomerization of the ring a great deal.

Williams: So it wasn't a straight cyclization of a peptide. Did you run the product on a column to see if the product was chemically different?

Pierschbacher: It was purified by reverse-phase HPLC. There was a major homogeneous peak of a single ring and there was a small peak of uncyclized material and some multimers that we excluded.

Williams: A previous Ciba Foundation Symposium was entitled 'Synthetic peptides as antigens' (1986). In that symposium it was shown that antigenic sites on proteins were frequently at exposed, mobile, β-turns, and RGD looks just like such a turn. Such exposed turns are the places where the antagonists for receptor sites may well be found. The mobility of these sites is a problem because many conformations can be visited, but the optical activity of the amino acids excludes some interesting possibilities. You should make not just a peptide from laevo amino acids but also a peptide from dextro amino acids.

Pierschbacher: I have done that, substituting the D isomer for L in each position of the RGDS sequence. D-Aspartic acid, for example, completely destroys activity, whereas D-arginine, at least with the fibronectin and vitronectin receptors, retains activity perfectly well. This is indicative of this sequence assuming a sharp β-turn for both the receptors. The collagen receptor, on the other hand, only sees the RGD sequence when it's twisted into the α-helix conformation, which is fairly linear. We've done some work with gamma 2-crystallin, which has an RGD sequence that is similarly linear, and the collagen receptor seems to see that. However, the receptor or an antibody should be able to take a floppy peptide and force it into the conformation it prefers. Then, just by mass action, receptor selectivity would be lost. That would result in a lower affinity because energy would have to be expended to accomplish this.

Williams: The more the mobility the lower the affinity.

Baron: What happens to the receptor once it binds its ligand and these cells are in excess ligand?

Pierschbacher: The receptor doesn't seem to be internalized. When cells move or round up to divide, patches of membrane get torn out and the receptor is left behind. Therefore the receptor is continually made at the cell surface. There also appears to be an internal pool of beta chain and the alpha chain seems to be the determining factor in terms of what gets to the cell surface.

Baron: You checked the specificity of the receptor for this ligand. Do all cell types express the receptor to the same degree?

Pierschbacher: Reagents that are reliable in terms of the alpha chain are only now becoming available. If you assay beta chains you find them scattered everywhere but the alpha chains seem to segregate in different combinations on different cell types. We are beginning to see a matrix adhesion receptor zipcode type of distribution, consistent with a hypothesis that would predict that cells can tell where they are on the basis of which receptors they express and which molecules in the matrix they are reading. It's a little premature at this point, though, to speculate too much about these possibilities.

Caplan: For these variant cells, have you analysed whether there is any difference between the ratio of alpha and beta receptors in the pool and in the membrane? Although you have selected for a cell which has up-regulated receptor, have you changed the balance of subunits at membrane locations?

Pierschbacher: The receptors are virtually all on membrane.

Osdoby: It has been suggested that as well as attaching to the cytoskeleton, fibronectin had endogenous kinase activity. Have you seen any evidence for kinase activity associated with your receptors?

Pierschbacher: Not in the receptor. In the fibronectin receptor beta chain there is a tyrosine that appears to be phosphorylated and seems to be homologous with the tyrosine that is phosphorylated in the EGF receptor. However, there is so little peptide of the fibronectin or vitronectin receptor inside the cell that I don't see room for any kinase activity.

Rodan: Dr Owen described (this volume) how, when bone marrow cells are put into chambers and make bone-like matrix, less fibronectin is found by cytochemical staining. We noticed that if you treat ROS 17/2.8 osteosarcoma cells with TGF-β the expression of several 'bone'-specific macromolecules is enhanced, but the fibronectin mRNA is reduced (Noda & Rodan 1987). Therefore there seems to be a link between bone cell maturation and the decrease in the attachment of these cells to fibronectin. This may be what you produce experimentally by adding large amounts of RGD.

Pierschbacher: I interpret the results to mean that there is a change in the balance of the receptors that are being ligated. Osteopontin apparently has an RGD receptor, and there could be a number of other receptors on the cell surface with an affinity for RGD. By putting these large amounts of peptide in the cultures, we could be tickling other receptors and transmitting other messages to the cells. We plan to transfect the fibronectin receptor into the cells to increase the amount of cell surface receptor in a way totally unrelated to the set of experiments described here. Then we shall ask whether the same type of cell changes in the cell will result. It may just be that by changing the adhesiveness of a cell to a different matrix component, you change the shape and that pulls all kinds of strings in the cell.

Prockop: Does bone have fibronectin?

Rodan: There is quite a lot in the early stages of differentiation, but mature bone contains less.

Termine: We have found similar results: fibronectin is expressed at high levels in the early stages of bone cell culture and then its expression goes down as matrix is produced.

Pierschbacher: The pool size does not appear to increase greatly. It's mostly the cell surface that increases. Therefore the ratio of cell surface to pool size would increase.

Martin: Was the continued presence of the RGD peptide necessary to maintain that morphological change in the cells?

Pierschbacher: It took about five months for us to obtain cells that were growing nicely in 3 mg per ml of inhibitory peptide. We have taken the peptide out and the cells continue to maintain the altered phenotype. We did karyotyping of both MG-63 cells and selected cells and found that they have a large number of unique chromosome markers in common, indicating that they have the same origin.

Krane: We have shown that monocyte precursors, such as U937 cells, increase their binding to fibronectin bound to plastic surfaces after incubation with $1,25(OH)_2D_3$ (Kroop et al 1987). $1,25-(OH)_2D_3$ induces binding to the 120 kDa cell-binding fragment. We also investigated collagen binding in solution (Polla et al 1987). We detect no binding at all in control U937 cells, but binding is also induced by $1,25-(OH)_2D_3$. Binding of labelled collagen is totally displaced by unlabelled collagen. Although there are very few binding sites induced per cell they are of high affinity. Thus in these cells there does appear to be a regulation of the putative receptors. The collagen-binding proteins do not seem to have the same electrophoretic characteristics as you describe for the fibronectin and vitronectin receptors, however. The induction of binding requires new protein synthesis, at least according to inhibition by cycloheximide.

Hauschka: In view of the predicted calcium-binding sites on the alpha chains of the fibronectin and vitronectin receptors, and recognizing that EDTA and other calcium depletion methods can be used to strip cells from surfaces on which they are grown, is it possible to study the calcium affinity of the alpha chain and show that it may account for the calcium-dependent part of the cell's mechanism for adhering to the substrate?

Pierschbacher: We have looked at isolated receptor in the presence and absence of calcium and it doesn't work without divalent cation. We have been trying to put in other metals that would substitute for calcium; we find some interesting activities.

Raisz: The differences between osteocytes and other cells include the vast number of gap junctions that they have in their processes with other cells. Has anybody looked at the gap junction to see if the fibronectin receptor is present?

Pierschbacher: The receptor is thought not to be there, but I would like to look at these cells to see what kind of gap junctions they have.

Krane: The gap junction protein has now been isolated and cloned and has been inserted into lipid bilayers to reproduce functional electrical channels with properties similar to those of intact gap junctions (Young et al 1987). One can thus analyse cells directly for this 27 kDa protein.

References

Ciba Foundation 1986 Synthetic peptides as antigens. Wiley, Chichester (Ciba Foundation Symposium 119)

Kroop SF, Polla BS, Krane SM 1987 Bone Miner Res 2 (Suppl 1):513

Noda M, Rodan GA 1987 Type beta transforming growth factor (TGFbeta) Regulation of alkaline phosphatase expression and other phenotype-related mRNAs in osteoblastic rat osteosarcoma cells. J Cell Physiol 133:426–437

Owen M, Friedenstein AJ 1988 Stromal stem cells: marrow-derived osteogenic precursors. In: Cell and molecular biology of vertebrate hard tissues. Wiley, Chichester (Ciba Found Symp 136) p 42–60

Peters DM, Mosher DF 1987 Localization of cell surface sites involved in fibronectin fibrillogenesis. J Cell Biol 104:121–130

Polla BS, Healy AM, Byrne M, Krane SM 1987 1,25-dihydroxyvitamin D_3 induces collagen binding to the human monocyte line U937. J Clin Invest 80:962–969

Young JD-E, Cohn ZA, Gilula NB 1987 Functional assembly of gap junction conductance in lipid bilayers. Demonstration that the major 27-KD protein forms the junctional channel. Cell 48:733–743

Expression of type I procollagen genes

Darwin J. Prockop, Karl E. Kadler, Yoshio Hojima, Constantinos D. Constantinou, Kenneth E. Dombrowski, Helena Kuivaniemi, Gerard Tromp and Bruce Vogel

Department of Biochemistry and Molecular Biology, Jefferson Institute of Molecular Medicine, Jefferson Medical College, Thomas Jefferson University, Philadelphia, Pennsylvania 19107, USA

Abstract. All of the type I collagen in connective tissue is the product of one structural gene for the proα1(I) chain and another for the proα2(I) chain of type I procollagen. An intriguing question therefore is how the expression of the two genes differs in mineralizing and non-mineralizing tissues. One approach that our laboratory has pursued to answer this and related questions is to develop a new system whereby one can examine the self-assembly of collagen fibrils *de novo* by controlled enzymic cleavage of procollagen to collagen under physiological conditions. The system has made it possible for the first time to define thermodynamic parameters for the self-assembly process. We are now using the system to define the normal kinetics for fibril formation. The results should make it possible to study the effects of other components of extracellular matrix on fibril assembly, including the effects of bone-specific components that initiate mineralization. A second approach has been to define mutations in type I procollagen genes that cause increased brittleness of bone. Over a dozen mutations in type I procollagen genes have been found in probands with osteogenesis imperfecta. One of the surprises has been that at least 25% of the probands with lethal variants of osteogenesis imperfecta have mutations in type I procollagen genes. Another surprise has been the observation that a number of the mutations are tissue specific in terms of their phenotypic manifestations even though the same abnormal proα chains are being synthesized in a variety of tissues.

1988 Cell and molecular biology of vertebrate hard tissues. Wiley, Chichester (Ciba Foundation Symposium 136) p 142–160

Type I collagen accounts for 80% or more of the protein content of bone, skin, tendons, ligaments and a variety of other connective tissues. All the type I collagen is the ultimate product of two structural genes for the precursor protein, type I procollagen; one for the proα1(I) chain and the other for the proα2(I) chain. One intriguing question, therefore, is how the expression of these two structural genes differs in mineralizing and in non-mineralizing tissues. At the moment we have little specific information with which to answer this question. However, we have a large amount of basic information about collagen biosynthesis that will be useful in formulating future experiments. Also, interesting data are being generated from two sources: a new system for studying assembly of collagen fibrils *de novo*; and a series of mutations in the structural genes for type I procollagen that cause heritable diseases of bone.

Biosynthesis of collagen

The biosynthesis of the procollagen molecule is a complex process involving at least eight post-translational enzymes that modify newly synthesized proα chains before they are secreted from cells (Fig. 1). At least three additional enzymes modify the procollagen molecule in the conversion of procollagen to collagen and are involved in the orderly assembly of collagen molecules into fibrils, either at the cell surface or some distance from the cell. (For detailed reviews, see Prockop & Kivirikko 1984, Cheah 1985, Prockop & Kuivaniemi 1986).

The pathway has several key features. One is that the hydroxylation of peptide-bound proline residues to hydroxyproline residues is essential for the three chains to fold into the unique triple-helical conformation of procollagen at body temperature. Another key feature is that the hydroxylation of proline residues to hydroxyproline residues, and the parallel hydroxylation of lysine residues to hydroxylysine residues, and the glycosylation of hydroxylysine residues, only occurs during the time when the protein polypeptide chains are non-helical. After the chains have folded into a triple-helical conformation, the processing by proline hydroxylase, lysine hydroxylase and the glycosyltransferases ceases. Another important feature of the pathway is that the association of two proα1(I) chains and one proα2(I) chain to form the heterotrimer of the protein is directed entirely by the conformation and binding properties of the C-terminal propeptides. The C-terminal pro-peptides of the proα chains associate and become disulphide cross-linked before the chains are hydroxylated sufficiently to fold into a triple-helical conformation at 37 °C. After chain association and disulphide linkage of the C-terminal propeptides, the folding of the triple helix begins with the formation of a nucleus in the C-terminal region of the α chain domain. The nucleus is then propagated from the C-terminus to the N-terminus of the protein. Another feature of the pathway is that transport of the protein from the rough endoplasmic reticulum to the Golgi apparatus and from the Golgi to the extracellular space is critically dependent on the protein's conformation. Folding of the chains into the triple-helical conformation can be prevented or delayed by inhibiting the hydroxylases, or by several other manipulations (see Prockop & Kivirikko 1984). If this occurs, the proα chains become disulphide-linked through the C-terminal propeptides, but the protein largely remains within the cisternae of the rough endoplasmic reticulum. If the inhibition of the hydroxylases is reversed, the protein is hydroxylated, it folds into the triple-helical conformation, and it is rapidly transported from the cell.

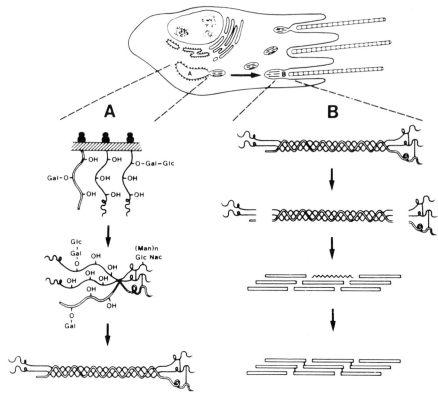

FIG. 1.   Fibroblast assembly of collagen fibrils. Frame A shows intracellular post-translational modifications of proα chains, association of C-propeptide domains, and folding into triple-helical conformation. Frame B shows the extracellular conversion of procollagen to collagen, self assembly of collagen monomers into fibrils, and cross-linking of fibrils. Reproduced with permission from Prockop & Kivirikko (1984).

Conversion of procollagen to collagen and assembly of collagen into fibrils

Extensive investigations in the past suggested that the formation of large collagen fibrils in tissues was largely a process of monomer self-assembly (Gross & Kirk 1958, Wood 1960, Na et al 1986). However, this was difficult to establish definitively because of experimental problems in dealing with the protein *in vitro*. The formation of collagen fibrils *in vitro* was examined by extracting collagen monomers from tissues at low temperatures and with solvents such as acetic acid, and then warming and neutralizing the solutions. Under these conditions, the monomer of the protein self-assembles into collagen fibrils that have the same cross-striated pattern as fibrils seen in tissues. Experiments with extracted collagen solutions, however, have a number of limitations. Most important among them is the fact that assembly of the monomer into fibrils occurs poorly at temperatures above 30–35 °C and in physiological buffers with low phosphate contents. Also, the kinetics of the reaction and the morphology of the fibrils formed are not easily reproduced with different collagen solutions.

We have recently developed a system for studying the assembly of type I collagen fibrils *de novo* by enzymic cleavage of an intermediate in the conversion of procollagen to collagen (Kadler et al 1987). The system has two protein components. One is pCcollagen, the partially processed form of type I procollagen that lacks N-terminal propeptides but contains C-terminal propeptides. The second component is procollagen C-proteinase, the enzyme that cleaves the C-propeptides from type I procollagen during the processing *in vivo*. The pCcollagen was purified to homogeneity by isolating type I procollagen from the medium of cultured human fibroblasts, cleaving the protein with partially purified procollagen N-proteinase, and then purifying the resulting pCcollagen by column chromatography. The procollagen C-proteinase was purified from the medium of organ cultures of chick embryo tendons by a five-step chromatographic procedure (Hojima et al 1985).

In a physiological buffer at 35–40 °C the pCcollagen did not assemble into fibrils or any readily detectable aggregates when examined in concentrations of up to 0.5 mg ml^{-1} (Kadler et al 1987). However, cleavage of pCcollagen to collagen with purified C-proteinase generated fibrils in a highly reproducible manner. The fibrils formed were visible by eye and were large enough to be separated from solution by centrifugation at 13 000 g for 4 min. With high concentrations of enzyme, the pCcollagen was completely cleaved in one hour, and turbidity was near maximal in three hours. Collagen continued to be incorporated into fibrils for over 10 h, which shows that turbidity assays are not entirely reliable for following fibril assembly.

The pCcollagen was uniformly labelled with [^{14}C]amino acids. Therefore, the concentration of soluble collagen, and hence the critical concentration for polymerization, could be determined directly. As expected, the critical concentration was independent of the initial pCcollagen concentration or the rate of cleavage. Also as expected, the critical concentration decreased with temperature between 29 °C and 41 °C. At 41 °C the mean critical concentration was 0.12 (SE 0.06) µg ml^{-1}. The temperature dependence of the critical concentration was consistent with the assumption that it reflected a simple equilibrium between monomer in solution and polymer in fibrils. Therefore, it was used to define the thermodynamic parameters of the system. All the thermodynamic parameters were similar to those for other protein self-assembly systems (Table 1). The data demonstrated that type I collagen fibril formation *de novo* is in fact a classical example of an entropy-driven self-assembly process similar to the polymerization of actin, flagella and tobacco mosaic virus protein.

TABLE 1   Thermodynamic parameters for systems of protein self-assembly[a]

Protein	ΔG at 37 °C (kJ mol^{-1})	ΔH (kJ mol^{-1})	ΔS (kJ K^{-1} mol^{-1}
G-ADP-actin		+42 to 63	
Flagella	−7.9	+423	+1.39
Tobacco mosaic virus protein	−42	+142	+0.58
Type I collagen de novo	−54	+234	+0.92

[a] For references see Kadler et al (1987).

Prospects for studying the interaction of collagen with other components of matrix

In our opinion, the system we have developed for studying the assembly of collagen fibrils *de novo* allows us to examine in detail the interaction of collagen with the other components found in many connective tissues — components such as proteoglycans, fibronectin and bone-specific components that may have critical roles in mineralization.

Our preliminary data suggest that the enzymic cleavage of procollagen to collagen by N-proteinase and C-proteinase (Fig. 2) provides a critical control point for determining the diameter and other morphological features of collagen fibrils. If, as now seems apparent, the assembly of collagen fibrils is essentially a crystallization process, the size of the fibrils formed should depend on the relative rates of two processes: the rate at which nuclei are generated and the rate at which the nuclei are propagated into fibrils. Therefore, the size of the fibrils formed should depend on the rate at which

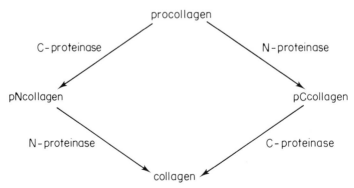

FIG. 2.   Schematic representation of the two pathways by which procollagen can be converted to collagen.

pCcollagen is cleaved in the initial phases of the reaction. Also, the process should depend on parameters such as the temperature of the reaction. In preliminary experiments (Miyahara et al 1984) the fibril diameter was markedly influenced by the rate at which collagen was generated *de novo*: the diameter of the fibrils formed could be varied from about 0.5 to 3 μm by altering the rate.

We also have preliminary results which indicate that fibril diameter is dramatically influenced by the temperature at which the reaction is carried out. In addition, we have confirmed preliminary impressions (Miyahara et al 1984) that the route by which the procollagen molecules are processed has a major influence (Fig. 2). If the processing occurs with cleavage of the C-propeptide and then the N-propeptide, the fibrils tend to be thin and irregular in shape. If the N-propeptide is cleaved first and then the C-propeptide is cleaved, larger and more rounded fibrils are formed (Hulmes et al 1988). This suggests that the processing of procollagen by N-proteinase and C-proteinase may have a major influence on the kinds of fibrils formed *in vivo*. Hence, they provide a basis for understanding the striking differences in morphology of type I collagen fibrils found in tissues as varied as bone, skin, tendon and cornea.

Mutations in type I procollagen genes that produce diseases of bone

One approach to establishing the role of type I collagen in the normal assembly and maintenance of bone is to search for mutations in genes for type I procollagen in patients with heritable disorders of bone such as osteogenesis imperfecta (OI). We and others initiated such a search about 10 years ago. At the time the work was begun, there was relatively little evidence that variants of OI were caused by defects in the synthesis or structure of type I procol-lagen or collagen. In fact, several observations suggested that mutations in type I procollagen genes would be rare in patients with OI. In particular, it was apparent that many patients with OI had marked brittleness of bone, but little change in a number of other tissues rich in type I collagen. Therefore, the common wisdom was that OI was likely to be caused, not by mutations in type I procollagen genes, but by mutations in genes that were bone specific and critical to the mineralization process. In spite of the initial misgivings, it is now apparent that one quarter or more of patients with OI have mutations in one or both of the two structural genes for type I procollagen (for reviews, see Byers & Bonadio 1985, Prockop & Kuivaniemi 1986).

Specific mutations in type I procollagen genes have now been defined in more than 14 variants of OI (Fig. 3). Clinically, OI is a highly heterogeneous disease. In the most commonly used classification (Sillence 1981) there are four major types of the disease with a number of subtypes of each. For consideration here, however, the variants can be simply categorized as non-

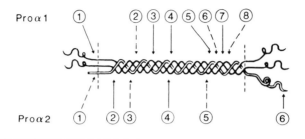

Pro α1

① Deletion 24 aas (Exon 6)   EDS VII

② ? → Cys   OI

③ Gly^{391} → Arg^{391} Lethal OI

④ Deletion Exons 24-26 (-88 aas)   Lethal OI

⑤ Gly^{748} → Cys^{748} Lethal OI

⑥ Gly^{904} → Cys^{904} Lethal OI

⑦ Gly^{988} → Cys^{988} Lethal OI

⑧ ? → Cys   OI

Pro α2

① Deletion 18 aas (Exon 6)   EDS VII

② Splice/Deletion Exon 11 (-18 aas)   EDS/OI

③ Deletion ~20 aas   OI

④ Splice Exon 28 (-18 aas)   Lethal OI

⑤ Deletion Exons 33 to 40 (-180 aas)   Lethal OI

⑥ Frameshift Deletion 4 bp   OI

FIG. 3. Schematic representation of mutations in type I procollagen genes in probands with osteogenesis imperfecta (OI) or Ehlers-Danlos syndrome (EDS) in which the mutation changes the structure of the protein at the sites indicated.

lethal or lethal, with lethal variants defined as those that produce death *in utero*, at birth, or within a few weeks of birth.

The lethal variants of OI with defined mutations in type I procollagen genes have several features in common. One is that the devastating effects of the mutations are largely traceable to the synthesis of an abnormal proα chain. Non-functioning alleles that produce no protein have milder effects. Another generalization is that, with a single exception, all the well-defined lethal variants are caused by a new or sporadic mutation in one allele for the proα1 or proα2 chain. The diseases, therefore, are dominant lethal in their manifestations. The mutations found in the lethal variants are of three types. One type consists of deletions from the gene itself. In one lethal variant, there is an in-phase deletion of exons 24, 25 and 26 from one allele of the proα1 chain

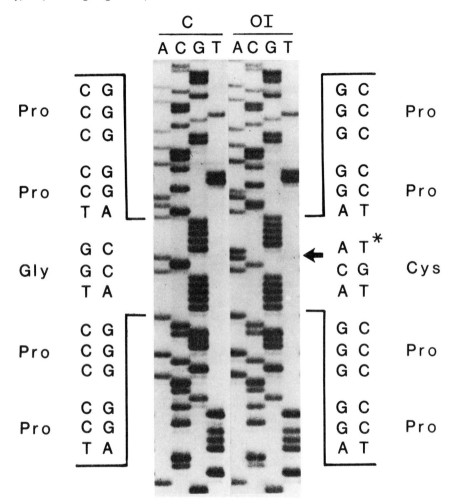

FIG. 4. DNA sequencing gel demonstrating the sequence of a normal cDNA from control fibroblasts (C) and a cDNA from fibroblasts from a proband with a lethal variant of osteogenesis imperfecta (OI). The sequence shown demonstrates the single base mutation (*) that converts the codon for glycine in amino acid position 748 of the α1 chain to a codon for cysteine. Reproduced with permission from Vogel et al (1987).

(Williams & Prockop 1983, Chu et al 1985, Barsh et al 1985). In a second lethal variant, there is another in-phase deletion that removes exons 33 to 40 from one allele of the proα2 chain (Willing et al 1987). A second type of lethal mutations consists of single base mutations that convert a glycine codon in a proα1(I) gene to a codon for another amino acid (Fig. 4). Two of the single base mutations convert a glycine codon to a codon for cysteine and the third converts a glycine codon to a codon for arginine (see Vogel et al 1987).

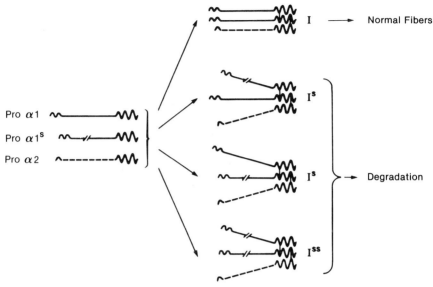

FIG. 5. Schematic representation of the phenomenon of 'protein suicide'. Because the deletion in the middle of the proα1(I) chain is in phase, the shortened proα1(I) chains associate with normal-length proα1(I) chains. The resulting trimers cannot fold into a stable triple helix at body temperature and rapidly become degraded. Reproduced with modification and permission from Prockop & Kivirikko (1984).

A third type of mutation found among lethal variants is an RNA splicing mutation. Only one variant of this type has been identified (Tromp & Prockop 1988). Here, the proband was a compound heterozygote with one non-functioning allele for the proα2 chain that was inherited from an asymptomatic father (de Wet et al 1983). The second allele for proα2 chain had a single base change that converted the last two nucleotides of intervening sequence 27 from -AG- to -GG- (Tromp & Prockop 1988). The level of protein expression from the allele was high (de Wet et al 1983). The single base change at the 3′ end of the intervening sequence apparently caused highly efficient splicing of the RNA so as to generate an mRNA that lacked only the codons found in exon 28. Several features of the RNA splicing mutation differ from RNA splicing mutations seen in globin and other genes (see Sharp 1987). Therefore, the observations raise the possibility that the processing mechanisms for RNA transcripts from type I procollagen genes are unusual.

The lethality of the seven variants is largely explained by a phenomenon we have termed 'protein suicide' (see Williams & Prockop 1983, Prockop & Kivirikko 1984). The phenomenon is based on the fact that association and disulphide linkage of the three proα chains in a procollagen molecule depend entirely upon the primary structure and conformation of the 250 or so amino acids found in each of the C-propeptides. If a mutation spares the structure of the C-propeptides, it has no effect on chain association and disulphide bonding. Therefore, structurally abnormal proα chains will associate with and become disulphide-linked with normal proα chains synthesized in the same cell. However, the presence of the abnormal proα chain frequently prevents the molecule from folding into a triple helix at 37 °C. The non-helical protein accumulates for a time in fibroblasts and then is degraded as it is slowly secreted (Fig. 5).

The seven lethal variants of OI all involve the phenomenon of protein suicide whereby synthesis of abnormal proα chains causes degradation of normal proα chains synthesized by the same fibroblast. The same phenomenon has previously been encountered with mutations in other systems and is usually referred to as trans-dominant defects caused by subunit interaction. With procollagen genes, the phenomenon appears to be particularly striking because so little of the total protein structure is essential for the subunit interaction. However, a large proportion of the 1014 amino acids found in the α chain domain of each proα chain is critical for correct folding (see below).

In some of the lethal variants, the phenomenon of protein suicide is augmented by the fact that some of the abnormal procollagen molecules are not degraded (Steinmann et al 1984). Instead, they are incorporated into fibrils. The abnormal primary structure of the proα chains apparently interferes with the normal packing and generates abnormal fibrils.

Among non-lethal variants of OI, mutations in type I procollagen genes have been identified in at least seven probands (Fig. 3). A variety of gene defects and phenotypic manifestations have been found. Therefore, only a few common features are apparent. There is a tendency, however, for the mutations to produce changes near the N and C termini of the proα chains. Those that alter the N-terminal region of the protein tend to prevent cleavage of the N-propeptide from the procollagen molecule. Apparently as a result, they produce the laxity of ligaments and tendons characteristic of Ehlers-Danlos syndrome rather than the brittleness of bone characteristic of OI.

One of the non-lethal variants is either a partial gene deletion or an RNA splicing mutation that causes synthesis of a proα1 chain lacking the 24 amino acids encoded by exon 6 (Cole et al 1986). Since the 24 amino acids include the peptide bond normally cleaved by the N-proteinase, the N-propeptide is not cleaved by the enzyme. Another non-lethal variant has a similar mutation: fibroblasts synthesize proα2 chains lacking the 18 amino acids encoded by exon 6 of the proα2 gene (Wirtz et al 1987). Again the N-propeptide is not cleaved by N-proteinase. A related non-lethal variant synthesizes proα2(I) chains lacking 18 amino acids encoded by exon 11 (Kuivaniemi et al 1988). Here the gene mutation is a deletion that removes the last five base pairs (bp) of intervening sequence 10, and 14 bp of exon 11. The deletion of 19 bp causes complete and apparently efficient splicing from exon 10 to exon 12 of the RNA transcript. Again, the RNA splicing mutation here differs from those seen in other genes (see Sharp 1987). At the protein level the in-frame deletion of 18 amino acids encoded by exon 11 prevents processing of the N-propeptide from both the proα1 chains and the abnormal proα2 chains in the same molecule. The explanation is that the deletion of amino acids that occurs 83 amino acid residues to the right of the cleavage site causes a shift in the registration of the three chains and thereby alters the conformation of the cleavage site.

Procollagen N-proteinase is unusual among proteinases in that it requires its substrate to have a native conformation. As a consequence, a mutation that alters the primary structure of a proα1 or proα2 chain at some distance from the cleavage site itself can make the procollagen molecule resistant to the enzyme. The patients in whom there is defective processing of the N-propeptide have extremely lax joint ligaments and other clinical manifestations generally classified as the type VII form of Ehlers-Danlos syndrome. Several other probands with the same clinical symptoms have a deficiency of N-proteinase itself (see Prockop & Kivirikko 1984). The reasons why failure to process the N-propeptide produces Ehlers-Danlos syndrome rather than OI have not been conclusively established. One possible explanation comes from the observation that persistence of the N-propeptide allows the formation of thin and irregular collagen fibrils. Such thin and irregular fibrils may be adequate, or nearly adequate, for apparently normal function of bones and other tissues. However, they may not provide adequate tensile strength for major ligaments and tendons in which collagen fibrils normally have large diameters.

The two non-lethal variants in which a new cysteine residue is introduced into the proα1 chain have not been fully defined (Fig. 6). Because they are non-lethal, it is generally assumed that they involve single base mutations that convert an amino acid residue other than glycine to a cysteine and, therefore, have less effect on the stability of the triple helix than the glycine-to-cysteine conversions that are lethal. This assumption, however, has not yet been verified.

One non-lethal variant contained a deletion of four base pairs coding for part of the C-propeptide of the proα2 chain (Pihlajaniemi et al 1984). The mutation is of special interest in relation to bone mineralization. The proband had a progressive form of OI and was homozygous for the deletion of four base pairs of coding sequence. The deletion caused a frame shift in reading of codons and, as a consequence, fibroblasts from the proband synthesized proα2 chains that were of the correct length but that had an abnormal sequence for the last 33 amino acids. The altered sequence of the last 33 amino acids, in turn, prevented the proα2 chains from associating with proα1 chains to form normal type I procollagen. Instead, the fibroblasts synthesized

Proα1 POINT MUTATIONS

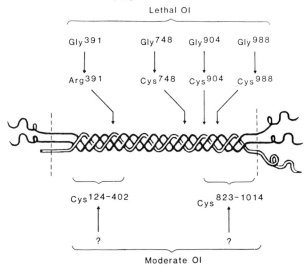

FIG. 6. Schematic representation of point mutations that have introduced a cysteine residue or an arginine residue into the proα1 chain of type I procollagen. The amino acid converted to a cysteine residue in the two moderate variants of OI is unknown.

only type I procollagen consisting of three proα1 chains (see Deak et al 1983). The procollagen consisted of three proα1 chains folded into a triple-helical conformation that was apparently normal at low temperatures and was normally processed by N-proteinase at 37 °C. The abnormal procollagen, however, partially unfolded at its C terminus at about 3 °C below the normal unfolding of type I procollagen. The results suggested, therefore, that the presence of the α2 chain in type I collagen contributes to the thermal stability of the C-terminal end of the molecule.

As discussed elsewhere (Prockop & Kuivaniemi 1986), the observations in the OI variant with the four base pair deletion in the proα2 chain appeared initially to have several broad implications for the synthesis and maintenance of normal bone. The proband's symptoms were primarily confined to bone. Therefore, it appeared that the presence of the α2 chain in the type I collagen molecule was not essential for the normal structure and function of most connective tissues, but was critical for bone. A second reason for interest in this variant was that the parents who were heterozygous for the same defect were found to have X-ray evidence of osteopenia even though they were asymptomatic and still in their thirties. Therefore, the results raised the possibility that the heterozygous state for the same mutation could be a cause of brittle bones late in life that would mimic osteoporosis. Recently, however, a proband whose fibroblasts also synthesized type I procollagen consisting entirely of proα1 chains was reported with the phenotype of Ehlers-Danlos syndrome rather than OI (Sasaki et al 1987). The reasons for the differences in phenotypes are not apparent. However, the gene alteration in the proband with Ehlers-Danlos syndrome has not been defined. It may be a mutation that shows tissue specificity in its expression. Alternatively, we cannot exclude the troubling possibility that the phenotype generated by a given mutation in a procollagen gene may vary from family to family or even from individual to individual, depending on other genetic factors or even environmental conditions.

Generalizations and questions about mutations in type I procollagen genes

We and others were initially concerned that the search for mutations of type I procollagen genes in individuals with OI and related heritable syndromes might yield only a few subjects in which such mutations were present. Instead, it is now apparent that a relatively large proportion of individuals with the diseases have mutations in type I procollagen genes. We ourselves, have found such mutations in about 25% of the fibroblast samples taken from some 60 lethal variants of OI. Therefore, we and others working in the field must address the question of why there are so many mutations in these genes. In effect, we must address the question of why the disease is largely linked to one set of genes.

One obvious reason for the large number of mutations is that the protein is a structural one. Hence, a mutation in any one of the four alleles that changes the primary structure of the chain can interfere with the complex process of fibril assembly in the extracellular matrix. A second and related reason is that much of the primary structure of the collagen molecule is essential for its normal function. The surface of the protein provides a series of hydrophobic and electrostatic sites that precisely direct fibril assembly and help to maintain the structure of the fibril. In effect, most of the 1014 amino acids in each α chain participate in the biological function of the protein. It is also apparent that every third residue in the repeating Gly-X-Y- sequence of the α chain domain must be glycine, the smallest amino acid, in order to fit into the compact central axis of the collagen triple helix. Hence, all of the 338 glycine residues in each of the two α chains are probably essential for the synthesis of a molecule that is stable at body temperature. These and related considerations about the normal assembly and function of procollagen and collagen suggest that the two structural genes may be particularly vulnerable sites for mutations that produce disease syndromes. Because some of these syndromes primarily affect bone, it is likely that the mutations causing them alter sites on the type I procollagen or collagen molecule that are essential for the normal formation of bone. Hence, detailed analyses of the mutated molecules should provide a means of defining precisely the important interaction sites.

Acknowledgements

The experimental work presented here was supported in part by research grants from the March of Dimes/Birth Defects Foundation and the National Institutes of Health (grant AR-38188).

References

Barsh GS, Roush CL, Bonadio J, Byers PH, Gelinas RE 1985 Intron-mediated recombination may cause a deletion in an α1 type I collagen chain in a lethal form of osteogenesis imperfecta. Proc Natl Acad Sci USA 82:2870–2874

Byers PH, Bonadio JF 1985 The molecular basis of clinical heterogeneity in osteogenesis imperfecta: mutations in type I collagen genes have different effects on collagen processing. In: Lloyd J, Scriver CR (eds) Genetic and metabolic disease in paediatrics. Butterworth, London p 56–90

Cheah KSE 1985 Collagen genes and inherited connective tissue disease. Biochem J 229:287–303

Chu M-L, Gargiulo V, Williams CJ, Ramirez F 1985 Multiexon deletion in an osteogenesis imperfecta variant with increased type III collagen mRNA. J Biol Chem 260:691–694

Cole WG, Chan D, Chambers GW, Walker ID, Bateman JF 1986 Deletion of 24 amino acids from the proα1(I) chain of type I procollagen in a patient with the Ehlers-Danlos syndrome type VII. J Biol Chem 261:5496–5503

Deak SB, Nicholls A, Pope FM, Prockop DJ 1983 The molecular defect in a nonlethal variant of osteogenesis imperfecta. Synthesis of proα2(I) chains which are not incorporated into trimers of type I procollagen. J Biol Chem 258:15192–15197

de Wet WJ, Pihlajaniemi T, Myers J, Kelly TE, Prockop DJ 1983 Synthesis of a shortened proα2(I) chain and decreased synthesis of proα2(I) chains in a patient with osteogenesis imperfecta. J Biol Chem 258:7721–7728

Gross J, Kirk D 1958 The heat precipitation of collagen from neutral salt solutions. Some rate-regulating factors. J Biol Chem 233:355–360

Hojima Y, van der Rest M, Prockop DJ 1985 Type I procollagen carboxyl-terminal proteinase from chick embryo tendons. Purification and characterization. J Biol Chem 260:5996–6003

Hulmes DJS, Kadler KE, Mould AP, Hojima Y, Chapman JA, Prockop DJ 1988 Pleomorphism in type I collagen fibrils produced by persistence of the procollagen N-propeptide. In preparation

Kadler KE, Hojima Y, Prockop DJ 1987 Assembly of collagen fibrils *de novo* by cleavage of type I pCcollagen with procollagen C-proteinase. Assay of critical concentration demonstrates that collagen self-assembly is a classical example of an entropy-driven process. J Biol Chem 262:15696–15701

Kuivaniemi H, Sabol C, Tromp G, Sippola-Thiele M, Prockop DJ 1988 A 19-base pair deletion in the proα2(I) gene of type I procollagen that causes in-frame RNA splicing from exon 10 to exon 12 in a proband with atypical osteogenesis imperfecta and his asymptomatic mother. J Biol Chem: in press

Miyahara M, Hayashi K, Berger J, Tanzawa K, Njieha FK, Trelstad RL, Prockop DJ 1984 Formation of collagen fibrils by enzymic cleavage of precursors of type I collagen *in vitro*. J Biol Chem 259:9891–9898

Na GC, Butz LJ, Carroll RJ 1986 Mechanisms of *in vitro* collagen fibril assembly. Kinetic and morphological studies. J Biol Chem 261:12290–12299

Pihlajaniemi T, Dickson LA, Pope FM et al 1984 Osteogenesis imperfecta. Cloning of a proα2(I) collagen gene with a frame-shift mutation. J Biol Chem 259:12941–12944

Prockop DJ, Kivirikko KI 1984 Heritable diseases of collagen. N Engl J Med 311:376–386

Prockop DJ, Kuivaniemi H 1986 Inborn errors of collagen. Rheumatology 10:246–271

Sasaki T, Arai K, Ono M, Yamaguchi T, Furuta S, Nagai Y 1987 Ehlers-Danlos syndrome. A variant characterized by the deficiency of proα2 chain of type I procollagen. Arch Dermatol 123:76–79

Sharp PA 1987 Splicing of messenger RNA precursors. Harvey Lect 81:1–31

Sillence DO 1981 Osteogenesis imperfecta. An expanding panorama of variants. Clin Orthop Relat Res 159:11–25

Steinmann B, Rao VH, Vogel A, Bruckner P, Gitzelman R, Byers PH 1984 Cysteine in the triple-helical domain of one allelic product of the α1(I) gene of type I collagen produces a lethal form of osteogenesis imperfecta. J Biol Chem 259:11129–11138

Tromp G, Prockop DJ 1988 Single base mutation in the proα2(I) collagen gene that causes efficient splicing of RNA from exon 27 to exon 29 and synthesis of shortened but in-frame proα2(I) chain. Proc Natl Acad Sci USA: in press

Vogel BE, Minor RR, Freund M, Prockop DJ 1987 A point mutation in a type I procollagen gene converts glycine 748 of the α1 chain to cysteine and destabilizes the triple helix in a lethal variant of osteogenesis imperfecta. J Biol Chem 262:14737–14744

Williams CJ, Prockop DJ 1983 Synthesis and processing of a type I procollagen containing shortened proα1(I) chains by fibroblasts from a patient with osteogenesis imperfecta. J Biol Chem 258:5915–5921

Willing MD, Cohn DH, Starman BJ, Byers PH 1987 Heterozygosity for a large deletion in the α2(I) collagen gene produces osteogenesis imperfecta type II. Clin Res 35:595 (abstr)

Wirtz MK, Keene DR, Glanville RW, Rao VH, Steinmann B, Hollister DW 1987 Ehlers-Danlos syndrome type VIIA: heterozygous deletion of 18 amino acids from a proα2(I) chain. Clin Res 35:213 (abstr)

Wood GC 1960 The formation of fibrils from collagen solutions. 2. A mechanism of collagen-fibril formation. Biochem J 75:598–605

DISCUSSION

Termine: In Pavia, at the recent international conference on osteogenesis imperfecta (OI), you showed that the proband with the deletion defect (proα2(I)) has siblings and family members who, while having the same collagen defect, are phenotypically variable. The involvement of other genes in the phenotypic expression of OI seems to be one of the most exciting aspects of current research in this disease.

Prockop: The heterogeneity with which the same defect expresses itself in different members of the same family is striking. One example is a boy whose mother has exactly the same gene defect. She has blue sclerae and she is short, but nothing else is apparently wrong; she does not have dislocated joints, or broken bones. In that same family we find that the defect was inherited as an autosomal dominant defect—half the people were affected. Some had broken bones, some had dislocated joints. There are at least five similar families. We don't know the explanation. The rest of the genetic background may make a difference. However, I find that hard to accept and I think there is a more complex biological explanation.

Baron: Is the protein suicide an intracellular or extracellular degradation, and does it result from collagenase acting on the molecule?

Prockop: In fibroblast cultures you can show that the non-helical protein builds up for a period of time and then slowly comes out into the culture medium, where it is rapidly degraded. It's hard to recover. The degradation may occur during the process of secretion through the membrane. When the chains are denatured and non-helical, any protease will digest them—you don't need a specific collagenase.

Ruch: Do you assume that gene amplification might exist? We have indirect evidence from cytophotometric and radioautographic data that during odontoblast terminal differentiation these cells become hyper-diploid and functional odontoblasts secrete increased amounts of collagen type I.

Prockop: I don't know of any evidence for gene amplification.

Ruch: Jaenisch (1983) produced a strain of mice that was no longer able to transcribe the α1 gene. E. Koller and K. Kratochwil (3rd Congress Tooth Morphogenesis and Differentiation, Alund, June 1987) removed tooth buds from these mice and grafted them in the anterior chamber of the eye. Normal collagen type I was produced.

Prockop: That's a complex experimental situation. Dr Jaenisch developed a mouse with retrovirus inserted into the first intervening sequence of the proα1 gene of type I procollagen (Schnieke et al 1987). The homozygous mouse died *in utero* at about Day 12. Jaenisch's experiment prompted Wouter de Wet and his colleagues (Rossouw et al 1987) to look for enhancers in the first intervening sequence of the proα1(I) gene. He found a powerful enhancer there. Other investigators have now found similar enhancers at the same location in other collagens. One explanation as to how the retrovirus inactivated the proα1(I) gene in the Jaenisch experiment is that it moved the enhancer 5 kb away from the promoter region. In my opinion, it has not been exclusively shown that the retrovirus causes permanent inactivation of the gene; it may be expressed in some tissues but not in others. It is surprising that the death of the homozygous embryo occurs as late as Day 12. This is an intriguing model that requires further investigation.

Rodan: Tissue-specific expression of a gene is not an uncommon phenomenon. The fact that the retroviral insertion does not prevent the expression of proα1 in dentine may indicate that control of the collagen expression in dentine is regulated in a different manner from other tissues. In what tissue does the gene not express *in utero*?

Prockop: At about 12 days *in utero* the mouse is a very small animal. It is difficult to look at all tissues in detail and the pathology of the homozygous mouse is complex. It begins with necrosis of the liver.

Krane: It is not necrosis. It is the failure to support haemopoiesis in the liver or bleeding due to blood vessel involvement (Löhler et al 1984).

Fleisch: You suggested that one should look for a gene defect in osteoporosis. Have you any data along these lines? Until now osteoporosis has not been thought to be caused by a defect in chemical structure.

Prockop: The basic data were our findings in OI. In addition, there are results from one or two scattered families. For instance, we studied a family with Michael Pope and Alan Nicholls here in England (Nicholls et al 1984). The child is homozygous for a defect in the gene for proα2(I) chains. The parents are heterozygous for the defect and while still in their thirties have evidence for osteopenia (Prockop et al 1985).

Fleisch: But this was a family with osteogenesis imperfecta.

Prockop: They were examined by X-ray and we had trouble getting a thorough work-up of their bone status. We and others are finding, as we look through the parents of children with OI, that a number of them have rather striking mutations. We don't have the full story yet: the parents are usually in their 30s, not in their 60s. In addition, I should point out that many individuals, particularly women, with mild osteogenesis imperfecta develop more extensive fractures after the menopause. If they do not have other symptoms of osteogenesis imperfecta, their bone status is difficult to distinguish from osteoporosis.

Termine: Are there no RFLP data in any of those families yet?

Prockop: No; RFLP data are difficult to obtain in osteoporosis because of the late onset of the disease. Generally you need three generations and rather large pedigrees to get conclusive results.

Caplan: Historically, we have used mutations to help us understand the normal process. I am interested in what controls fibril diameter of the same basic gene product. Perhaps the rate of cleavage of the N-terminal propeptide might control fibril diameter. Is there any evidence for this?

Prockop: People have been extracting collagen and reprecipitating fibrils for decades and have done some important work. In the system we are now using we can control the critical variables much better than before. We can show that it works as expected: it is a process of crystallization.

Caplan: Is this a question of the amount of processing enzyme or is there a cell structural component? Osteoblasts do business as sheets of cells that have many side-to-side contacts. The coordination of these sheets to produce large fibrils may affect how the cells are arranged as well as how much type I collagen is synthesized.

Prockop: We think of it as a process that can be modulated at several different levels. Rate of cleavage is one; the order in which the propeptides are cleaved is another; temperature seems to have a dramatic effect; and cell surfaces seem likely to be important. We can't study such effects as yet.

Veis: There are probably several different mechanisms by which fibril diameters are regulated *in vivo*. One can describe mechanisms such as those Dr Prockop has discussed which are intrinsic to the collagen itself or others which relate to the interaction of collagen with many other macromolecules. One can control fibril diameter *in vitro* by any one of a dozen or so manipulations, such as the ones discussed here. The question is how does the regulation operate *in vivo* to control temporal–spatial relationships?

Prockop: A basic problem in our system is that we produce crystal-like, spiky structures that don't look like fibrils. We realize that when a monomer polymerizes *in vivo* there is modulation by many other components of the matrix.

Canalis: I share your views about osteoporosis, but because it is a heterogeneous disease and because collagen is regulated at multiple post-translational steps, do you think that defects at multiple levels could explain the heterogeneity of the disease?

Prockop: In osteogenesis imperfecta we thought we would find defects in processing enzymes, but the evidence is dramatically against this. Bryan Sykes in Oxford and P. Tsipouras in Connecticut looked through 42 families in England with dominantly inherited OI. RFLP studies revealed that in essentially every family, the disease was co-inherited with an allele for either the proα1(I) or proα2(I) chain of type I procollagen (Sykes 1987).

Canalis: But is OI more homogeneous than osteoporosis?

Prockop: We thought so initially and we can't exclude the idea that there are multiple causes of osteoporosis. We are dealing with a biological system that does have multiple gene products, so there could be many genes at fault. However, mutations in genes for structural proteins are by far the most likely to produce dominant phenotypes.

Beyond that, all of us in the field have been amazed at how much of the structure of type I procollagen must be conserved to produce a normal phenotype. The results demonstrate dramatically that it is a highly valuable gene–protein system. I think no one would have predicted in advance that a single substitution of cysteine for glycine would have been sufficient to destabilize the whole collagen molecule and to produce a lethal phenotype. The effect of the single cysteine substitution on the cooperative unfolding of the molecule is in itself a surprising observation.

References

Jaenisch R, Harbers K, Schnieke A et al 1983 Germline integration of Moloney leukemia virus at the *Mov*13 locus leads to recessive lethal mutation and early embryonic death. Cell 32:209–216

Löhler J, Timpl R, Jaenisch R 1984 Embryonic lethal mutation in mouse collagen-1 gene causes rupture of blood-vessels and is associated with erythropoietic and mesenchymal cell death. Cell 38:597–607

Nicholls A, Osse G, Schloon HG et al 1984 The clinical features of homozygous alpha2(I) collagen deficient osteogenesis imperfecta. J Med Genet 21:257–262.

Prockop DJ, Chu ML, de Wet W et al 1985 Mutations in osteogenesis imperfecta leading to the synthesis of abnormal type I procollagens. Ann NY Acad Sci 460:289–297

Rossouw CMS, Vergeer WP, du Plooy SJ, Bernard MP, Ramirez F, de Wet WJ 1987 DNA sequences in the first intron of the human proα1(I) collagen gene enhance transcription. J Biol Chem 262:15151–15158

Schnieke A, Dziadek M, Bateman J et al 1987 Introduction of the human pro-alpha 1(I) collagen gene into pro-alpha 1(I)-deficient mov-13 mouse cells leads to formation of functional mouse human hybrid type I collagen. Proc Natl Acad Sci USA 84:764–768

Sykes B 1987 Genetic cracks in bone disease. Nature 330:607–608

Phosphoproteins from teeth and bone

Arthur Veis

Connective Tissue Research Laboratories, Department of Oral Biology, Northwestern University, 303 E. Chicago Ave, Chicago, Illinois 60611, USA

Abstract. The anionic non-collagenous proteins of the extracellular matrices of bone and dentine have been proposed as important participants in the mineralization of these tissues. Phosphorylated protein components have been implicated as mediators of the specific nucleation of the growth of hydroxyapatite crystals on the matrix collagen fibrils. However, the phosphoproteins of bone and dentine are quite different and it is difficult to postulate a common mechanism for the nucleation. If there is a common mechanism some particular domain on each molecule is likely to be involved. Therefore, we have initiated studies of the domain structure of the most highly phosphorylated dentine protein, phosphophoryn. At least three sequence domains with different character have been found, and studies are under way to relate these domains to the antigenic, collagen-binding and calcium ion-binding properties of phosphophoryn. The collagen- and calcium ion-binding regions appear to be localized within the same domains.

1988 Cell and molecular biology of vertebrate hard tissues. Wiley, Chichester (Ciba Foundation Symposium 136) p 161–177

In vertebrates the mineralized tissues have important structural and metabolic functions. Bone is perhaps the most complex tissue, serving both as a labile reservoir for calcium ions and as a key element in the mechanical-chemical transduction system. Bone is continually growing and remodelling in response to both external mechanical stresses and internal hormonal and metabolic signals, but it must maintain its structural integrity. This integrity is ensured by the cross-linking of the collagen fibre array and the deposition of mineral crystals within the matrix. The mechanical, structural role is paramount in most other mineralized tissues in vertebrates and lower organisms. Although there are many differences in organization and composition of the mineralized tissues in diverse organisms, there do appear to be some common principles used in their bioassembly. The main feature seems to be the inclusion of a crystalline phase within a preformed organic matrix, producing the necessary rigidity for the structure. Most of our work in this area has been designed to analyse the mechanisms governing the placement of the crystals within the matrix.

The extracellular matrices of the mineralized tissues are constructed from three distinct components: a structural protein phase which defines the over-

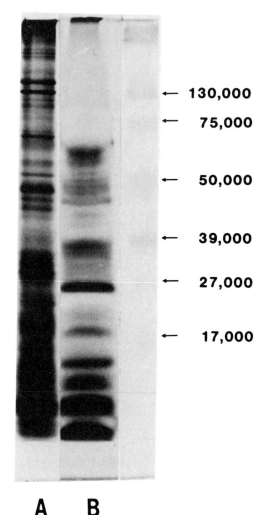

$\leftarrow$ 130,000

$\leftarrow$ 75,000

$\leftarrow$ 50,000

$\leftarrow$ 39,000

$\leftarrow$ 27,000

$\leftarrow$ 17,000

A   B

FIG. 1.   Gel electrophoretic patterns of the total EDTA-extractable proteins of rat bone and rat incisor dentine; 5–15% acrylamide gradient, 0.1% SDS, mercapto-ethanol reduced. A, rat incisor dentine; B, rat bone. Arrows represent positions of globular protein molecular mass standards (unpublished work of B. Sabsay & C.B. Wu). These gels are silver stained. Several components in each lane do not stain well with silver, therefore these patterns do not show the complete distribution of protein components.

all shape and structure of the tissue; the crystalline phase deposited within or upon the structural protein framework; and, finally, a set of non-structural anionic macromolecules that interact with both structural protein and mineral components. One can classify the mineralized tissues in several ways, but two

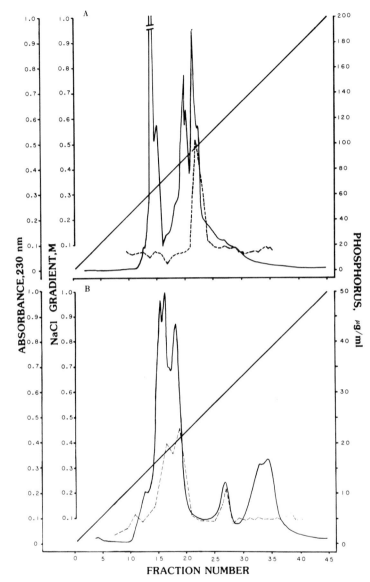

FIG. 2. Anion exchange chromatography of total EDTA extracts of rat bone and rat incisor dentine on Bio Gel TSK DEAE-5-PW, in 0.05 M Tris HCl, pH 8.2 with NaCl gradient from 0.0 to 1.0 M. A, rat incisor dentine (Rahima & Veis 1988); B, rat bone (M. Rahima & A. Veis, unpublished work); solid line, absorbance; dashed line, phosphorus content; diagonal solid line, NaCl gradient.

are particularly appealing. One can specify the nature of the mineral (e.g. apatite or calcite), or the nature of the organic structural component (e.g.

TABLE 1 Amino acid analyses of the principal non-collagenous extracellular matrix proteins of bone and dentine[a]

	Bone proteins				Dentine proteins			
	Osteopontin (Oldberg et al 1986)	Osteonectin (Termine et al 1981a)	Bone phosphoprotein (Lee & Glimcher 1981)	Fish scales (Sauk et al 1984)	Bovine molar phosphophoryn (Stetler-Stevenson & Veis 1983)	Rat incisor phosphophoryn (Rahima & Veis 1988)	Rat incisor phosphophoryn (Butler et al 1981)	
							Fraction	
Amino acid							I	II
Lysine	58	71	20	35	44	11	13	25
Histidine	45	44	19	28	5	7	8	14
Arginine	33	28	35	35	3	6	3	27
Hydroxylysine	—	—	—	—	—	—	—	—
Hydroxyproline	—	—	—	—	—	—	—	—
Aspartic acid	147	130	165	120	447	363	375	258
Threonine	44	59	40	60	7	18	12	37
Serine and phosphoserine	153	42	128	117	426	481	491	284

Glutamic acid	184	148	237	205	14	44	31	140
Proline	56	72	52	55	6	8	6	37
Glycine	32	65	81	88	27	31	31	65
Alanine	75	58	80	98	6	15	13	36
Cysteine	—	6	2	1	—	4	—	—
Valine	49	69	46	50	3	2	4	20
Methionine	—	11	2	6	—	—	—	—
Isoleucine	23	37	14	20	3	2	4	12
Leucine	61	90	34	47	4	4	4	23
Tyrosine	17	24	20	11	2	3	3	16
Phenylalanine	21	44	19	19	2	2	2	12
(O-Phosphoserine)			(62)	(55)	(404)			
(O-Phosphothreonine)			(25)	(25)				

[a] Data expressed as residues per 1000 amino acid residues.

collagen or chitin). Bone and dentine are usually considered together because they have the same structural protein, type I collagen, as their principal constituent, and they have hydroxyapatite crystals of similar size distributed in essentially the same way within their collagenous matrices. They differ sharply, however, in function and in their content of anionic matrix proteins. These differences provide the focus for this discussion.

General characteristics of the non-collagenous proteins

The non-collagenous proteins (NCP) of bone and dentine constitute 5 to 10% of the total matrix proteins (Veis 1984). The majority of these NCP cannot be extracted without prior or concomitant demineralization, but are readily solubilized at neutral pH during demineralization. A small portion of the NCP is tightly bound to the collagen and cannot be removed without degradation of the collagen. Some serum proteins, such as the α_2-HS glycoprotein, are present in the neutral pH extracts (Dickson et al 1975), but the majority of the most prominent proteins are bone or dentine specific.

The large number of protein components present, and their wide range of molecular masses, is illustrated in Fig. 1, which compares the SDS–polyacrylamide gel electrophoresis of the total EDTA extracts of rat bone and rat dentine. Anion exchange chromatography of the same extracts (Fig. 2) demonstrates that the majority of these proteins are anionic in character. The phosphorus analyses reported in Fig. 2 show that several phosphorylated proteins are prominent constituents of the extracts. Table 1 is a collection of amino acid analyses reported for the principal bone and dentine matrix proteins. These data show that the tissue-specific bone and dentine proteins have quite different compositions. The dentine proteins are rich in aspartic acid and are highly phosphorylated on serine, whereas the bone proteins generally have more glutamic acid than aspartic acid, and have only a few phosphorylated serine and threonine residues.

Some of the bone and dentine phosphoproteins have been reported to share the properties of binding to collagen, binding calcium ions, and binding to the surfaces of hydroxyapatite crystals (Termine et al 1981a,b, Stetler-Stevenson & Veis 1986, 1987). These data have resulted in hypotheses that the phosphoproteins have some particular influences on the mineralization process. However, because of the marked differences in the bone and dentine proteins, it is difficult to generalize about their role as a common mechanism. Indeed, it has been suggested that one protein may serve to regulate more than one step during mineralization (Veis & Sabsay 1983).

A recurring theme in modern protein chemistry is the frequent existence of several functional and structural domains in a single molecule. This has led us to suggest that only certain domains in each type of matrix protein are important in the regulation of mineralization, and that one should search for these domains rather than for overall similarities among the anionic bone and dentine matrix proteins.

Phosphophoryn

Until recently, our laboratory has focused its attention on the dentine system and on the potential role of its most highly phosphorylated proteins, the phosphophoryns (DiMuzio & Veis 1978a). This emphasis was chosen for two basic reasons. First, the dentine system is inherently more simple than bone because remodelling and calcium ion homeostatic processes are not important in dentine — it is primarily a system for the deposition and maintenance of the mineral phase. Second, phosphophoryns are remarkable in their high content of aspartic acid and phosphoserine; those two residues constitute almost 90% of the bovine phosphophoryn molecule (Table 1). A phosphophoryn molecule can bind to the surface of a collagen fibre and retain its capacity to bind an enormous number of calcium ions (Stetler-Stevenson & Veis 1986). Thus, a phosphophoryn could act as the factor regulating the specific initiation and localization of mineral onto collagen fibre surfaces. We have begun an examination of rat incisor and bovine molar phosphophoryns (RIPP and BPP, respectively) to determine whether they have specific domains that might fulfil this role.

Phosphophoryn sequence domains

Phosphophoryns were isolated and purified from bovine molar and rat incisor dentines by well-established techniques (Stetler-Stevenson & Veis 1983, Tsay & Veis 1985). It was important for us to establish that a single protein was present. This was most readily accomplished with the BPP, because bovine molar dentine contains only one phosphophoryn. The situation is more complex for the RIPP because several species of phosphophoryns are present (DiMuzio & Veis 1978a, b, Butler et al 1981). Because only a single component was found in the BPP, we concentrated on its analysis. RIPP was studied in the same way, but only for confirmation of the BPP data.

Two methods were used to examine the structure of the BPP: partial acid hydrolysis and limited trypsin digestion. The BPP contains two residues of tyrosine per molecule. These were labelled with ^{125}I (Stetler-Stevenson & Veis 1986) and the product was used as the substrate for tryptic digestion. The digestion was followed by gel electrophoresis and by gel filtration chromatography as a function of the digestion time. As shown in Fig. 3, the labelled BPP appeared as a single component at zero time. With increasing time of digestion a small decrease in molecular mass of the main peak was seen, together with a decrease in the density of labelling, while three low molecular mass peptides appeared. HPLC gel filtration chromatography confirmed

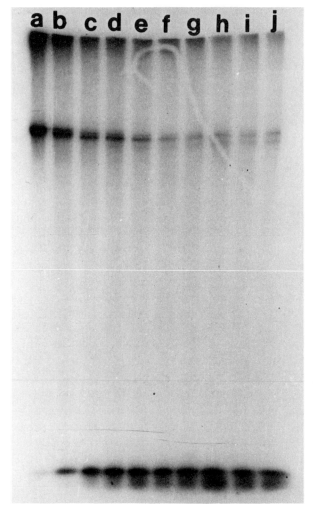

FIG. 3. Autoradiogram of gel electrophoretic analyses of the trypsin digestion of BPP as a function of digestion time. The ^{125}I-BPP was digested with trypsin at a 10:1 w/w ratio. a, BPP substrate; b, zero time after addition of trypsin, then trypsin inhibitor; c–j, 5, 10, 20, 30, 45, 60, 90 and 120 min (W.G. Stetler-Stevenson & A. Veis, unpublished work).

these results and permitted recovery of the large molecular mass component that remained after several hours of digestion. Of the three small peptides, only two were ^{125}I-labelled. The high molecular mass digest fragment and a mixture of the small peptides were tested for collagen- and calcium ion-binding properties. The peptides were inactive in both assays, whereas the high molecular mass component retained, undiminished, both of these

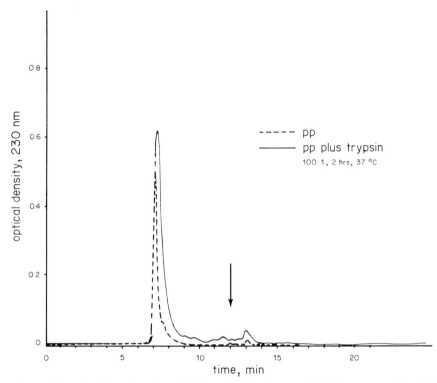

FIG. 4.   Trypsin digestion of rat incisor phosphophoryn. Gel filtration on Bio Gel DEAE 5-PW before and after 120 min of digestion using 100:1 RIPP to trypsin at 37 °C and pH 7.5. The arrow marks the elution position of trypsin. After digestion the major peak is reduced slightly in molecular size and a few small peptides appear.

properties. After a limited two hour digestion, the high molecular mass component was devoid of tyrosine, reduced in its content of hydrophobic amino acids, but markedly enriched in aspartic acid and phosphoserine residues.

Similar results were found after the digestion of RIPP (Fig. 4). The calcium chloride precipitate of RIPP was passed over a Bio-Gel DEAE 5-PW column and the PP fraction was collected. This was then fractionated on a Zorbax GF-250 gel filtration column eluted isocratically with 0.1 M sodium orthophosphate, pH 7.5. The main peak that emerged just after the void volume was collected. This was then re-chromatographed. The protein emerged as a single sharp band just after the void volume. However, on SDS-polyacrylamide electrophoresis it proved to contain two closely migrating bands, not seen with silver staining, but heavily blue-stained with Stains All. The two components were in the $M_r \approx 90\,000$ range. The important point was that neither silver nor Stains All staining revealed components of lower

FIG. 5. A schematic representation of the types of sequence domains in bovine molar phosphophoryn. (P)Ser, phosphoserine residue.

molecular mass. After trypsin digestion under the same conditions as used for the BPP, the high molecular mass component was still the main peak, but several lower molecular mass fragments were present. Thus, the preliminary RIPP data confirm the behaviour of the BPP.

The model that emerges from the trypsin digestion data is of a molecule with a large central region rich in the acidic amino acid residues and nearly free of apolar, hydrophobic residues. The tyrosine-containing apolar peptides released by trypsin must be in terminal domains of the molecule.

The nature of the acidic regions of the molecule was examined by taking advantage of the fact that sequences containing aspartyl residues are particularly labile to hydrolysis in dilute acetic acid, releasing free aspartic acid. An analysis of the other amino acids and peptides freed in this process can then provide information on the distribution of the aspartic acid and yield peptides suitable for sequencing. This strategy is being applied to the BPP acidic domain. The data are too detailed to present here. In summary, after 48 hours of hydrolysis 43% of the total aspartic acid and 14% of the serine residues were released as free amino acids. These data agree well with the early study by Krippner & Nawrot (1977). Alanine and glycine were released in lesser amounts, together with other peptides, some of which contained only serine residues. We concluded that within the acidic region there must be major domains of $(Asp-Asp)_x$, some of $(Asp-Ser[P])_y$ and some of $(Ser[P]-Ser[P])_z$, where Ser [P] denotes a serine residue which may or may not be phosphorylated.

The overall view of the phosphorphoryns which emerges from these studies is depicted in Fig. 5.

Domain function

It is too early to discuss the functions of any of the sequence domains of the phosphophoryns until further detailed studies have been carried out. However, there are some preliminary observations which appear to be important. Rahima & Veis (1987) have examined the reactivity of a number of phosphophoryns from various species to the antibody prepared against RIPP. It was found that the phosphophoryns of every species examined were cross-reactive, and it was argued that the antigenic epitope was strongly conserved during evolution. However, the high molecular mass acidic region of RIPP recovered after digestion by trypsin did not show reactivity with the antibody. Thus, the conserved antigenic region must be in one of the terminal domains of the molecule. Stetler-Stevenson & Veis (1986) found that the collagen-binding property of phosphophoryn was ionic strength dependent and, probably, electrostatic in nature.It seems likely that one of the domains $(Asp)_n$, $(Ser)_m$ or $(Asp\text{-}PSer)_o$ is involved. One or more of these must also be involved in the calcium ion binding.

Oldberg et al (1986) have recently published the complete sequence of rat bone sialoprotein (osteopontin). Glutamic acid (or glutamine) and aspartic acid (or asparagine) are present in near equal amounts (48 and 49 residues per molecule of 295 total residues, respectively). However, there is a unique region of $(Asp)_9$, residues 90–98. Osteopontin binds to hydroxyapatite surfaces (Poser & Price 1979) as well as to rat osteosarcoma cells. This aspartic acid-rich region in osteopontin may be related in function to similar regions in bovine molar or rat incisor phosphophoryn. Sequences for the other bone-specific proteins are not available but, as seen in Table 1, all are rich in aspartic acid and glutamic acid, with variable amounts of serine.

When the possibility of a repeating (Asp-Ser) sequence in PP was first discovered (Krippner & Nawrot 1977), we synthesized a repeating dipeptide $(Glu\text{-}Ser)_{15}$ with which to examine its calcium ion-binding properties before and after phosphorylation, and to compare those with the behaviour of PP and dephosphorylated PP (Lee & Veis 1980). Phosphorylation within the repetitive dipeptide appeared to have two effects. It increased the apparent binding constant for calcium ions tenfold and promoted the affinity of the carboxylate groups such that all binding sites (carboxylate and phosphate) had the same binding constant. Conversely, partial dephosphorylation of BPP halved the high affinity calcium ion-binding constant. These data suggest that the repetitive Asp-PSer domains within the phosphophoryn are of particular importance in determining the calcium ion-binding properties of the PP, and that phosphorylated regions within the other mineralized extracellular matrix proteins may similarly regulate the properties of those macromolecules.

Acknowledgements

All of the work described in this manuscript has been supported by NIH grant RO1-DE01374. I am pleased to acknowledge that support. The previously unpublished work described has been carried out with my colleagues, Drs B. Sabsay, W.G. Stelter-Stevenson and M. Rahima in this laboratory.

References

Butler WT, Bhown M, DiMuzio MT, Linde A 1981 Noncollagenous proteins of dentin. Isolation and partial characterization of rat dentin proteins and proteoglycans using a three-step preparative method. Collagen Res 1:187–199

Dickson IR, Poole AR, Veis A 1975 Localization of plasma α_2-HS glycoprotein in mineralizing human bone. Nature (Lond) 256:430–432

DiMuzio MT, Veis A 1978a Phosphophoryns—major noncollagenous proteins of rat incisor dentin. Calcif Tissue Res 25:169–178

DiMuzio MT, Veis A 1978b The biosynthesis of phosphophoryns and dentin collagen in the continuously erupting rat incisor. J Biol Chem 253:6845–6852

Krippner RD, Nawrot CF 1977 The distribution of aspartic acid residues in bovine dentin phosphoprotein. J Dent Res 56:873

Lee SL, Glimcher MJ 1981 Purification, composition and ^{31}P NMR spectroscopic properties of a noncollagenous phosphoprotein isolated from chicken bone matrix. Calcif Tissue Int 33:385–394

Lee SL, Veis A 1980 Cooperativity in calcium ion binding to repetitive carboxylate–serylphosphate polypeptides and the relationship of this property to dentin mineralization. Int J Pept Protein Res 16:231–240

Oldberg Å, Franzén A, Heinegård D 1986 Cloning and sequence analysis of rat bone sialoprotein (osteopontin) cDNA reveals an Arg-Gly-Asp cell-binding sequence. Proc Natl Acad Sci USA 83:8819–8823

Poser JW, Price PA 1979 A method for decarboxylation of γ-carboxyglutamic acid in proteins. Properties of the decarboxylated γ-carboxyglutamic acid protein from calf bone. J Biol Chem 254:431–436

Rahima M, Veis A 1988 Two classes of dentin phosphophoryns, from a wide range of species, contain immunologically cross-reactive epitope regions. Calcif Tissue Int 42:104–112

Sauk JJ, Cocking-Johnson D, Cervenka VA, Van Kampen CL 1984 Noncollagenous phosphoprotein derived from teleostean fish-scales. Biochim Biophys Acta 798:199–203

Stetler-Stevenson WG, Veis A 1983 Bovine dentin phosphophoryn: composition and molecular weight. Biochemistry 22:4326–4335

Stetler-Stevenson WG, Veis A 1986 Type I collagen shows a specific binding affinity for bovine dentin phosphophoryn. Calcif Tissue Int 38:135–141

Stetler-Stevenson WG, Veis A 1987 Bovine dentin phosphophoryn: calcium ion binding properties of a high molecular weight preparation. Calcif Tisue Int 40:97–102

Termine JD, Belcourt AB, Conn KM, Kleinman HK 1981a Mineral and collagen-binding proteins of fetal calf bone. J Biol Chem 256:10403–10408

Termine JD, Kleinman HK, Whitson SW, Conn KM, McGarvery ML, Martin GR 1981b Osteonectin, a bone-specific protein linking mineral to collagen. Cell 26:99–105

Tsay TG, Veis A 1985 Preparation, detection and characterization of an antibody to rat α-phosphophoryn. Biochemistry 24:6363–6369

Veis A 1984 Bones and teeth. In: Piez KA, Reddi AH (eds) Extracellular matrix biochemistry. Elsevier, New York, p 328–374

Veis A, Sabsay B 1983 Bone and tooth formation. Insights into mineralization strategies. In: Westbroek P, De Jong EW (eds) Biomineralization and biological metal accumulation. D Reidel, Dordrecht, Holland, p 273–284

DISCUSSION

Martin: Quite a lot is known about phosphorylation site sequence determinants in other phosphorylated substrates. Is there anything about the sequence around the phosphorylated serines that is peculiar to this protein and might indicate involvement of a specific kinase?

Veis: The serines occur in two kinds of sequences: (Asp-Ser) $_X$ and (Ser) $_Y$. In both the polyserine and Asp-Ser sequences over 90% of the serines are phosphorylated. They are all in acidic regions. We are investigating which of the kinases might act specifically by using inhibitors for the various kinds of kinases. We have already found that odontoblasts and osteoblasts contain every kind of apo-kinase. We are trying to find out which ones work most specifically on newly synthesized phosphophoryn as the substrate and at what stage in the secretory process phosphorylation occurs.

Rodan: Are you saying that the distribution of phosphophoryn in human dentine changed your views on its possible role in calcification?

Veis: I imagined that the distribution of phosphophoryn in dentine would be fairly uniform. We have shown that this is not so (Rahima et al 1988). There is a heightened concentration of phosphophoryn in the region immediately surrounding the odontoblastic processes. Apparently the phosphophoryn is extruded from the odontoblastic processes and may precipitate with calcium. Moreover, the human dentine has a lower content of phosphophoryn than rat dentine. A continuously erupting tooth, which has a very short life time for the mineralized phase, has a higher content of phosphophoryn than the mature permanent tooth. A number of studies (e.g. Masters 1985) have shown *in situ* degradation in human teeth. If you take mature human tooth material rather than rat incisors, the molecular mass distribution of phosphoproteins on gels presents as a broad smear rather than sharp bands (Takagi & Veis 1984). We have to look more carefully at the potential role of the phosphophoryns in dentine: their action may extend well beyond the process of mineral deposition.

Butler: If you treat phosphophoryn with alkaline phosphatase you can remove 40 or 50% of the phosphate, but one cannot readily remove it all. Can you account for this observation on the basis of its proposed structure?

Veis: We initially thought that phosphophoryns would have an extended structure. That is not the case. The intrinsic viscosity relative to the chain length shows that the protein is folded. Even when you suppress the ionization with high salt concentrations, the molecule has some internal folding which could be of beta structure type. In 0.5 M potassium chloride, circular dichroism studies demonstrate that phosphophoryn is still not a completely random chain. If you put in 10 mM calcium ions, you can increase the amount of beta structure. In spite of all our efforts during purification, we may not have removed every calcium, magnesium, or other divalent ion. Alternatively, there may be some non-ion-mediated structure present in the intact molecule.

Butler: Dr Y. Takagi has shown that mantle dentine has little or no phosphoprotein. There also appears to be none in reparative dentine. How do you explain these results?

Veis: Secondary dentine is that which occurs after the tooth is mature. There is a continued deposition of dentine along the way. When we looked at that material with our antibody we saw that phosphophoryn was there. If you have an injury to the tooth which kill the cells you get reparative dentine which stains just like mantle dentine. The reparative dentine interfaces between the dentine and pulp. Secondary dentine is a normal process of dentine formation, albeit at a slower rate than the primary dentine. Like the primary dentine, the secondary dentine can be classified as intertubular.

Slavkin: You presented evidence that collagen plus the bovine phosphophoryn enhances calcium-45 binding to the surfaces of collagen fibrils. You suggested that the epitopes for the antibody were located at either the N-terminus or the C-terminus, but clearly not in the other regions. How do you prepare the material for immunolocalization?

Veis: That is a good question. In other words, our antibody may not be picking up what is in the calcium-binding domain. If there is degradation in the human tooth you would say that, even though we are not seeing it, there could still be the very anionic part. Analysis of the calcium and phosphate ion concentration in the regions between and surrounding the tubules shows that it follows the same kind of distribution seen by immunostaining: there is less calcium phosphate in the centre between two tubules than there is at the outer edge. Electron microscopy showed that the calcium phosphate precipitation is outside, not inside the tubule.

Slavkin: Your cells didn't light up with your antibody?

Veis: Not very much, but there is some staining there. We have looked, at the electron microscopic level, with colloidal gold and antibody labelling and we can clearly see the phosphophoryn being made in the endoplasmic reticulum in localized patches. We are working on the pathway for secretion.

Slavkin: The *in vitro* approach is potentially very elegant. If you used the immunoperturbation approach with a fab fragment to the dentine phosphoprotein (DPP), in the collagen phosphoprotein-binding assay that you showed, you should be able to identify the epitopes.

Veis: We know where the regions containing the epitopes are. We are now isolating those limited tryptic peptides and we shall sequence them. Then we shall be able to construct good probes for the intact phosphophoryn molecule.

In contrast to the work by Dr Slavkin, when we do the cell-free translation we obtain a higher molecular mass material. In organ culture experiments we clearly have as the first biosynthetic product a material of higher molecular mass than the extracellular matrix phosphoprotein. Therefore there is a pro form of the molecule and we need to investigate its processing, just as for the procollagen/collagen system.

Weiner: Much of the rationale for work over the last 20 years on non-collagen proteins in bone and dentine was based on the observation that skin type I collagen does not mineralize, whereas bone type I does. We now know that the three-dimensional fibril structure is different in these two tissues. Shouldn't we focus on this difference as being a possible basic explanation for the difference in behaviour of the two tissues?

Veis: Phosphophoryn binds very well to the fibrils from skin collagen. The question is whether there is something else in the mixture of proteins in the extracellular matrix of mineralizing tissues which modulates the specificity of binding. Our studies so far do not show much specificity. The phenotypic expression of this phosphoprotein only comes when the tissue is mature, when odontoblasts mature, and then it makes true dentine. You never see phosphophoryn in the skin. If you could transpose the gene into the skin you would mineralize the matrix.

Glimcher: Highly purified non-skeletal collagens and purified decalcified bone collagen, almost all of which is type I, calcify *in vitro* when exposed to metastable solutions of Ca-P, or *in vivo* when implanted in the animal from which the collagen was isolated. Proteins other than collagen cannot induce calcification under the same conditions and only native collagens with about a 700 Å axial periodicity can nucleate the Ca-P crystals *in vitro*. Reconstituted native collagen fibrils (700 Å axial period) calcify when placed in non-skeletal locations, such as muscle and peritoneum, *in vivo*. This is convincing evidence that native collagen fibrils, at least type I, regardless of the source, may contain nucleation sites for Ca-P crystals.

As Dr Weiner knows from his own and previous studies of turkey and other avian species, calcification of type I collagen of some tendons occurs during normal development. Type I collagen from tendons and other soft tissues calcifies in some pathological conditions in humans, and in experimentally induced pathologies in animals.

Some questions remain unanswered. Are all type I collagens as effective as bone, dentinal or cemental type I collagen? Are there structural and/or chemical differences (including post-translational modifications) which cause these type I collagens to be more effective nucleators than those from normally unmineralized tissues?

There is evidence that the nature and distribution of the cross-links in bone collagen fibrils produces an architecture that is more efficient as a heterogeneous nucleator than that in most soft tissue collagens. We have tested the putative role of phosphoproteins by *in vitro* calcification experiments. Phosphoproteins are absent from uncalcified tissues, but are found in all mineralized tissue and in pathologically calcified human and animal tissue *in vivo*. We demonstrated a direct relationship between the concentration of phosphoproteins in collagen–phosphoprotein complexes and the efficacy of their calcification *in vitro*. When preparations of chick bone collagen were treated with wheat germ acid phosphatase to remove any phosphoproteins remaining after decalcification, the collagen did calcify *in vitro*, although more slowly than phosphorylated bone collagen. Therefore, bone collagen fibrils may be necessary but not sufficient as nucleation substrates. In addition to specific post-translational modifications which change the architecture of the collagen fibrils, synthesis of phosphoproteins (and possibly other macromolecules that bind to specific locations in the fibril) may be needed to produce an efficient system for calcification *in vivo*. We are doing experiments to test this hypothesis.

Weiner: I think the fine detailed structures of collagen are very important in understanding the mineralization process. However, if you look at the whole world of biomineralization and not just the Phylum Chordata, the view that Dr Veis presents, namely that anionic glycoproteins are important, is substantiated. In every mineralized tissue in which crystal growth is controlled you find the same group of acidic proteins and there's good evidence to say that they are actively involved in the mineralization process. We have to understand a lot more about collagen to understand details of bone mineralization and we should not focus solely on either the non-collagenous proteins or the collagen.

Veis: That was exactly where we started: looking to see how the collagen of dentine differed from the collagen of soft tissues. Type I collagen is particular to bone and dentine; there is very little of any of the other collagen types. All the soft tissues and basement membranes have a wide range of other collagen types. So there must be something about type I, probably the $\alpha2$ chain, that has some peculiar binding to or some peculiar interaction with the anionic non-collagenous proteins.

Krane: The primary structure of collagen is not the whole picture. There are chemical differences between skin and bone collagen; at least in the human (Pinnell et al 1971), the pattern of hydroxylation and the hydroxylysine glycosides are very different. The first genetic disease that we described showed nothing the matter with bone collagen, but the skin and tendon hydroxylation were very defective (Pinnell et al 1972).

Veis: I agree with that. Dentinogenesis imperfecta (DI) type II is a dentine disorder in which mineralization is inhibited. In osteogenesis imperfecta you still get mineralization of the bone, but in DI type II there is no dentine mineralization, or only a very limited amount. What's missing in DI is the phosphoprotein and what seems to be maximized is the type I collagen. When mineralization is missing there seems to be no barrier to the continued secretion of collagen by those cells. Therefore, the pulp cavity in a very young person with DI type II is closed up completely. This means that there is both a biosynthetic regulatory role and a mineralization regulatory role for the non-collagenous proteins in the mineralized tissue structure.

Butler: I think your last point is well taken. An outstanding feature of the dentine phosphoprotein is its moderately high affinity for large amounts of calcium. This property would be ideal for mineral nucleation or regulation of mineralization in some other fashion.

Williams: In a nuclear magnetic resonance (NMR) study of rat incisor dentine phosphoprotein by Cookson et al (1980) it was shown that it is a mobile stretched-out polymer in aqueous solution. It binds calcium strongly to its phosphate and carboxylate groups and could not be expected to act in an epitaxial way. The interaction of the protein with a calcium mineral surface could not be a simple spatial matching, but it may function by generating a series of conformers which match stages in the build-up of a calcium mineral nucleus.

A second possible function for such a protein is to assist diffusion of ions in a restricted matrix, much as ions can diffuse in an ion-exchange matrix. The diffusion of ions on a polyelectrolyte surface is a well-known observation.

Veis: I understand what you are saying about the polyelectrolyte nature of the phosphophoryn and the territorial nature of the ion binding at these low ion-binding constants. We've done our calcium binding studies in 0.5M potassium chloride, which would swamp out these polyelectrolyte effects in most other polyelectrolytes of this charge density. Even then we still get calcium ion uptake and a conformational transition into a regular structure. *In vivo* it is probable that the molecule always has its calcium burden once it gets outside the cell, unless it's secreted out in packets with very little calcium ion on it. Therefore, I do not think the protein is stretched out.

Williams: It doesn't have to be stretched out in any particular way. All that matters is that if you put a calcium on to it at one point then every other calcium in a chain knows about it, and one of them will leave.

Veis: We looked at the calcium binding of $(Ser-Glu-Ser-Glu)_{15}$ versus $(Ser-Asp-Ser-Asp)_x$ sequences in phosphorylated and dephosphorylated phosphophoryn (Lee & Veis 1980). They show different binding affinities.

It is a system in which a local high concentration of calcium ions is induced. The calcium ions certainly are not 'piped' to a particular site along the backbone, and are free to redistribute among equivalent groups. The conformation of the phosphophoryn in calcium ion-containing solutions is not that of a random chain.

References

Cookson DJ, Levine BA, Williams RJ, Jontell M, Linde A, Bonard BD 1980 Cation binding by the rat incisor dentine phosphoprotein: a spectroscopic investigation. Eur J Biochem 110:273–278

Lee S, Veis A 1980 Cooperativity in calcium ion binding to repetitive carboxylate–serylphosphate polypeptides and the relationship of this property to dentine mineralization. Int J Pept Protein Res 16:231–240

Masters PM 1985 *In vivo* decomposition of phosphoserine and serine in noncollagenous protein from human dentin. Calcif Tissue Int 37:236–241

Pinnell SR, Fox R, Krane SM 1971 Human collagens—differences in glycosylated hydroxylysines in skin and bone. Biochim Biophys Acta 229:119–122

Pinnell SR, Krane SM, Kenzora JE, Glimcher MJ 1972 Heritable disorder of connective-tissue—hydroxylysine-deficient collagen disease. N Engl J Med 286:1013–1020

Rahima M, Tsay T-G, Andujar M, Veis A 1988 Localization of phosphophoryn in rat incisor dentin using immunocytochemical techniques. J Histochem Cytochem 36:153–157

Takagi Y, Veis A 1984 Isolation of phosphophoryn from human dentin organic matrix. Calcif Tissue Int 36:259–265

Non-collagen proteins in bone

John D. Termine

National Institute of Dental Research, National Institutes of Health, Bethesda, Maryland 20892, USA

Abstract. The non-collagen proteins of bone are a complex set of molecules that arise from local or exogenous sources. Because bone mineral is an excellent adsorbent, many circulatory and/or cell surface proteins bind to bone, where they may have immediate or subsequent effects. These include the α_2-HS-glycoprotein from blood and the potent growth factors TGF-β, PDGF, IGF-1, FGF-a and -b, and IL-1, derived from both bone and non-bone cells. Furthermore, bone cell membrane proteins such as alkaline phosphatase may be cleaved from the cell surface and entrapped in the bone matrix. Bone is enriched in a variety of enzymes and their inhibitors by similar adsorption processes. Even osteocalcin, a bone cell product, is adsorbed to bone via mineral-binding (Gla) groups. The bone sialoproteins (BSP-I or osteopontin and BSP-II) also bind to the mineral via acidic groups. Because of this phenomenon it is difficult to distinguish whether a given protein's presence in bone is advantageous or merely fortuitous. The bone matrix proper consists of type I collagen and other osteoblast products such as osteonectin (a phosphorylated glycoprotein) and small proteoglycans (PG-I and/or PG-II) which are incorporated into bone collagen fibrils. These proteins may have additional roles in tissue morphogenesis and/or differentiation.

1988 Cell and molecular biology of vertebrate hard tissues. Wiley, Chichester (Ciba Foundation Symposium 136) p 178–202

Healthy cortical and trabecular bones consist of an extracellular organic framework (matrix) spatially arranged in a highly organized, tightly layered (lamellar) pattern and embedded (mineralized) with tiny crystals of a basic calcium phosphate called hydroxyapatite. It is the organization of this mineralized matrix which gives human bone its unique structural and biomechanical properties. Less efficiently organized bone matrix, such as the loose woven bone of the fetus, the focalized lesions of diseased bone found in Paget's disease and the hypercellular, woven bone of individuals with severe forms of osteogenesis imperfecta (a genetic, 'brittle bone' disease of children), is poor in structure and biomechanically unsound, despite being fully mineralized.

Normal bone-forming cells are programmed to produce a healthy bone matrix via the synthesis and secretion of a set of protein molecules that are arranged into a precise three-dimensional network within the bone extracellular space. The major product of the bone cell is a fibrous protein, type I collagen, that polymerizes via lateral and staggered longitudinal aggregation to form the architectural motif of bone tissue. The forming collagen fibrils interact with selected non-collagen bone proteins to make a heteromolecular complex which constitutes a basic building structure for bone. These hetero-molecular matrix units are then organized into lamellar sheaths of (eventually mineralized) tissue, thereby defining the mass of skeleton. The main job of all bone-forming (osteoblast) and bone-resorbing (osteoclast) cells throughout life is to maintain skeletal mass through the selective removal of older, and faithful restoration of new, bone matrix units. These are then mineralized (i.e. they accumulate calcium phosphate crystals) by non-vital, physicochemical mechanisms.

The main molecular differences between the bone matrix units of healthy lamellar bone and those of less biomechanically sound woven bone are that the latter generally contain lower proportions of certain non-collagen proteins (such as osteonectin, bone proteoglycan, osteocalcin) (Conn & Termine 1985). This may reflect significant differences in protein metabolism within the osteoblasts producing these forms of bone tissue: several non-collagen bone proteins can be severely reduced in amount in congenital bone diseases such as osteogenesis imperfecta (e.g., Termine et al 1984). It is quite probable that the expression of non-collagen proteins varies significantly during normal osteoblast differentiation, growth and maturation. It has been suggested that non-collagen bone proteins facilitate several aspects of bone biology, including cell differentiation and growth, cell recognition and attachment, organized matrix production, and modulation of bone resorption processes regulating calcium homeostasis. The non-collagen proteins may therefore be key players in bone physiology and metabolism. In this regard, osteocalcin (bone Gla protein), a low molecular mass non-collagen bone protein, has proved invaluable as a non-invasive marker of bone metabolism and disease (e.g., see Price et al 1980).

In this article I shall briefly review current knowledge of the non-collagen proteins of bone. Detailed considerations of many aspects of bone protein function only briefly considered here are available elsewhere (e.g. see Termine 1986).

Non-collagen proteins in bone: general considerations

Two-dimensional electrophoretic analyses of the non-collagen proteins of bone suggest that there may be well over 200 of them (Delmas et al 1984). Many of these proteins are identified as originating from plasma or other non-bone sources (Table 1). The inorganic component of bone is primarily hydroxyapatite, which is widely distributed in nature. In bone it is rich in sodium, carbonate and other ions, very high in surface area, and an excellent adsorbent material, especially for electronegative ligands: in preparative biochemistry, hydroxyapatite chromatography is often used to purify proteins, acidic carbohydrates, phospholipids and nucleic acids. Therefore, the

TABLE 1   Circulating proteins found in bone

a$_1$-Acid glycoprotein	FGF-a[a] (2)	IgE
Acid phosphatase (1)	FGF-b[a] (2)	IgG
Albumin	GC-Globulin	IgM
Alkaline phosphatase[a]	α_2-HS-Glycoprotein[b] (3,4)	IL-l[a] (5)
Antithrombin III	Haemopexin	β_2-Microglobulin[a]
a$_1$-Antitrypsin	Haptoglobulin	PDGF[a] (2)
Apo-I-lipoprotein	Haemoglobin	Transferrin
Cholinesterase	IGF-l[a]	TFG-β[a] (6)
Collagenase[a]	IgA	

[a] Also produced locally in bone tissue.
[b] Particularly enriched in bone.

References: (1) K.H. Lau et al 1987 J Biol Chem 262:1389–1397; (2) P. Hauschka et al 1986 J Biol Chem 261:12665–12674; (3) B.A. Ashton et al 1976 Calcif Tissue Int 22:27–33; (4) J.M. Mbuyi et al 1982 Calcif Tissue Int 34:229–231; (5) S. Hanazawa et al 1987 Calcif Tissue Int 41:31–37; (6) S. Sayedin et al 1986 J Biol Chem 261:5693–5695.

bone mineral has enormous potential to adsorb and concentrate exogenously derived materials (e.g. the α_2-HS-glycoprotein of plasma). Such adsorption is often advantageous to bone as a living system. For example, bone contains a plethora of potent growth factors, many of which apparently originate from both local and exogenous cell sources (Centrella & Canalis 1985). These factors include both forms of transforming growth factor β (TGF-β_1 and β_2), platelet-derived growth factor (PDGF), insulin-like growth factor 1 (IGF-1), the acidic and basic forms of fibroblast growth factor (FGF-a and -b), and interleukin 1 (IL-1), all of which have profound effects on bone-forming cells. The capacity of bone to regenerate on injury may depend greatly on the presence of these materials. Similarly, the incorporation of various enzymes and their inhibitors in bone (Table 1) may serve a useful function in maintaining and/or altering the matrix prior to bone remodelling. For example, most of the bone non-collagen proteins are gradually degraded after they are deposited in the tissue. Consequently, entrapment of macromolecules — such as growth factors and enzymes — in bone may be critical to even its most essential physiological functions.

Osteoblasts, like most cells, incorporate several of their gene products within their plasma membranes. Alkaline phosphatase is the best known protein of this type in bone. Many different roles have been suggested for this enzyme in bone physiology, but the true function of alkaline phosphatase remains unclear. This may be because this protein can display different properties depending on its immediate environment. Like all cell surface proteins, alkaline phosphatase in bone can be cleaved from its membrane loci and drift to the mineralized matrix space, thereby confounding descriptive data taken from composite (e.g. whole tissue) experimental systems. A similar situation may exist for the bone sialoproteins (see below). However, cDNAs encoding human bone alkaline phosphatase have now been cloned and its sequence has been completely determined (Weiss et al 1986). These new data provide a solid base from which to investigate alkaline phosphatase expression with greater precision and experimental accuracy. Similarly, cDNAs encoding the most abundant exogenous non-collagen protein of human bone, the α_2-HS-glycoprotein, have been cloned and both its A and B chains have been fully sequenced (Lee et al 1987). Perhaps future experiments using this new information will discover whether this plasma protein has a physiological role in bone or is merely a fortuitous constituent of the tissue. Recent data, obtained *in vitro*, on bone resorption (Lamkin et al 1987) point to the former of these two possibilities.

It is easy to see why bone protein biochemistry has advanced so slowly over the years. In fact, the early work in bone (see Herring 1972) represents an even more remarkable achievement because of these difficulties. Investigators were faced with both technical problems associated with the presence of the bone mineral phase, and substantial perplexity about the origin of the proteins studied. Many of the difficulties were recently overcome by using sequential dissociative extraction procedures (Termine et al 1981a) that selectively remove proteins either unprotected by, or loosely associated with the mineralized bone compartment. This then facilitates the enrichment and eventual purification of proteins tightly bound in the mineralized bone matrix proper (Termine et al 1981a). Such proteins are the focus of the remainder of this article. Interestingly, one protein isolated and purified by these procedures was initially described as a 24 000 dalton phosphoprotein (Termine et al 1981a) and is, in fact, the amino propeptide of the $\alpha1(I)$ procollagen chain (Table 2). This acidic, non-helical (and mostly non-collagenous in overall amino acid sequence) protein is cleaved from the procollagen $\alpha1(I)$ chain and retained as a phosphorylated protein bound within the mineralized bone matrix. Thus, procollagen (I) itself is a phosphorylated protein as synthesized by osteoblastic cells.

Bone sialoproteins

One of the first bone proteins to be characterized was a glycoprotein, rich in sialic acid, initially described by Geoffrey Herring in the early 1960s (see Herring 1972). Because of its origin, acidic nature and overall composition, Herring named this 23 000 dalton species as the bone sialoprotein. We now know that this molecule is a proteolytic degradation product of an 80 000 dalton (as determined on SDS gels) parent molecule (Table 2) which is approximately 50% protein and 50% carbohydrate, and is rich in sialic acid (12–14% of total protein) and O-linked oligosaccharides.

In fact, bone is reported to contain two sialoproteins that account for 5–6% of the total bone non-collagen protein and are products of completely dif-

TABLE 2 Non-collagen protein products of bone cells

Constituent	Other names used	Molecular mass (SDS gels)	N-terminal sequence[a]	Unmodified or core protein
A. Major species				
Osteonectin (1,2,3,4)	SPARC, BM-40, 43 kDa protein	45 kDa	APQQEALPDE- (human)	32 kDa (cDNA)
Proteoglycan I (5)	PG-I, PG-Sm-1	240 kDa	DEEASGADTS- (human)	45 kDa (Chond-roitinase ABC)
Proteoglycan II (5,6,7)	PG-II, PG-Sm-2, proteodermatan sulphate	120 kDa	DEASGIGPEV- (human)	36 kDa (cDNA)
Sialoprotein I (8,9,10)	Osteopontin, 44 kDa phosphoprotein	80 kDa	IPVKQADSGE- (human)	32 kDa (cDNA)
Sialoprotein II (9,11,12)	Bone sialoprotein	80 kDa	FSMKNLHRRV- (human)	—
Osteocalcin	Bone Gla protein (BGP)	10 kDa	YLYQWLGAPV- (human)	6 kDa (cDNA)

Matrix Gla protein	MGP	14 kDa (bovine)	Y$\underline{E}$SHESLESY- (bovine)	9 kDa (protein sequence)
24K phosphoprotein (13)	α1 (I) Procollagen amino propeptide[b]	24 kDa	EEEGQEEGQE- (bovine)	—
B. Others				
19 kDa Protein (14)	—	19 kDa		
4 kDa, 12 kDa, 30 kDa, 60 kDa Phosphoproteins (15,16)	—	4–60 kDa		

[a] $\underline{E}$: Gla residue.
[b] Produced by proteolytic cleavage of pNα1(I) chains.

References: (1) R. W. Romberg et al 1986 Biochemistry 25:1176–1180; (2) K. Mann et al 1987 FEBS Lett 218:167–172; (3) F. Kuwata et al 1985 J Biol Chem 260:6993–6998; (4) J. Engel et al 1987 Biochemistry 26:6958–6965; (5) L. W. Fisher et al 1983 J Biol Chem 256:6588–6594; (6) T. Krusius & E. Ruoslahti 1986 Proc Natl Acad Sci USA 83:7683–7687; (7) A. Franzén & D. Heinegård 1984 Biochem J 224:47–58 and 59–66; (8) C. W. Prince et al 1987 J Biol Chem 262:2900–2907; (9) A. Franzén & D. Heinegård 1985 Biochem J 232:715–724; (10) M. Somerman et al 1987 J Bone Min Res 2:259–265; (11) L. W. Fisher & R. Kinne 1987 J Biol Chem 262:10206–10211; (12) L.W. Fisher et al 1983 J Biol Chem 258:12723–12727; (13) L.W. Fisher et al 1987 J Biol Chem 262:9702–9708; (14) H. Mardon & J.T. Triffitt 1987 J Bone Min Res 2:191–199; (15) A. Uchiyama et al 1986 Biochemistry 25:7572–7583; (16) S. Lee & M.J. Glimcher 1981 Calcif Tissue Int 33:385–394.

ferent genes (Fisher et al 1987). One of these, lower in sialic acid content ($\approx 5\%$) is a minor component of bovine and human bone but a substantial part (50–60%) of the total sialoprotein pool in rat bone. This low sialic acid component co-elutes with the major (human, bovine) bone sialoprotein during both size and ion-exchange chromatography and, therefore, was initially called bone sialoprotein I (BSP-I). This molecule, however, is identical to an acidic, 44 kDa rat bone phosphoprotein (Table 2). (The apparent difference in molecular mass (see Table 2) arises from different methods (SDS gels and sedimentation equilibrium) used to characterize the protein.) cDNAs encoding this bone glycoprotein have been cloned and its primary sequence has been determined (Oldberg et al 1986). Because the protein contains an internal, fibronectin-like Arg-Gly-Asp (RGD) cell-attachment sequence, it was renamed osteopontin. Analysis of the unmodified 'core' protein of rat osteopontin by cDNA methods reveals its size to be $\approx 32\,000$ daltons. Osteopontin serves as an attachment protein for gingival fibroblasts *in vitro*, and is typical of vitronectin-like cell-attachment proteins (Table 2). The protein is localized to both bone and non-bone (kidney, nervous system) tissue (C.W. Prince & M.P. Mark, unpublished data). Similarly, mRNAs encoding the protein are widely distributed (L.W. Fisher & M.F. Young, personal communication). It is interesting that both osteopontin and BSP-II (see below) closely resemble cell surface glycoproteins in overall composition. They may well originate as osteoblast surface constituents which, like alkaline phosphatase, eventually become lost from the plasma membrane to be entrapped in the mineralized bone matrix space, binding to the bone mineral via their electronegative (acidic) constituent groups.

The major human and bovine bone sialoprotein is more like the species originally studied by Herring and is called bone sialoprotein II (BSP-II, Fisher et al 1987). Rabbit bone contains essentially all of its bone sialoprotein II protein as a constituent core protein of a keratan sulphate proteoglycan (Table 2) rather than as a 'normal' sialoprotein. That is, rabbit bone utilizes, almost exclusively, the protein core, usually fated to be a true sialoprotein in other species, as a substrate with which to construct (by post-translational modification) a chemically different keratan sulphate proteoglycan constituent. Calf and human bone, conversely, contain no keratan sulphate proteoglycan. Similar post-translational selective modifications (partial glycosaminoglycan attachment, primarily as keratan sulphate) have been noted for cell surface glycoproteins in other systems. What is unusual about this one is its apparent species specificity. cDNAs encoding human bone sialoprotein II have now been obtained and are being fully characterized (L.W. Fisher & M.F. Young, personal communication).

Bone proteoglycans

The bone proteoglcyans were among the earliest bone components to be studied, primarily through their constituent glycosaminoglycan chains. These acidic polysaccharides (mostly chondroitin sulphate in bone) were first called 'ground substance', based on their affinity for histochemical dyes, and then 'mucopolysaccharides' because of their chemical composition and properties. It was then discovered that in almost all tissues, acidic sulphated polysaccharides are attached to core proteins, and the composite macromolecules were eventually called proteoglycans. The polysaccharide chains of proteoglycans are now called glycosaminoglycans and in bone are repeating disaccharides of glucuronic acid and N-acetylgalactosamine sulphate. The glycosaminoglycan chains are attached to their protein cores (synthesized in the usual way within the rough endoplasmic reticulum) via a series of complex reactions in the cell's Golgi apparatus.

The best-studied proteoglycan is that of cartilage and consists of a long, central core protein to which are attached N- and O-linked oligosaccharides, keratan sulphate and chondroitin sulphate, with an overall composition of about 5% protein, 95% carbohydrate. The bone matrix proteoglycans are much smaller in overall size, richer in protein (25 to 40%) and their chondroitin sulphate chains are larger (average $M_r \approx 40\,000$) than those of their cartilage counterparts (Fisher 1985, Heinegård et al 1986). However, most connective tissues, including cartilage, also contain similar small proteoglycans and in cartilage there are as many small proteoglycan molecules as there are larger ones.

Fetal bone contains two small proteoglycans, PG-I and PG-II (Table 2), which are products of completely different genes (Fisher et al 1987). PG-I is generally richer in carbohydrate, which is probably contained in two attached glycosaminoglycan chains (Fisher 1985). PG-I is reported to be absent from adult bone and may be selectively removed from the tissue by proteolysis after its deposition in the bone matrix space. cDNAs encoding human bone PG-I have recently been cloned and are being fully characterized (L.W. Fisher & M.F. Young, personal communication). As expected, PG-I mRNAs are widely distributed in different connective tissues.

The smaller bone proteoglycan, PG-II, contains a single glycosaminoglycan chain attached (via O-glycosidic linkage) to serine residue 4, as measured from the amino terminus. cDNAs encoding the bone PG-II protein core have already been described (Day et al 1986) and its mRNA is widely distributed amongst connective tissues. The amino acid sequence of bovine bone PG-II core protein (A.A. Day & M.F. Young, personal communication) is almost identical to that of its human placental counterpart (Table 2), suggesting that the primary sequence of the PG-II protein is highly conserved. However, proteolytic cleavage studies suggest that different tissues may post-translationally modify PG-II in different ways. For example, PG-II molecules from other tissues contain dermatan sulphate rather than the chondroitin sulphate found in bone. This involves activation of a specific epimerase converting some glucuronic acid residues to iduronic acid during glycosaminoglycan biosynthesis. Interestingly, bone cells synthesize both the PG-I and PG-II molecules in culture as dermatan sulphate species (B. Charrier, P. Gehron Robey, personal communications). Bone cells also synthesize a large (600 000 M_r) chondroitin sulphate proteoglycan with a core protein of $\approx$350 000 daltons, and a 400 000 dalton heparan sulphate proteoglycan *in vitro* (P. Gehron Robey, personal communication). These latter proteoglycans may represent pericellular constituents expressed more definitively in developmentally 'younger' bone tissue: analysis of the distribution of PG-I and PG-II in connective tissue suggests that their tissue-specific expression may be developmentally regulated. This is of particular interest because the small proteoglycans are thought both to interact with growing collagen fibrils in a spatially precise manner and to regulate the growth, maturation and subsequent interactions of collagen within the connective tissues (for more detail see Ciba Foundation 1986).

Bone Gla-containing proteins

Vitamin K modulates a particular post-translational modification of proteins, first found for the blood-clotting factors, which entails the γ-carboxylation of glutamic acid side-chains. The γ-carboxylated glutamyl (called Gla) residues facilitate low-affinity calcium binding via their vicinal dicarboxy side-chain groups. Bone contains two proteins with this particular modification (Table 2): osteocalcin (or bone Gla protein, BGP), and matrix Gla protein (MGP). The two proteins have some sequence homology, but are the products of different genes (Price 1987). Furthermore, osteocalcin and MGP are synthesized by different rat osteosarcoma cell lines (Price 1987). MGP is also a cartilage protein and is found in bone at an earlier developmental stage than osteocalcin, which usually appears later in bone embryogenesis (Price 1987).

   Speculation that the Gla-containing proteins regulate bone mineralization via their Gla residues has been at the fore of studies aimed at elucidating their function. Warfarin-treated animals (in which γ-carboxylation reactions are blocked) contain minimal (1–2%) levels of non-γ-carboxylated BGP and MGP, thereby proving that retention of these proteins in bone is mediated by hydroxyapatite bonding to their (constituent) Gla residues. However, the bones of warfarin-treated animals seem perfectly normal in all aspects. Rats treated with warfarin show growth plate closure, but this is an effect on cartilage rather than on bone and has not been attributed to any one of the various affected Gla-proteins present in such animals.

   Osteocalcin is perhaps the best studied of all the bone proteins. Its complete precursor sequence is known from cDNA analysis (Pan & Price 1985) and its full secreted protein sequence has been determined for many animal species. It has become the best indicator of bone turnover. Nevertheless, its function in bone is completely unknown, a refrain that by now is depressingly familiar to bone enthusiasts. Human bone contains almost no osteocalcin and human serum BGP levels are 5 to 10% of those in the calf and the rat. Again, species variation seems operative here and caution must be used in extrapolating data from one animal to all others. It was fortunate that calf rather than human bone was first used to purify osteocalcin, otherwise its significance to bone metabolism might not have been evident.

It is striking that osteocalcin biosynthesis seems to be regulated by one, and only one, vitamin/hormone, $1,25\text{-}(OH)_2$ vitamin D_3, which greatly enhances the nuclear transcription and eventual secretion of BGP from bone cells. This hormone's main bone-related action *in vivo* is to increase bone resorption, yet osteoclasts have no receptors for $1,25\text{-}(OH)_2$ vitamin D_3. While it is possible that $1,25\text{-}(OH)_2$ vitamin D_3 acts on presumptive osteoclast precursors, it should be stressed that osteoblasts are replete with functional receptors for this calcium-regulating hormone. The general effect of $1,25\text{-}(OH)_2$ vitamin D_3 on the synthesis of matrix constituents such as collagen and osteonectin by bone cells in culture is either no response or strong suppression. The same cells, however, greatly increase their osteocalcin biosynthesis on stimulation with $1,25\text{-}(OH)_2$ vitamin D_3, as noted above. Such behaviour is inconsistent with an overall anabolic effect and suggests that osteocalcin expression may correlate with bone formation–resorption coupling rather than with mineralization as hypothesized earlier. In this regard, osteocalcin may be critical to normal bone turnover and metabolism.

Osteonectin

The most abundant non-collagen protein produced by human and bovine osteoblasts is osteonectin (Table 2). This phosphorylated glycoprotein represents $\approx 2\%$ of the total protein of developing bone for cows, pigs, dogs, rabbits, chickens, sheep, monkeys, and humans, but not apparently for rats (Termine et al 1981a, 1981b, Fisher et al 1987, Table 2). In calves the osteonectin content of woven bone is only half that of lamellar bone (Conn & Termine 1985). Rat bone remains fairly woven in character until the animals approach their adult weight, which might explain, at least in part, why osteonectin is not as prominent as in other mammals. Anyway, as in the study of osteocalcin, it is also fortunate that calf bone was chosen for the first purification of osteonectin, because of this variance.

Studies on fetal (Termine et al 1981b) and adult (Romberg et al 1985) bovine bone show that osteonectin binds both calcium ions and hydroxyapatite strongly ($K_d \approx 10^{-8}$). It also binds tightly to native collagen (Termine et al 1981b, Romberg et al 1985). In solution, osteonectin is a potent inhibitor of hydroxyapatite crystal growth and replication (Romberg et al 1985, Table 2). However, bound to collagen in the solid state, the protein promotes calcium phosphate deposition from metastable solutions (Termine et al 1981b). These activities represent the same function displayed differently in operationally distinct experimental systems. Both sets of data indicate high affinity interaction with hydroxyapatite crystal growth sites. In solution, excess osteonectin saturates a limited number of such sites (exhibited as crystal seeds in the solid phase) and prevents their further kinetic growth. In the solid state, osteonectin seems to replicate a mineral or crystal growth site on its topological surface (exhibited as a solid phase) and the ions (which are in excess in solution) then become activated to accumulate at that site and grow into a true mineral phase. Both events are kinetic in nature and act through the common mechanism of an osteonectin–mineral interface. The former represents saturation inhibition, the latter catalytic-stimulation of growth.

cDNAs encoding bovine (Young et al 1986) and human (L.W. Fisher & M.F. Young, personal communication) bone osteonectin have been cloned and characterized. By sequence comparison it is now apparent that osteonectin is identical to a protein of the mouse parietal endoderm called SPARC (an acronym for secreted protein, acidic and rich in cysteine) (Mason et al 1986). Osteonectin protein has also been found in human platelets, where it is part of a calcium phosphate-rich secretion granule and is released by thrombin (Stenner et al 1986). Osteonectin is also identical to a 43 kDa endothelial cell protein not present in steady-state tissue, but whose synthesis is stimulated in isolated cells by conditions of cell culture (Sage et al 1986). A similar observation has been made for most fibroblast cells *in vitro* (Table 2). Finally, osteonectin is identical to the BM-40 glycoprotein of the EHS (Englebreth-Holm-Swarm) basement membrane tumour, where it is thought to maintain protein–protein interactions in the extracellular matrix space (Dziadek et al 1986). Indeed, osteonectin mRNA seems widely distributed in developing tissues (Young et al 1986, Mason et al 1986, Holland et al 1987).

However, osteonectin protein is not ubiquitous and is substantially more abundant (100- to 1000-fold) in bone than in almost all non-bone sources. Osteonectin is expressed in non-bone cells primarily during rapid proliferation and differentiation and is temporally activated in human tissues when they undergo profound growth and/or remodelling processes (U. Wewer & J.D. Termine, unpublished data). In normal non-bone tissues, osteonectin expression is almost always deactivated during steady-state or maintenance conditions. It is of considerable interest that bone, the major producer of osteonectin, is a tissue undergoing constant remodelling throughout life and regenerates spontaneously on injury. The physiological effects of the timed (non-bone) and continual (bone, certain tumour cells) expression of osteonectin are not yet known, but elucidation of the mechanisms regulating its expression should provide valuable insights into its potential biological functions.

Osteonectin is encoded by a large and complex gene containing about 10 expressing (exon) domains (D.M. Findlay & M.F. Young, unpublished data).

One of these (exon 3) seems to vary in sequence (but not in chemical character) between species, while the remainder seem highly conserved throughout mammalian evolution. The bovine protein has a 17 amino acid signal peptide and the secreted molecule consists of 287 amino acid residues (M.E. Bolander & J.D. Termine, unpublished data). The bone matrix protein can be divided into four domains: an acidic amino terminus characterized by an abundance of vicinal acidic amino acids, a likely site for hydroxyapatite binding; a cysteine-rich region, a probable area for extensive disulphide bonding in this tightly folded macromolecule; and two remaining domains, one hydrophilic and one aromatic, each containing a potential high affinity calcium-binding loop (M.E. Bolander & J.D. Termine, unpublished data). In fact, the molecule is multi-faceted in character and from a structural point of view has potential for a number of biological activities. Future work with osteonectin should prove biochemically and biologically interesting.

Acknowledgements

The author is grateful to his colleagues, L.W. Fisher, P. Gehron Robey and M.F. Young, for their excellent suggestions throughout the preparation of this paper. Their insights were helpful and invaluable.

References

Centrella M, Canalis E 1985 Local regulation of skeletal growth, a perspective. Endocrinol Rev 6:544–551

Ciba Foundation 1986 Functions of the proteoglycans. Wiley, Chichester (Ciba Found Symp 124)

Cohn KM, Termine JD 1985 Matrix protein profiles in calf bone development. Bone (NY) 6:33–36

Day AA, Ramis CI, Fisher LW, Gehron Robey P, Termine JD, Young MF 1986 Characterization of bone PG-II cDNA and its relationship to PG-II mRNA from other connective tissues. Nucl Acids Res 14:9861–9867

Delmas PD, Tracy RP, Riggs BL, Mann KG 1984 Identification of the noncollagenous proteins of bovine bone by two-dimensional electrophoresis. Calcif Tissue Int 36:308–316

Dziadek M, Paulsson M, Aumailley M, Timpl R 1986 Purification and tissue distribution of a small protein (BM-40) extracted from a basement membrane tumor. Eur J Biochem 161:455–464

Fisher LW 1985 The nature of the proteoglcyans of bone. In: Butler WT (ed) The chemistry and biology of mineralized tissues. Ebsco, Birmingham, p 188–196

Fisher LW, Hawkins GR, Tuross N, Termine JD 1987 Purification and partial characterization of small proteoglycans I and II, bone sialoproteins I and II and osteonectin from the mineral compartment of developing human bone. J Biol Chem 262:9702–9708

Heinegård D, Franzén A, Hedbom E, Sommarin Y 1986 Common structures of the core proteins of interstitial proteoglycans. In: Functions of the proteoglycans. Wiley, Chichester (Ciba Found Symp 124) p 69–82

Herring GM 1972 The organic matrix of bone. In: Bourne GH (ed) The biochemistry and physiology of bone. Academic Press, New York, vol 1:127–189

Holland PWH, Harper SJ, McVey JH, Hogan BLM 1987 In vivo expression of mRNA for the Ca^{++}-binding protein SPARC (osteonectin) revealed by in situ hybridization. J Cell Biol 105:473–482

Lamkin MS, Colclasure C, Rodrick M et al 1987 Three forms of BRP-2 (bone resorptive proteins) from human cancer ascites fluid and their relationship to human serum alpha-2-HS-glycoprotein. Calcif Tissue Int 41:171–175

Lee CC, Bowman BH, Yang F 1987 Human α_2-HS-glycoprotein: the A and B chains with a connecting sequence are encoded by a single mRNA transcript. Proc Natl Acad Sci USA 84:4403–4407

Mason IJ, Murphy D, Munke M, Francke U, Elliott RW, Hogan BLM 1986 Developmental and transformation-sensitive expression of the SPARC gene on mouse chromosome 11. EMBO (Eur Mol Biol Organ) J 5:1831–1837

Oldberg Å, Franzén A, Heinegård D 1986 Cloning and sequence analysis of rat bone sialoprotein (osteopontin) cDNA reveals an arg-gly-asp cell-binding sequence. Proc Natl Acad Sci USA 83:8819–8823

Pan LC, Price PA 1985 The propeptide of rat bone γ-carboxyglutamic acid protein shares homology with other vitamin K-dependent protein precursors. Proc Natl Acad Sci USA 82:6109–6113

Price PA, Parthemore JG, Deftos LJ 1980 New biochemical marker for bone metabolism. J Clin Invest 66:878–883

Price PA 1987 Vitamin K-dependent bone proteins. In: Cohn DV et al (eds) Calcium regulation and bone metabolism: basic and clinical aspects. Elsevier Science Publishers, Amsterdam, vol 9:419–425

Romberg RW, Werness PG, Lollar P, Riggs BL, Mann KG 1985 Isolation and characterization of native adult osteonectin. J Biol Chem 260:2728–2736

Sage H, Tupper J, Bramson R 1986 Endothelial cell injury in vitro is associated with increased secretion of an M_r 43,000 glycoprotein ligand. J Cell Physiol 127:373–387

Stenner DD, Tracy RP, Riggs BL, Mann KG 1986 Human platelets contain and secrete osteonectin, a major protein of mineralized bone. Proc Natl Acad Sci USA 83:6892–6896

Termine JD 1986 Bone proteins and mineralization. In: Kuhn K, Krieg T (eds) Rheumatology; connective tissue: biological and clinical aspects. Karger, Basel, vol 10:184–196

Termine JD, Belcourt AB, Conn KM, Kleinman HK 1981a Mineral and collagen-binding proteins of fetal calf bone. J Biol Chem 256:10403–10408

Termine JD, Kleinman HK, Whitson SW, Conn KM, McGarvey ML, Martin GR 1981b Osteonectin, a bone-specific protein linking mineral to collagen. Cell 26:99–105

Termine JD, Gehron Robey P, Fisher LW et al 1984 Osteonectin, bone proteoglycan and phosphophoryn defects in a form of bovine osteogenesis imperfecta. Proc Natl Acad Sci USA 81:2213–2217

Weiss MJ, Henthorn PS, Lafferty MA, Slaughter C, Raducha M, Harris H 1986 Isolation and characterization of a cDNA encoding a human liver/bone/kidney-type alkaline phosphatase. Proc Natl Acad Sci USA 83:7182–7186

Young MF, Bolander ME, Day AA, Ramis CI, Gehron Robey P, Yamada Y, Termine JD 1986 Osteonectin mRNA: distribution in normal and transformed cells. Nucl Acids Res 14:4483–4497

DISCUSSION

Glimcher: Many of the proteins in bone, dentine and other calcified tissues have now been shown to be phosphorylated, and several that were formerly considered to be distinct proteins are now thought to be the same protein.

The most commonly measured (or postulated) property of these non-collagenous proteins is their ability to bind to apatite crystals. However, apatite binds large numbers of proteins and is used to purify proteins wholly unrelated to bone. The fact that a particular protein in bone binds to certain preparations of apatite crystals *in vitro* does not mean that it does so *in vivo* to the exclusion of other proteins. Even if it is bound to the crystals, its biological function is not necessarily related to that fact.

As Arthur Veis has pointed out, when EDTA or HCl extracts of bone or dentine, or the fractions obtained by molecular sieving and ion-exchange chromatography, are subjected to SDS-PAGE, many bands are obtained. It would have been impossible to isolate and purify the individual phosphoproteins without rhodamineB which specifically stains only phosphoproteins. We were eventually able to isolate 16 homogeneous phosphoprotein components which we now think are derived from at least two genomically distinct precursors (Uchiyama et al 1986). Five of the major components were shown to contain significant amounts of carbohydrates, including sialic acid.

Although we were aware of the previous isolation of rat bone sialoproteins by Franzén & Heinegård (1985) and Oldberg et al (1986), there was no evidence that these two sialoproteins were phosphoproteins. This blinded us to the fact that the amino acid compositions of their proteins were very similar to those of our chicken bone phosphoproteins. As he will describe (p 203), Bill Butler recognized the relationship between the rat bone sialoproteins and the 44 kDa rat bone phosphoprotein that he and his colleagues isolated. From the analyses of amino acid and phosphate content we conclude that our phosphoproteins can be put into two groups: one with low Ser(P) and Thr(P) contents and the other with much higher Ser(P) and Thr(P) contents. These groups probably correspond to the sialoprotein I and sialoprotein II proteins (Oldberg et al 1986). Now we have the unfortunate situation that one protein or a single group of proteins has several names: osteopontin (sialoprotein), rat 44 kDa phosphoprotein, and (chicken) bone phosphoprotein.

The three investigators propose quite different functions for the three proteins. What do we call these proteins now? Clearly osteopontin ('bone bridge') is a catchy term and likely to remain fixed in the literature, certainly more so than '44 kDa phosphoprotein' or simply '(chicken) bone phosphoproteins', despite the fact that there is, as yet, little or no evidence that osteopontin's function is to bind bone cells to the mineral phase. In addition, with all due respect to John Termine for his fine work in discovering the protein in bone which he named osteonectin, we now have an osteonectin which is really not a 'nectin' in the sense that fibronectin is. If, as John proposes, osteonectin links the collagen fibrils to the mineral phase, and if at least one function of osteopontin is to bind to bone cells, osteonectin is really an osteopontin and osteopontin is better defined as an osteonectin!

We propose that the chicken phosphoproteins facilitate mineral nucleation and also function as a bridge, linking the collagen to the mineral phase—that is, as an osteopontin but in a different manner from the other proposed osteopontins [sialoproteins of Franzén & Heinegård (1985) and the osteonectin (really an osteopontin) of Termine]. Let's not create further confusion by giving any more names to new protein components until we are sure what they do!

Termine: I would agree. We have some evidence, and so does Helene Sage, that the endothelial cell and the skin fibroblast handle these proteins differently. Helene Sage found that a 43 kDa protein (osteonectin, SPARC or BM-40) is dominant in proliferative cultures and reduced in confluent ones. The opposite is true in bone cells. Furthermore, the collagen type I amino propeptide is phosphorylated in bone but not in fibroblasts. Therefore, the bone cell seems to specialize in phosphorylating proteins; almost every protein that we have isolated has at least one phosphorylated residue.

Veis: Dr Martin mentioned earlier that the kinase activity in these tissues is high. Therefore, a lot of proteins will be phosphorylated without a highly specific mechanism in a particular case.

Urist: Dr Termine, you described these non-collagenous proteins in different species. How does the amount of osteonectin as a fraction of the total non-collagenous proteins differ in human and bovine marrow-free cortical bone?

Termine: In adult bone, as well as in dentine, all the non-collagenous proteins are degraded and it is not possible to discern how many epitopes have thus been conserved. Therefore, one has to look at developing human and bovine bone at essentially the same stages, i.e. when they are forming lamellar tissue. At that point the two species are almost identical: the major difference is that bovine bone has substantial levels of osteocalcin, about 10 to 20 times as much as in human bone or serum.

Urist: But is there 10% non-collagenous protein in adult bone? What percentage of that is Gla protein (osteocalcin) and what percentage is osteonectin?

Termine: There is only about 10% non-collagenous protein in adult bone. I don't think those measurements would have any meaning.

Hauschka: Adult bovine bone contains about 2.5 μg of osteocalcin per mg of dry bone, which represents 10–20% of the total non-collagenous protein. In human bone the osteocalcin concentration is about 10-fold lower, based on our direct radioimunoassay measurements. Human osteocalcin was reported to have only two Gla residues rather than the usual three (Poser et al 1980). This may reduce the affinity of osteocalcin for bone matrix and explain why there is less in human bone than in other species.

There is exciting evidence linking osteocalcin with chemotactic recruitment of bone-resorbing cells. However, the local concentration of osteocalcin in bone matrix cannot yet be correlated with resorption rate. For example, osteocalcin occurs at the highest levels in the oldest, densest regions of adult bone which have a long half-life *in vivo*. Osteocalcin is much less abundant in the matrix of rapidly remodelling embryonic bones. In developing a working hypothesis we must also recognize that resorption rate depends on both the abundance of osteoclasts and the metabolic activity of the resident osteoblasts.

Termine: I don't think osteocalcin gets into the matrix *per se*. I think it has a tendency to bind apatite and, therefore, the more you mineralize the more you will find. That is consistent, because those are the areas that will be resorbed.

Raisz: Dr Termine, did you say that osteonectin production was related to rapid growth? I think the rapidly dividing preosteoblast layer might have more osteonectin than a mature osteoblast.

Termine: No. As soon as osteonectin expression is turned on in a bone cell, it's on at a very high level and stays as such in bone. In non-bone tissues it is turned on at a high rate during growth and proliferation and then is down-regulated during steady-state tissue maintenance.

Urist: I ask this because the quantities of osteonectin being reported are so large as to represent the major non-collagenous protein found in the adult human matrix.

Termine: Osteonectin has to act with collagen in the matrix, and it is the dominant bone protein in the forming bone of all species except the rat. In a bone cell culture you see type I collagen and a lot of osteonectin being made and secreted into medium.

Rodan: The function that a non-collagenous protein fulfils may not be unique; another protein may do the same thing.

Termine: I would agree with that.

Canalis: If the function of these non-collagenous proteins is unknown but antibodies are available, why hasn't anybody attempted neutralization studies?

Termine: It took a long time to characterize these proteins but they have now been cloned. The best approach to studying their function might be to use classical molecular biology. Neutralization studies would require the introduction of the antibody into the cell, and that is difficult. However, these things will be done eventually.

Prockop: What molecular biology approach do you suggest?

Termine: Antisense DNA, mutations, transgenic mice: those experiments will be done in the next few years.

Prockop: The antisense approach appeared more promising about three years ago than it does now. There are technical problems.

Termine: There are newer ways of getting around them which make this approach more enticing.

Prockop: From the study of collagen in transgenic mice it seems important that most of the mutations found in patients with osteogenesis imperfecta are dominant, because of the phenomenon of 'protein suicide' and other effects. Therefore, one can probably produce an abnormal phenotype without inhibiting expression of the endogenous genes. A number of exciting experiments are in progress in several laboratories. We ourselves have been making transgenic mice with mutated collagen genes. The potential of the method is huge, but it is a complex biological system. Only 15% of the eggs that are microinjected produce viable pups (Brinster et al 1985).

One needs to establish a causal relationship between the injection of a mutated collagen gene and an abnormal phenotype. A simple approach, which you and others have tried, is to make collagen fibrils, add the test protein and see if it enhances mineralization.

Termine: I'm comfortable with that sort of approach, but biologists want to study more biologically relevant functions. This may best be done by looking at gene mutations, because you can't get precise information from inexact or dirty experiments.

Canalis: If the neutralization study does not work, it doesn't mean anything because the antibody may be non-neutralizing. But if it happens to work, you could detect changes in various biological functions.

Termine: What biological functions for the non-collagenous proteins would you like to see measured in a bone cell? The cell is not expressing a hormone nor performing a function in the same way that a bone endocrine system does. The main role here is to make an extracellular matrix that is very complex and then mineralizes. Therefore, you can't get a truly negative control and we don't know what endpoints to choose.

Weiner: I have always been intrigued by the observation that in adult bone the non-collagenous proteins are highly degraded. In enamel this occurs at the same time as a late-stage crystal growth phase. Might something analogous be happening in bone?

Termine: I would speculate that the more you break down these proteins, the higher is the chance of the post-maturation, 'crystal replication' form of mineralization. Dr Chambers has shown (this volume) that the osteoclast doesn't like to deal with unmineralized organic matrix. If matrix degradation allows bone site crystallization to reach a point where the outermost surface becomes fully mineralized, that should be the place where resorption would occur. Anything that the cells or the system would do to enhance that would thus modulate the resorptive response. So post-synthetic, long-term matrix degradation may have a real biological function.

Veis: In mature bone, where there is non-collagenous matrix protein degradation, is there evidence for collagen degradation? These non-collagenous proteins are trapped in the mineral phase. Does the degradation occur in that phase or is it cellularly mediated?

Termine: The mechanism is not known.

Veis: Have the non-collagenous proteins gone down-hill?

Glimcher: The phosphoproteins have. We have evidence that the 16 or more phosphoprotein components isolated from chicken bone are mostly degradation products of two precursor proteins. The maximum number of precursors is four.

John Termine mentioned that information on the degradation of bone collagen with ageing and maturation might come from an examination of the highest density bone, such as that recovered using density centrifugation of powdered adult chicken bone (Bonar et al 1983). The most dense bone is the most heavily mineralized and the oldest bone substance in the bone. The EDTA-insoluble residue recovered from these high density particles after decalcification was principally collagen, as it was in bone particles from less dense bone. There was no evidence (including by electron microscopic examination of the particles) that this collagen was degraded.

Although we did not measure the recovery of collagen, qualitative recoveries suggest there could only have been a small amount of degraded collagen in the original mineralized sample. It is possible that a significant amount of collagen was degraded but was rapidly removed from the tissue and therefore not measured by this method.

There are two major ways in which the high density bone could end up at maturity with so much more mineral: (1) collagen content remains the same but more mineral is packed in and between the collagen fibrils; (2) some collagen is degraded, resorbed and rapidly removed from the bone, and the volume originally occupied by these fibrils is now used to accommodate the additional mineral. The latter suggestion is not unlikely in view of the degradation (proteolysis) and removal of the enamel proteins, which commences early after the onset of mineralization and progresses rapidly with maturation until over 99% of the protein matrix has been removed (Strawich & Glimcher 1985, Glimcher & Levine 1966, Glimcher et al 1964, Robinson et al 1977). We are now exploring the possibility that a similar process occurs in ageing bone.

Canalis: We have studied collagen turnover. It varies tremendously, depending on the system. If you pre-label intact bones and study the hydroxyproline release you find that the release into the culture medium is very small: 10% at 24 hours. If you do the same experiment in isolated osteoblast-rich cells, at 24 hours the amount of radiolabelled hydroxyproline is 80%. In both systems most of the product is not trichloroacetic acid precipitable, suggesting that the collagen is fully degraded. The organ culture model is useful because most of the newly synthesized collagen is laid normally.

Termine: Would that be the same in the fetus or a neonate compared to a rat past the stage of maturity?

Canalis: We have not done that experiment.

Krane: It was shown earlier by Avioli & Prockop (1967) and by Harris & Sjoerdsma (1966) that when PTH is given to animals there is significant degradation of mature collagen. This may even be preferential degradation of old collagen. The cell systems don't answer that question. You still have to look at the animal, and that's difficult.

Slavkin: How do these cells prevent themselves from mineralizing? Do a large number of these non-collagenous proteins function as inhibitors of biomineralization, rather than all of them participating in crystal nucleation and growth?

Termine: I think the cells prevent themselves mineralizing by staying alive. When they die in culture, you get dystrophic mineralization. I prefer to think of these proteins having a more dynamic normal function in cells.

Glimcher: When a protein is described as binding calcium it is often assumed that it must be related to the initiation of calcification. This is not necessarily true. The factors that influence whether or not calcium ions, calcium phosphate clusters or other anions, such as carbonate, and their calcium complexes can participate in the initiation or nucleation of a solid phase of Ca-P from a metastable solution of Ca and P_i have been described previously (Glimcher 1976, Glimcher & Krane 1968).

A major requirement is that the bond between the calcium ion and the protein is such that the bound calcium ions or complexes remain able to react with additional ions such as phosphate or carbonate. Bound calcium ions can be rendered unreactive by being complexed as a clathrate or chelate, for example by reaction with EDTA. Because of this ability to sequester calcium ions, EDTA dissolves Ca-P and $CaCO_3$ crystals. Any ligand, protein, etc. which binds calcium strongly and/or forms a complex such as a chelate or clathrate, and thus makes the ions unreactive to phosphate or carbonate ion, will *not* be capable of participating in the formation of mineral crystals of phosphorus or carbonate, and indeed, it will tend to inhibit such a nucleation reaction.

Heterogeneous nucleation of Ca-P or $CaCO_3$ by a protein matrix is a more complicated problem than the simple binding of calcium ions or calcium ion complexes. The 'lattice' parameters of the specific side-chain groups which bind the appropriate mineral ions must be complementary to the lattice parameters of the particular crystal to be nucleated. That is, the less the disregistry between the stereochemical disposition of the reactive side-chain groups of the protein in the nucleation site and the appropriate planes of the crystal to be nucleated, the more potent the heterogeneous nucleator will be. If the disregistry exceeds a certain amount, nucleation will not occur. The closer the nature of the bonds between the nucleation site and the crystal to be nucleated and the intercrystalline bonds of the crystal to be nucleated, the better the nucleator, as well. Therefore a protein or other macromolecule which not only binds significant numbers of calcium ions, but binds them in such a way that they remain reactive towards phosphate and carbonate ions, may still fail to facilitate nucleation, and may actually inhibit it, because of the stereochemistry and site location in the tissue. For example, proteoglycans, which bind considerable amounts of calcium ions that are reactive before binding, are thought to inhibit the deposition of calcium ions. Although the role of bound Ca^{2+} in heterogeneous nucleation is complex, it seems reasonable to first examine the potential of protein-bound calcium ions to participate in nucleation by studying the reactivity of the protein-bound calcium. Sandra Lee did this by ^{31}P nuclear magnetic resonance (NMR) spectroscopy (Lee et al 1983) of the dentine phosphoprotein described by Lee & Veis (1980). She used solutions containing only calcium or inorganic orthophosphate (P_i) ions—i.e., single-species solutions and *not* the metastable solutions of Ca^{2+} and P_i used for nucleation experiments. She recorded the ^{31}P NMR signals from: (1) the phosphoprotein in solution; (2) the phosphoprotein plus calcium ions in solution; (3) the phosphoprotein plus P_i ions in solution; (4) the phosphoprotein, followed by addition of calcium ions and then addition of P_i solution.

When the calcium ions were added to the phosphoprotein, the phosphoprotein (phosphate ester) peak was almost completely eliminated, indicating that the Ca^{2+} ions were bound to the phosphate groups. When the P_i ions were added to the protein, there was no effect on the phosphoprotein peak (phosphate ester) but a separate P_i peak appeared. This indicated that there was no significant interaction between the protein and the P_i ions. Where P_i ions were added to the phosphoprotein to which Ca^{2+} was already bound, there was a marked shift of the P_i peak as well as the obliteration of the phosphoprotein peak due to calcium binding. This indicated that the protein-bound Ca^{2+} had interacted with the P_i ion to form a ternary complex—the phosphoprotein-bound calcium ions were still capable of further reactions with other ions. It does *not* mean that calcium ions bound to dentine phosphoproteins can or do participate in heterogeneous nucleation of a Ca-P solid phase: as already mentioned, many other chemical and stereochemical requirements must be met. This experiment does show that calcium ions bound to phosphophoryn are potentially capable of being involved in nucleation and the system warrants further study.

Termine: There are high affinity and low affinity calcium-binding sites in bone proteins and these are often conformation sensitive. As Ken Mann has shown, high affinity calcium-binding proteins, such as calmodulin, appear not to be involved in nucleation or crystal growth processes.

Williams: Nucleation is an intriguing problem. A solution of inorganic salts may be saturated but crystallization won't occur until the energy barrier of nucleation is overcome. The protein can act here in one of two ways. It can facilitate crystallization by reducing the energy barrier of nucleation. Alternatively, it could bind calcium and phosphate on the nucleus and prevent further growth.

Weiner: We have studied how a protein from mollusc shell nucleates *in vitro* (Addadi et al 1987). It is analogous in some ways to the bone sialoprotein. There are two components: the protein moiety which is aspartic acid-rich and adopts the β-sheet conformation; and associated sulphated polysaccharides which are covalently attached to the protein. We showed that these two parts of the macromolecule act cooperatively to induce crystal nucleation. The sulphate groups could act as concentrators of calcium ions from the solution and create a flux of Ca^{2+} into and over the ordered site. The carboxylate groups do not envelop an individual calcium ion, but are arranged in a two-dimensional sheet. The 30 or 40 calcium ions necessary for nucleation have to be tied down long enough for the carbonate ions to come into the system and then the nucleation event takes over. This study might have ramifications for the role of proteoglycans in mineralization—not as inhibitors of mineralization but working to enhance mineralization in conjunction with an ordered substrate.

Williams: Drs P.V. Hauschka, J. Triffit and B.A. Levine and Mr R. Meats and I have examined the proton NMR spectrum of rabbit osteocalcin (bone Gla protein) in the presence and absence of calcium ions. The conformation of this protein, as is also true of calmodulin, is very different with and without calcium. Without calcium, osteocalcin in free solution is not well structured except around the disulphide bridge (Fig. 1), characterized by a typical β-bend which results in a through-space α-CH to α-CH nuclear Overhauser enhancement (nOe) in the NMR spectrum. Elsewhere there is little evidence for order and all the aromatic residues (Y and F) flip freely and the tyrosines have normal pKa values.

When it binds calcium ions the protein folds (Fig.1). The short β-turn (residues 26–29 within the disulphide cross-link) remains unaffected, but there is clear evidence from nOe data for new β-structures and for some increase in helical elements. Circular dichroism studies also provided evidence for increased helical elements. Most striking is the formation of a hydrophobic cluster around a tyrosine (Tyr-42) and the single phenylalanine (Phe-46). The cluster was also shown to include several assigned amino acids: a valine (Val-36), two leucines, one alanine (Ala-32), one proline and residues that could be aspartate or asparagine. Tyrosine-42 was observed to have a high pK_a and to become immobile. Interactions with one edge of its aromatic ring differ from those with the other.

The local structure is very rigid and could easily be a signalling system as well as a calcium-binding molecule. All the carboxylic acids, groups GL, D and E, are together and form a surface to which calcium can bind (Fig.1). Thus the calcium binds the two helices together while the induced bends generate a hydrophobic core, F,Y,V. Hauschka & Carr (1982) had reported previously that the helical stretches exist in this state.

The general impression of the calcium-bound structure is a plane of very polar, frequently charged residues which could match the surface of apatite. Some other part of the helical surface may actually resemble the structure of collagen. The structure of the calcium-free protein does not directly match that of bone, but the protein is always in the calcium form in serum. The suggestion

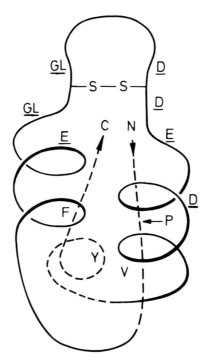

FIG. 1. (Williams). The structure of osteocalcin as determined by nuclear magnetic resonance spectroscopy. The single letters refer to normal amino acids. GL, γ-carboxy glutamate (see text); C and N, the termini.

arises that crystals and metal-free biopolymers do not have epitaxial relationships but rather that they have cooperative energetics of interaction, the protein being stabilized in a folded form as it stabilizes the calcium nucleus or lattice.

Termine: The N-terminus of osteonectin is positioned near to the C-terminus. On binding Ca^{2+}, this N-terminus takes the form of an α-helix which resembles the structure you show for osteocalcin. Thus it may not be necessary to form a β-pleated sheet to provide a double face of carboxylate sites in adjacent arms of the protein. A double α-helix may also suffice.

Williams: Osteocalcin also has an hydroxyproline in some sequences. It is suitably situated to become part of the binding face to apatite and it is on a three or four amino acid stretch which is collagen-like.

Krane: Is that structure only in the bovine and not in the human osteocalcin?

Hauschka: Hydroxyproline occurs at sequence position 9 in osteocalcin from nine of the 13 species studied so far. The proteins from human, mouse, chicken and swordfish do not contain hydroxyproline.

Urist: Are you suggesting that osteocalcin is part of the nucleation site?

Williams: No. I know what calcium does to osteocalcin. The calcium could obviously be part of the surface of a crystal. Looking at the other side of the molecule, not the calcium-binding side, we see that it has a very rigid structure. It could be some signalling device, for example a receptor surface.

Termine: How does the structure of osteocalcin compare with those of other Gla-containing proteins in serum? There are many clotting factor proteins that have the same motif and similar sequences to osteocalcin. A few have been studied but not with this level of sophistication.

Williams: I have no idea.

Fleisch: How would osteocalcin look intracellularly? If it is in a different form, antibodies which react with extracellular protein might not react with intracellular osteocalcin, and *vice versa*.

Williams: Intracellularly this protein will be completely unfolded.

Rodan: When osteocalcin binds to apatite, would the hydrophobic parts be exposed?

Williams: It's not hydrophobic. The protein is quite hydrophilic.

Urist: How does it compare with citrate in its affinity for calcium?

Williams: I haven't actually measured the binding constants. Osteocalcin binds more than one calcium ion. The first binding constant would be higher than for citrate and the last one might be similar to that for citrate.

Hauschka: The K_d values for Ca^{2+} binding to osteocalcin vary from 0.8mM for the chicken protein (Hauschka & Gallop 1977, Hauschka & Carr 1982) to about 3mM for bovine osteocalcin (Price et al 1977).

Veis: Models like Dr Williams' structure of osteocalcin are exciting. I would guess that most molecules with this kind of extracellular matrix function will have a specific face or regions with a similar set of properties, whether they are more or less ionic than those for osteocalcin. When we look for proteins that are mineral nucleators or binders to hydroxyapatite, we could speculate that they will have a beta-like display of the carboxyl groups or phosphate groups. When you make the β-sheet from phosphophoryn, two different folding arrangements are possible. One results in the β-sheet having all the phosphates on one side and the other gives a phosphate-Asp on each side.

Nijweide: Osteocalcin may have a function in osteoclast resorption. What is the structure like at low pH? It may be that the structure differs in the resorption cavity where there is a lower pH.

Williams: If we put protons on the molecule instead of calcium, it starts to fold. It doesn't give precisely the same tight fold as with calcium, but it does give a fold with protons at about pH 4.

Nijweide: What sort of pH do you need for the protein to exchange calcium for protons?

Williams: Calcium falls off below about pH 6. Below pH 4 there would be very little calcium and protons would be bound.

Nijweide: So it may be that in the resorption cavity, at low pH, the molecule is in its active form and stimulates the removal of the osteoclast.

Weiner: The behaviour of osteocalcin on binding calcium is identical to that of the β-sheet proteins. Acidic proteins only fold into the β-sheet conformation in the presence of calcium ions. An important implication of that observation is that in self-assembling systems, where these acidic glycoproteins need to find their specific location before they start functioning, they must first be loaded up with calcium ions. Thus a matrix which is getting itself organized to nucleate and calcify is actually calcium-loaded before nucleation.

Williams: There is a conserved arginine in the middle of the protein and we have yet to locate it exactly in the structure. It may be that this residue responds to the phosphate on the surface of bone.

References

Addadi L, Moradian J, Shay E, Maroudas NG, Weiner S 1987 A chemical model for the cooperation of sulfates and carboxylates in calcite crystal nucleation: relevance to biomineralization. Proc Natl Acad Sci USA 84:2732–2736

Avioli LV, Prockop DJ 1967 Collagen degradation and the response to parathyroid extract in the intact rhesus monkey. J Clin Invest 46:217–224

Bonar LC, Roufosse AH, Sabine WK, Grynpas MD, Glimcher MJ 1983 X-ray diffraction studies of the crystallinity of bone mineral in newly synthesized and density fractionated bone. Calcif Tissue Int 35:202–209

Brinster RL, Chen HY, Trumbaue ME, Yagle MK, Palmiter RD 1985 Factors affecting the efficiency of introducing foreign DNA into mice by microinjecting eggs. Proc Natl Acad Sci USA 82:4438–4442

Chambers TJ 1988 The regulation of osteoclastic development and function. In: Cell and molecular biology of vertebrate hard tissues. Wiley, Chichester (Ciba Found Symp 136) p 92–107

Franzén A, Heinegård D 1985 Isolation and characterization of two sialoproteins present only in bone calcified matrix. Biochem J 232:715–724

Glimcher MJ 1976 Composition, structure and organization of bone mineralized tissues and the mechanism of calcification. In: Greep RO, Astwood EB (eds) Handbook of physiology. Am Physiol Soc, Washington DC vol 7:25–116

Glimcher MJ, Levine PT 1966 Studies of the proteins, peptides and free amino acids of mature bovine enamel. Biochem J 98:742–753

Glimcher MJ, Krane SM 1968 The organization and structure of bone, and the mechanism of calcification. In: Ramachandran GN, Gould BS (eds) Treatise on collagen. Academic Press, New York vol 78:68–251

Glimcher MJ, Friberg UA, Levine PT 1964 The isolation and amino acid composition of the enamel proteins of erupted bovine teeth. Biochem J 93:202–210

Harris ED, Sjoerdsma A 1966 Effect of parathyroid extract on collagen metabolism. J Clin Endocrinal & Metab 26:358–359

Hauschka PV, Gallop PM 1977 Purification and calcium-binding properties of osteocalcin, the gamma-carboxyglutamate-containing protein of bone. In: Wasserman RH et al (eds) Calcium binding proteins and calcium function. Elsevier Science Publishers, Amsterdam, p 338–347

Hauschka PV, Carr SA 1982 Calcium-dependent alpha-helical structure in osteocalcin. Biochemistry 21:2538–2547

Lee SL, Veis A 1980 Studies on the structure and chemistry of dentin collagen–phosphophoryn covalent complexes. Calcif Tissue Int 31:123–134

Lee SL, Glonek T, Glimcher MJ 1983 ^{31}P nuclear magnetic resonance spectroscopic evidence for ternary complex formation of fetal phosphoprotein with calcium and inorganic orthophosphate ions. Calcif Tissue Int 35:815–818

Oldberg Å, Franzén A, Heinegård D 1986 Cloning and sequence analysis of rat bone sialoprotein (osteopontin) cDNA reveals an Arg-Gly-Asp cell binding sequence. Proc Natl Acad Sci USA 83:8819–8823

Poser JW, Esch FS, Ling NC, Price PA 1980 Isolation and sequence of the vitamin K-dependent protein from human bone: under-carboxylation of the first glutamic acid residue. J Biol Chem 255:8685–8691

Price PA, Otsuka AS, Poser JW 1977 Comparison of gamma-carboxyglutamic acid-containing proteins from bovine and swordfish bone: primary structure and Ca^{2+} binding. In: Calcium binding proteins and calcium function. Wasserman RH et al (eds) Elsevier Science Publishers, Amsterdam, p 333–337

Prince CW, Oosawa T, Butler WR et al 1987 Isolation, characterization, and biosynthesis of a phosphorylated glycoprotein from rat bone. J Biol Chem 262:2900–2907

Robinson C, Fuchs P, Weatherell JA 1977 The fate of matrix proteins during the development of dental enamel. Calcif Tissue Res 22(suppl):185–190

Strawich E, Glimcher M 1985 Synthesis and degradation *in vivo* of a phosphoprotein from rat dental enamel. Biochem J 230:423–433

Uchiyama A, Suzuki M, Lefteriou B, Glimcher MJ 1986 Isolation and chemical characterization of the phosphoproteins of chicken bone matrix: heterogeneity in molecular weight and composition. Biochemistry 25:7572–7583

General discussion II

Osteopontin: structure and biological activity

Butler: As Dr Glimcher pointed out earlier (p 191), there are at least four names for this protein: osteopontin; bone sialoprotein I; 44 kDa bone phospho-protein; and 2ar. The name 2ar was given to the protein coded by a cDNA clone isolated from a mouse transformed epidermal cell line by Smith & Denhardt (1987). The cDNA sequence revealed that 2ar was the murine counterpart to rat osteopontin (Oldberg et al 1986). As we reported (Prince et al 1987), osteopontin (from rat) was found to be a 44 kDa protein by sedimentation equilibrium. However, it runs anomalously on SDS-polyacrylamide gel elec-trophoresis (PAGE): by using SDS-PAGE we obtained a 75 kDa molecular mass on 5–15% gradient gels, whereas with 15% gels we obtained a more accurate molecular mass of 45 kDa. Osteopontin is rich in the acidic amino acids, containing about 36% glutamic and aspartic acid residues. It contains about 16% carbohydrate, including 7% sialic acid. From the mannose content we predicted that it contained one N-glycoside. We also predicted five or six 0-glycosides, based on *N*-acetylgalactosamine values and on β-elimination experiments, quantifying the losses of serine and threonine. From β-elimination experiments and other qualitative methods we estimated that the molecule contained 12 phosphoserine residues and one phosphothreonine.

Oldberg et al (1986) reported the cDNA sequence, which showed good agreement with our data. Using the N-terminus from our studies (Prince et al 1987), and their cDNA sequence, we can determine that osteopontin contains 301 amino acids. If one adds the carbohydrate and phosphate moieties, the actual molecular mass can be calculated to be about 41 500 Da. From the cDNA sequence, Oldberg et al (1986) predicted one N-linked oligosaccharide, in agreement with our conclusion. They also found a string of nine aspartic acids which they suggested might be involved with hydroxyapatite binding. Osteopontin has an Arg-Gly-Asp sequence (Oldberg et al 1986) involved in binding to receptors on osteoblast-like osteosarcoma cells and subsequent attachment and spreading. Martha Somerman has also demonstrated that the protein is a cell attachment and spreading factor with receptors on certain fibroblasts (Somerman et al 1987). Thus we know of at least two cell types that have receptors for osteopontin and there may be others. One important ques-tion is: what are the target cells for osteopontin and in which cells does this protein act?

In our laboratory, Manuel Mark studied the biosynthesis and secretion of osteopontin by immunolocalization experiments (Mark et al 1987a,b, 1988).

These indicated that osteopontin is synthesized by osteoblasts, osteocytes and preosteoblasts (i.e. fibroblast-shaped cells) in an osteogenic region that are not located against bone (Mark et al 1987a,b). In the calvaria of an 18-day-old rat embryo, the preosteoblasts, the osteoblasts and the osteoid are immunopositive. A few other cells outside bone also produce osteopontin: hypertrophic chondrocytes, as well as certain sensory cells and neural cells in the inner ear and brain, are immunopositive for this protein (Mark et al 1988). Therefore, osteopontin is not totally bone specific.

However, the major cell types and the major tissue involved are still bone: osteopontin is an important bone protein. Every bone tissue that we have examined is immunopositive. In addition, we may have detected osteopontin in serum, although the results are not yet conclusive. The protein appears to be degraded by kidney but we don't find evidence for synthesis in kidney. Thus, in certain kidney cells localization is in lysosomes whereas in bone cells the antigen is concentrated in the Golgi zone (Mark et al 1987b, 1988). Thus it could be taken up from the circulation and subsequently degraded by kidney. Osteopontin is stimulated modestly (by a factor of 1.5–2) when ROS 17/2.8 cells are incubated with 1,25-$(OH)_2$ vitamin D_3 (Prince & Butler 1987). As Dr Rodan described earlier, the mRNA level is increased several-fold.

What is osteopontin's function? Unlike Dr Glimcher, I think that the focal point of activity of the phosphorylated bone protein is related to cells and not necessarily to a function in extracellular mineralization. To summarize the main points, osteopontin is made early, is secreted into osteoid, and has an activity for attachment of cells to substratum. Therefore, the cells may be making their own attachment activity and secreting it into the matrix. Osteopontin persists in the tissue, because it is entrapped in the matrix after mineralization. One important observation by Martha Somerman was that if dishes were plated with collagen type I, and then with fibronectin, cell attachment was enhanced over that with collagen alone. However, if osteopontin was added to the collagen no additional cell attachment was achieved. The conclusion is that, unlike fibronectin, which has several binding domains including one for collagen, osteopontin does not have a collagen-binding domain. The question therefore arises as to how osteopontin is attached to osteoid and fulfils the cell attachment function. We are designing experiments to answer that question.

Martin: Is there any evidence to suggest that osteopontin might bind to osteoclasts?

Butler: No, as yet there are no data with which to answer that important question.

Raisz : In your slide of rat bone (not shown) there were many cells that did not stain for osteopontin.

Butler: That slide showed immunofluorescence of cells in an 18-day-old rat calvaria, where no calcification has yet taken place. In this early stage of development a lot of cells are immunostained with antibodies to osteopontin and those are presumably preosteoblasts. You are right that there are also a number of cells which do not stain.

Raisz: If you look at that kind of tissue under the light microscope you see a condensing mesenchyme in which you don't see much qualitative difference in histology. It looks like a mass of cells that are about to start making bone and yet your results look as though there are more than one class of cells. Do you have more evidence?

Butler: No, but that's a very interesting point.

Slavkin: In the cDNA sequence for osteopontin was there a transmembrane motif? In the immunolocalization experiments, have you tried to unmask the possible antigens by a hyaluronidase pretreatment? Is the image you saw the only immunostaining or might there be more?

Butler: The cDNA sequence does not indicate a transmembrane domain in this protein. It's a very small, hydrophilic protein. In general, in appropriate tissues that were negative, the immunolocalization experiments were done before and after hyaluronidase treatment. For example, no immunostaining was unmasked in cartilage after hyaluronidase treatment.

Slavkin: Is the protein pericellular, lying on the surface of the plasma membrane but not physically part of the plasma membrane?

Butler: In electron microscope immunolocalization experiments (Mark et al 1987b) we can detect reactivity in the Golgi apparatus of osteoblasts but not in a pericellular region. The majority, of course, is found extracellularly in bone.

Termine: How specific is the antibody? When Dick Heinegård first studied osteopontin he had what he thought was a specific antibody but then discovered some contaminating antibodies to the other sialoprotein. Proteoglycan contaminants or type I collagen determinants are common. If you overload a gel and do a Western blot, do you see any traces of potential contaminant?

Butler: Using competition ELISAs we can show no cross-reactivity with other bone and dentine proteins, except possibly with the low phosphorylated dentine phosphoprotein. We have not done Western blots successfully.

Rodan: We have done immunoblots of ROS 17/2.8 conditioned media with Dr Butler's antibody and visualized a single, relatively broad band.

Termine: I meant in bone matrix, because those cells don't exactly reproduce everything that's in the matrix.

Veis: Is the bone sialoprotein II also in the osteoid?

Termine: I think it looks much the same as other proteins in bone.

Veis: Suppose you took more mature bone, extracted the osteoid around the cells with guanidine HCl, and then demineralized after that, would you expose a lot more? Is the bone sialoprotein II moved into the mature bone mineralized matrix to the same extent as osteopontin, or does it stay where it is, surrounding the cells?

Butler: All the studies have been done with a developing bone system, not a mature bone system. Cutting sections from mature bone presents technical problems.

References

Mark MP, Prince CW, Oosawa T, Gay S, Bronckers ALJJ, Butler WT 1987a Immuno-histochemical demonstration of a 44-kD phosphoprotein in developing rat bones. J Histochem Cytochem 35:707-715

Mark MP, Prince CW, Gay S et al 1987b A comparative immunocytochemical study on the subcellular distributions of a 44kDal bone phosphoprotein and bone γ-carboxyglutamic acid (Gla)-containing protein in osteoblasts. J Bone Miner Res 2:337-346

Mark MP, Prince CW, Gay S, Austin RL, Butler WT 1988 44kDal bone phosphopro-tein (osteopontin) antigenicity at ectopic sites in newborn rats: kidney and nervous tissues. Cell Tissue Res 251:23-30

Oldberg Å, Franzén A, Heinegård D 1986 Cloning and sequence analysis of rat bone sialoprotein (osteopontin) cDNA reveals an Arg-Gly-Asp cell binding sequence. Proc Natl Acad Sci USA 83:8819-8823

Prince CW, Butler WT 1987 1,25-dihydroxyvitamin D₃ regulates the biosynthesis of osteopontin, a bone-derived cell attachment protein, in clonal osteoblast-like osteosarcoma cells. Collagen Rel Res 7:305-313

Prince CW, Oosawa T, Butler WT et al 1987 Isolation, characterization and biosynth-esis of a phosphorylated glycoprotein from rat bone. J Biol Chem 262:2900-2907

Smith JH, Denhardt DT 1987 Molecular cloning of a tumor promoter-inducible mRNA found in JBG mouse epidermal cells: induction is stable at high, but not at low, cell densities. J Cell Biochem 34:13-22

Somerman MJ, Prince CW, Sauk JJ, Foster RA, Butler WT 1987 Mechanism of fibroblast attachment to bone extracellular matrix: role of a 44 kilodalton bone phosphoprotein. J Bone Miner Res 2:259-265

Polypeptide growth factors in bone matrix

Peter V. Hauschka, *Theresa L. Chen and Athanasios E. Mavrakos

Harvard School of Dental Medicine and Children's Hospital Medical Center, 300 Longwood Avenue, Boston, Massachusetts 02115, USA and *California Biotechnology Inc., 2450 Bayshore Parkway, Mountain View, California 94043, USA

Abstract. The presence of many types of polypeptide growth factors in the mineralized extracellular matrix of bone is now well established. These factors are generally referred to as bone-derived growth factors (BDGFs), and are similar, or possibly identical, to the following species; platelet-derived growth factor (PDGF); acidic and basic forms of fibroblast growth factor (aFGF, bFGF); transforming growth factor β (TGF-β); and insulin-like growth factor 1 (IGF-1). Several osteoinductive factors, such as bone morphogenetic protein (BMP) and osteogenin, a skeletal growth factor (SGF), and osteoblast-derived BDGFs, have also been identified. Complete description of the biological functions of these BDGFs which are relevant to bone will ultimately require specific bioassays involving specific cell types *in vitro*, as well as *in vivo* animal implant models. Studies with primary rat osteoblast-like cells exposed either to mixed BDGFs, pure TGF-β, or heparin-purified PDGF, aFGF, or bFGF from bovine bone have shown a general dose-dependent mitogenic effect. Phenotypic changes which accompany the BDGF-induced wave of proliferation include: decreased osteocalcin secretion and a reduction in 1,25-$(OH)_2$ vitamin D_3-stimulated osteocalcin synthesis; reduced alkaline phosphatase specific activity; decreased cyclic AMP responsiveness to parathyroid hormone (PTH); and increased collagen synthesis. Bone exhibits the most complex spectrum of growth factor activities of any tissue yet described. In bovine bone powder free of blood and cartilage contamination, the volume concentration of mitogens is up to 20 times greater than that in serum. Bone cells and other indigenous cell types must be considered as possible sources of the BDGFs, in addition to sequestration from blood. Mechanisms for the unmasking or release of BDGFs from the mineralized matrix that result in local action on osteoblasts, endothelial cells, and other target cells are undoubtedly important for the development and maintenance of bone tissue.

1988 Cell and molecular biology of vertebrate hard tissues. Wiley, Chichester (Ciba Foundation Symposium 136) p 207–225

Polypeptide growth factors appear to play an important role in the development and growth of osseous tissue. Bone is unique because of its abundant mineralized extracellular matrix which may sequester growth factors and modulate their biological action through complex modes of release and presentation to responding cells. Extracellular matrix accounts for about 90% of the total weight of compact bone and is composed of microcrystalline calcium phosphate resembling hydroxyapatite (60%) and fibrillar type I collagen (27%). The remaining 3% of bone consists of minor collagen types and other proteins including osteocalcin, osteonectin, matrix γ-carboxyglutamic acid (Gla) protein (MGP), phosphoproteins, sialoproteins and glycoproteins, as well as proteoglycans, glycosaminoglycans and lipids. Osteoblasts are principally responsible for biosynthesis of this complex matrix. Additional compositional and biological complexity is imposed by: (1) matrix adsorption of numerous plasma proteins; (2) osteoclastic remodelling of calcified cartilage during endochondral bone development; (3) the capillary network with its associated endothelial cells and basement membranes; and (4) haemapoietic marrow- and blood-borne cells including monocytes, the apparent precursors of multi-nucleated osteoclasts which resorb bone, and other leukocytes which may regulate resorption via monokines and lymphokines.

The effects of polypeptide growth factors on bone are necessarily divided into two areas of study: exogenously produced endocrine factors which act on specific bone target cells; and endogenously produced 'local' factors with possible autocrine or paracrine action. Fundamental work with fetal rat calvaria showed that epidermal growth factor (EGF), fibroblast growth factor (FGF), platelet-derived growth factor (PDGF), and insulin-like growth factor 1 (IGF-1) generally stimulated cellular proliferation. EGF and FGF depressed formation of the collagenous extracellular matrix, while PDGF and IGF-1 (somatomedin C) stimulated collagen synthesis. EGF, PDGF and transforming growth factors (TGF-α and β) may also play a role in bone resorption, as they stimulate calcium release by a prostaglandin-mediated mechanism. Many of these effects have been reviewed by Canalis (1985).

Endogenous or 'local' growth factors of possible relevance include: (1) an 11 kDa protein produced by cultured fetal rat calvaria. This was originally called bone-derived growth factor (BDGF) but is now known to be β_2-microglobulin (Canalis et al 1987); (2) TGF-β, a 25 kDa protein demonstrated in the same organ culture system (Centrella & Canalis 1985); (3) cartilage-derived factor (CDF), an 11 kDa somatomedin C-like substance; (4) cationic cartilage-derived growth factor (CDGF), a 19 kDa FGF-like protein (Sullivan & Klagsbrun 1985); (5) human skeletal growth factor (hSGF), 83 kDa and 9–12 kDa forms isolated from human bone matrix (Farley & Baylink 1982); (6) bone morphogenetic protein (BMP) an 18.5 kDa bovine protein with bone-inducing activity (Urist et al 1984); (7) osteogenin, a heparin-absorbable osteoinductive factor from bovine bone (Sampath et al 1987), and (8) cartilage induction factors from bovine bone (CIF-A, now recognized to be identical to TGF-β, and the related CIF-B (Seyedin et al 1987)). Other local factors affecting bone cells include: interleukin-1; macrophage-derived growth factors (MDGF) (Rifas et al 1984) and two forms of PDGF (Shimokado et al 1985); endothelial cell-derived growth factor (ECDGF); polypeptide factors from neoplastic cells; the chemoattractant activity of PDGF (Deuel et al 1982); and a chemoattractant acting on mesenchymal cells (Lucas et al 1986). (For additional references to these factors see Hauschka et al 1986.) A wide range of assay systems involving different target cells and various bioresponses have been used in the study of these factors. Therefore, it is difficult to compare the findings and assess whether different names have been assigned to the same proteins.

Extraction and resolution of bone-derived growth factor activities

The informative technique of chromatography on heparin-Sepharose, developed by Klagsbrun and colleagues, permits the classification of growth factors on the basis of their empirically defined affinity. This procedure has been applied to the isolation and partial purification of several bone-derived growth factors (Hauschka et al 1986). For a quantitative survey of growth factors in bone matrix, the bioassay must be relatively non-discriminatory in its response to various BDGFs resolved by heparin-Sepharose. The quiescent BALB/c 3T3 fibroblast bioassay has been widely used, and is known to respond in a quantifiable, dose-dependent manner to a wide spectrum of competence and progression factors, except for TGF-β (Hauschka et al 1986).

The total mitogenic activity of bone is determined by EDTA extraction, dialysis, and titration in the 3T3 assay. Fetal calf bone yields up to 570 growth factor units (GFU) per g dry bone powder, where 5 GFU/ml causes half-maximal mitogenic stimulation of quiescent 3T3 cells (Hauschka et al 1986). For rat and mouse bone, the yield ranges between 300 and 1200 GFU/g and is typically two- to three-fold greater than for bovine material (R. Bevilacqua, unpublished work; C. Glass, unpublished work). The concentration of total growth factor activity in bone is at least 1000 GFU/ml. This is about 100 times higher than the levels required to cause maximal mitogenic stimulation of BALB/c 3T3 fibroblasts, capillary endothelial cells and osteoblasts (Hauschka et al 1986). Because the cells of bone are not continuously proliferating, the matrix must play an important role in sequestering and masking these factors. Extraction of BDGFs requires demineralization, but this is not adequate evidence to establish the mineralized extracellular matrix as the precise location of the BDGFs. In fact, some or all of these activities could be associated with the cells normally present in bone. Marrow elements, cartilage, and periosteum have been rigidly excluded by our preparation procedures, and the pre-extraction hypotonic washes of the finely ground bone powders (<150 μm particle diameter) thoroughly remove trapped blood and many cytosolic components without releasing appreciable growth factor activity.

The variety of growth factors in bone matrix is displayed in the heparin-

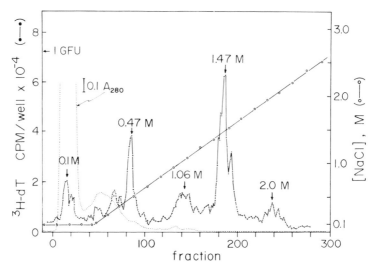

FIG. 1. Heparin-Sepharose affinity chromatography of growth factors extractable by 0.5 M EDTA from 50 g of fetal calf mandible power. 15 500 GFU were applied to an equilibrated 1.6 × 9.5 cm column of heparin-Sepharose. A linear gradient (○—○) of 0.1 M NaCl to 3 M NaCl in Tris buffer, pH 7.0, was applied. Absorbance at 280 nm of the effluent flow (......) was recorded. The molarity of NaCl corresponding to the position of each growth factor peak is indicated. Growth factor activity (●—●) was measured by assay of 10 µl aliquots on quiescent BALB/c 3T3 cells as described (Hauschka et al 1986), and the [³H] c.p.m. equivalent to 1 GFU (72 400) is indicated on the ordinate. (Reproduced with permission from Journal of Biological Chemistry).

Sepharose chromatogram of the EDTA extract of fetal calf mandible, where at least six separate peaks of activity are evident (Fig. 1). Previous evidence (Hauschka et al 1986) documents the occurrence of a PDGF-like factor (eluting at 0.45 M NaCl), an aFGF-like factor (1.1 M), and two bFGF-like factors (1.5 M and 1.7 M). The other peaks of mitogenic activity in Fig. 1 are awaiting postive identification. The most interesting variation in the BDGF elution pattern is the dramatic increase in the bFGF-like peak (1.7 M) found in adult bovine femur, compared with the relative paucity of this activity in fetal calf mandible. This factor elutes from heparin-Sepharose at the same position as CDGF (Sullivan & Klagsbrun 1985), and its abundance may stem from differences in bone maturity or developmental origin (membranous mandible versus endochondral femur). Although it is estimated that up to 100 000-fold purification is required to bring the BDGFs to homogeneity, the heparin methodology provides some hope that molecular characterization will eventually be achieved.

Bone is unique among tissues in the variety of polypeptide growth factors which it harbours (Table 1). In contrast, cartilage and brain exhibit a very

TABLE 1   Concentrations of growth factors in bone powder

Growth factor	Bovine bone[a] (ng/g dry bone)	Human bone[c] (ng/g dry bone)
TGF-β	400[b]	460
PDGF	50	67
aFGF	0.5–12	nd
bFGF	40–80	nd
IGF-1	nd	85
EGF	nd	0
SGF	nd	1260

nd, not determined; 0, not detectable.
[a] Estimated as described by Hauschka et al (1986).
[b] Seyedin et al (1987).
[c] Mohan et al (1987).

simple growth factor pattern on heparin-Sepharose, with only one or two principal peaks of activity. Although the exposure of a single factor to proteolytic degradation, or aggregation of a factor with different 'carrier' proteins (eg. Pfeilschifter et al 1987), might account for the multiple peaks in bone, this is an unlikely explanation for the diversity of factors in bone. It is now clear that the variety of BDGFs represent different types of polypeptide growth factors according to the previously defined criteria (Hauschka et al 1986). It is indeed possible that bone matrix contains factors in addition to the BDGFs of Table 1 and Fig. 1. Screening of extracts and heparin-Sepharose chromatograms with other specific bioassays will eventually answer this question. Other factors could also have been overlooked because of the requirement for a second, synergistic factor separated during heparin-Sepharose chromatography.

Origins of BDGFs

Several indigenous bone cell types are capable of contributing to the complex mixture of growth factors in bone matrix referred to as BDGFs. These include osteoblasts, osteocytes, endothelial cells and macrophages. As well as being locally produced, the BDGFs indicated in Table 1 and other factors mentioned previously may be sequestered from circulating blood after originating in extraosseous sources, such as platelets and brain. Pathologically derived factors, such as those from tumours, could also lodge in the bone matrix. The origin of the BDGFs, however, has no bearing on their presence in the matrix or their potential regulatory actions on bone cells.

Reports that growth factors were produced in bone organ culture (Canalis 1985, Centrella & Canalis 1985, Canalis et al 1987) or by U2-Os osteosarcoma cells (Betsholtz et al 1986) prompted the investigation of MG63 human

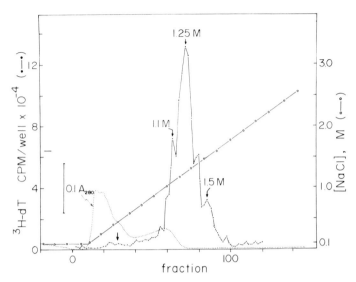

FIG. 2. Growth factor activity associated with the cell layer of MG63 human osteosarcoma cells after chromatography on heparin-Sepharose. 10^9 cells were extracted by hypotonic shock followed by 1 M NaCl and the dialysed extract was resolved as described in Fig. 1. The 25 000 GFU recovered consists of aFGF-like (1.1 M) and bFGF-like (1.5 M) activities, as well as a mitogen with intermediate affinity for heparin (1.25 M). There is no evidence of PDGF (arrow at 0.45 M NaCl) production or association with the MG63 cell layer.

osteosarcoma for growth factor synthesis. This cell line has osteoblastic properties but does not express the v-*sis* oncogene, and thus might not be expected to make a PDGF-like substance. Heparin-Sepharose chromatography of factors extracted from the cell layer of confluent MG63 cells showed no PDGF-like activity, but there was abundant activity eluting between 1.1 M and 1.5 M NaCl and analogous to aFGF and bFGF (Fig. 2). Parallel studies with primary rat osteoblast-like cells have shown a similar pattern of several FGF-like factors after heparin chromatography (A.E. Mavrakos, unpublished work). Thus the osteoblast remains a likely candidate for biosynthesis of at least several of the polypeptide growth factors found in bone matrix.

Response of osteoblasts to various growth factors of bone

Many polypeptide growth factors appear to exert a mitogenic effect on differentiated osteoblasts (Farley & Baylink 1982, Rifas et al 1984, Centrella & Canalis 1985, Hauschka et al 1986, Jennings et al 1986, Pfeilschifter et al 1986, Mohan et al 1987). Most of the BDGFs are generally active on mesodermally derived cells, but several factors have apparent target cell selectivity for stimulation of fibroblasts (F), chondrocytes (C), osteoblasts (O) and capillary endothelial cells (E) including: hSGF, O and C, but not F (Farley & Baylink 1982); PDGF-like factor, F and O, but not E (Hanks et al 1986, Hauschka et al 1986); MDGF, O and C, but not F (Rifas et al 1984); FGF-like factors, F, E and O (Hauschka et al 1986); and TGF-β, O but not F (Centrella & Canalis 1985, Jennings et al 1986, Pfeilschifter et al 1986).

While mitogenesis is often the focus of interest with growth factors, the potential hormonal action by which specific differentiated functions of non-proliferating cell populations may be regulated is of equal importance. Growth factors modulate such osteoblastic functions as collagen and non-collagen protein synthesis, alkaline phosphatase activity, and prostaglandin-dependent bone resorption (reviewed by Canalis 1985). However, the exact relationship of growth factor-induced changes in these functions to the state of osteoblastic proliferation has not yet been established.

Quiescent, confluent cultures of rat osteoblasts exposed for 48 hours to crude bone extracts, or to isolated BDGF peaks from heparin-Sepharose, or to fresh 5% serum show enhanced incorporation of [^{3}H]thymidine (4- to 6-fold), increased DNA and protein content (1.5- to 2-fold) and increased collagen biosynthesis relative to untreated controls. The concentration at which the maximal mitogenic effect of each BDGF component is attained varies from 0.4 to 20 GFU/ml, with PDGF- and aFGF-like activities appearing somewhat more potent towards osteoblasts than the bFGF-like BDGFs (Hauschka et al 1986). BDGFs in this study caused a very rapid decline in the important ability of rat osteoblast-like cells to respond to parathyroid hormone (PTH) with a burst of cyclic AMP synthesis (Fig. 3). Whether this is due to transient loss of the PTH receptor or to other events is unclear. Recovery does occur after 48 h (Fig. 3). Alkaline phosphatase activity, an important osteoblastic marker, is found to decrease about 3-fold, coincident with a mitogenic stimulus (Fig. 3). Similarly, the specific bone matrix protein osteocalcin is secreted at a reduced rate by mitogenically stimulated osteoblasts (Table 2), regardless of whether the stimulus is from pure TGF-β, bFGF or mixed BDGFs. These changes are perhaps to be expected from the general inverse relationship between proliferation rate and expression of differentiated function.

Of particular relevance to bone may be the enhanced protein kinase activity and the intracellular Ca^{2+} mobilization which typically accompanies growth factor stimulation of cells. Such stimulation could provide for the special needs of mineralizing tissues involving calcium phosphate mineral deposition and the prolific biosynthesis and secretion of phosphorylated proteins into the extracellular matrix.

In an initial exploration of the growth factor receptors of osteoblast-like cells, binding of ^{125}I-aFGF (bovine) to monolayer cultures has been measured at 19 °C under conditions where internalization is minimal (Tables 3 and 4).

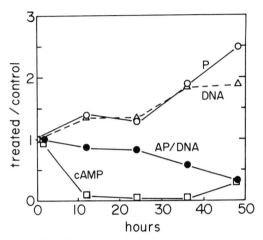

FIG. 3.   Effect of mixed bovine bone BDGFs on the phenotypic properties of rat
osteoblast-like cells. Confluent monolayers were treated with mixed BDGFs (4 GFU/
ml containing activities in the proportions shown in Fig. 1 and Table 1) for periods of
up to 48 h. Total protein (○) and DNA (△) increased 2- to 2.5-fold compared to
control cells at the same time points, in agreement with the mitogenic effect of this
BDGF dose. Alkaline phosphatase activity per cell (●) decreased 3.2-fold. The
stimulation of cyclic AMP synthesis by 10-min treatment with PTH (2 U/ml) dropped
rapidly from a control value of 54-fold to only 2.7-fold after 24 h of BDGF exposure
(□); this property appears to partially recover by 48 h.

TABLE 2   Effects of mitogenic stimulation on osteocalcin production by rat osteoblast-
like cells

	Osteocalcin secretion[a]	
Treatment	+ 1,25-(OH)$_2$D$_3$	− 1,25-(OH)$_2$D$_3$
Experiment 1		
Control	100	23
BDGF mix (5 GFU/ml)	24	15
Experiment 2		
Control	100	10
TGF-β (2 ng/ml)	37	7
TGF-β (20 ng/ml)	9	5
Experiment 3		
Control	100	20
bFGF (2.5 ng/ml)	35	10

Experiment 1. Confluent monolayers were treated for 48 h in fresh Eagle's Minimal Essential
Medium α-MEM + 1% fetal calf serum, with or without BDGF, and medium was harvested.
Experiments 2 and 3. Confluent monolayers were treated for 18 h in Bone Fluid Medium-serum,
with or without growth factor, and medium was harvested.

[a] Expressed relative to control without growth factor (100% secretion in presence of 1 nM
1,25-(OH)$_2$ vitamin D$_3$ alone).

TABLE 3   Properties of the FGF receptor in osteoblast-like cells

Species	Cell type	No. of receptors per cell	K_d (pM)
Rat	1° fetal calvarial osteoblasts	20 000	40
Mouse	MC3T3 osteoblasts	32 000	53
Human	MG63 osteosarcoma	10 000	26

Results of Scatchard analysis of binding data for cell monolayers at 19 °C using ^{125}I-aFGF and protocols kindly provided by Olwin & Hauschka (1986).

Three types of osteoblast-like cells have reasonable numbers (10 000–32 000 sites/cell) of high affinity receptors (K_d = 20–60 pM) when compared to fibroblasts and MM14 myoblasts (Olwin & Hauschka 1986). Another common property is the ability of bFGF to displace ^{125}I-aFGF from its osteoblastic receptor equally as well as unlabelled aFGF (Table 4). Thus the properties of the osteoblastic aFGF receptor are compatible with the action of aFGF (and perhaps bFGF) at their local concentrations in bone matrix (about 60 pM to 10 nM, calculated from Table 1). The well-known effect of heparin on enhancing aFGF activity has not yet been examined in this system.

Endothelial cell growth factor activity of BDGFs

Stimulation of bovine capillary endothelial (BCE) cell proliferation in the system established by Folkman et al (1983) showed potent actions of the aFGF-like (1.1 M) and bFGF-like (1.5 M, 1.7 M) factors from bone (Hauschka et al 1986). As expected, the PDGF-like (0.45 M) material did not enhance BCE cell proliferation. Dose–response curves for the three active BDGFs are substantially similar to those for HGF, the endothelial cell growth factor from bovine hypothalamus which is the normal positive control in this

TABLE 4   Competition between aFGF and bFGF for the osteoblastic FGF receptor

Unlabelled FGF (pM)	aFGF (% ^{125}I-aFGF bound)	bFGF
0	100.0	100.0
50	82.9 ± 6.0	84.9 ± 0.8
250	64.3 ± 6.4	63.0 ± 5.8
1000	43.1 ± 2.2	41.4 ± 2.4
5000	32.2 ± 5.6	24.7 ± 4.4
30 000	22.5 ± 4.4	16.6 ± 3.3

[125]I-aFGF (52 pM) was mixed with various concentrations of unlabelled bovine aFGF or human bFGF and allowed to bind to cells for 1 h at 19 °C following Olwin & Hauschka (1986). Data are mean ± SEM for the three osteoblastic cell types shown in Table 2.

LOCAL FACTORS AND BONE REMODELLING

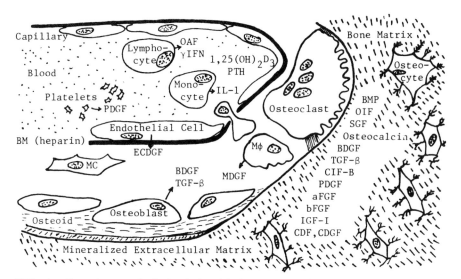

FIG. 4. Complex topological relationship between the indigenous bone cell types and the various growth factors and other effector proteins of bone. An osteoclast leads a capillary into dense bone, forming a typical 'cutting cone'. Growth factors in plasma and those produced by blood-borne cells and capillary endothelial cells thereby gain intimate access to the bone matrix. Factors trapped in the matrix may be released by the mining behaviour of the osteoclast, causing local stimulation of osteoblastic proliferation (coupling).

assay. The ability of capillary endothelial cells to respond to BDGFs may be of great biological significance. Nutrient artery ingrowth and neovascularization of developing bone could require this type of matrix-derived stimulus. Poor vascularization, as in osteoradionecrosis, could stem from a defective stimulus. Perhaps a mechanism exists, analogous to the 'coupling phenomenon' for osteoblasts (Howard et al 1981), which links endothelial cell proliferation to osteoclastic resorption on a local scale.

Local control of bone formation by BDGFs in the extracellular matrix

The exquisite balance between osteoclastic resorption and osteoblastic formation of bone is currently understood in terms of local regulation or 'coupling' (Howard et al 1981), in which both the extracellular matrix and local growth factors are believed to play a central role. Access of resident osteoclasts and recruited monocytes to the bone surface destined for resorption is apparently modulated by osteoblasts (Rodan & Martin 1981). Within the locus of biochemical signals emanating from an established osteoclastic resorption site, local proliferation and differentiation of osteoblasts is promoted and new

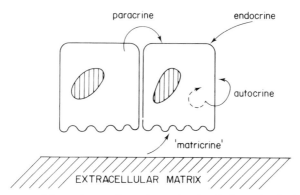

FIG. 5. Demonstrable pathways for mitogenic stimulation of cells by polypeptide growth factors include *endocrine*, *paracrine* and *autocrine* routes. The existence of growth factors bound to components of the extracellular matrix such as collagen, hydroxyapatite and heparan sulphate could allow direct stimulation by a '*matricrine*' route.

bone is formed. Capillary ingrowth involving endothelial cell proliferation in haversian bone is also coupled to resorption by the osteoclastic 'cutting cone' (Fig. 4). Candidates for providing these signals include growth factors, the collagenous extracellular matrix, and the calcium phosphate mineral phase itself.

Osteoclastic resorption can solubilize polypeptide growth factors from bone matrix, and an example of active TGF-β release by osteoclasts has been provided by Pfeilschifter et al (1986). Clarification of the mechanism of matrix resorption, including the kinetics of release of H^+, collagenases, other proteinases, and heparinases (eg. Baird & Ling 1987), will ultimately enhance our understanding of growth factor liberation. Which BDGFs can be liberated from matrix or exposed without being proteolytically damaged? Must these factors act on target cells in a soluble form? Could BDGFs function as effectors if presented in an adsorbed state on heparan sulphate, collagen, hydroxyapatite, or other matrix elements? In distinction to endocrine, paracrine and autocrine routes for growth factor stimulation, cellular interaction with matrix-adsorbed factors could be called 'matricrine' (Fig. 5). Osteoblasts in bone are generally prevented from making direct contact with the calcium phosphate (hydroxyapatite) mineral phase by a layer of protein-rich non-mineralized osteoid. Direct mineral contact resulting in mitogenic stimulation may occur where either osteoblasts migrate into freshly denuded resorption lacunae, or osteoid is degraded by osteoblastic collagenase. Osteoblast-mediated exposure of hydroxyapatite has also been suggested as a prerequisite for osteoclastic degradation (Chambers & Fuller 1985).

The extracellular matrix apparently plays a critical role in controlling the distribution and presentation of growth factors. First, the matrix provides a temporally extended storage depot for local factors awaiting response by cells which could arrive on the scene long after the producing cell has departed or ceased to function. For example, CDGF produced during endochondral osteogenesis could be zonally concentrated and released at a later time. Second, the regulation of extracellular matrix biosynthesis by growth factors may actually allow modulation by a feedback loop involving adsorption of the factors to new matrix. The steadily increasing number of related polypeptide sequences in each growth factor class (currently: PDGF, ≥5 forms; FGF, 5 forms; TGF-β, ≥3 forms; EGF, 13 forms; IGF, 3–4 forms) raises the question of bone-specific types. Might there be sequence variations or proteolytic modifications which confer matrix affinity, stability, and/or target cell specificity on the BDGFs? What BDGF signals are reserved for actual growth stimulation, in contrast to the maintenance and remodelling functions in bone? Clearly, with six or more potential osteoblastic mitogens at large in bone matrix, there must be a hierarchy of control and differential response. These mechanisms will remain elusive until the entire bone lineage ranging from stem cells to osteoblasts to osteocytes is fully accessible to *in vitro* study.

Systemic responses of the skeleton to endocrine signals do not explain the complex patterns of bone growth and remodelling which are exquisitely sensitive to local physical stresses. In a dense, labyrinthine mineralized tissue such as bone, with many barriers to the free diffusion of circulating hormones, it is an exciting prospect that these local responses may be regulated by the reservoir of specific growth factors in the matrix.

Acknowledgements

We thank S. Doleman and G. Hintsch for expert technical assistance, and C. Glass, R. Bevilacqua and K. Choi for providing additional data. Drs B.B. Olwin and S.D. Hauschka kindly supplied the ^{125}I-aFGF for osteoblast receptor studies. Supported by grants AR38349, DE05408, DE08235, and CA37391 from the National Institutes of Health, DHHS.

References

Baird A, Ling N 1987 Fibroblast growth factors are present in the extracellular matrix produced by endothelial cells *in vitro*: implications for a role of heparinase-like enzymes in the neovascular response. Biochem Biophys Res Commun 142:428–435
Betsholtz C, Johnsson A, Heldin C-H et al 1986 cDNA sequence and chromosomal localization of human platelet-derived growth factor A-chain and its expression in tumour cell lines. Nature (Lond) 320:695–699
Canalis E 1985 Effect of growth factors on bone cell replication and differentiation. Clin Orthop Relat Res 193:246–263
Canalis E, McCarthy T, Centrella M 1987 A bone-derived growth factor isolated from rat calvaria is beta$_2$ microglobulin. Endocrinology 121:1198–1200
Centrella M, Canalis E 1985 Transforming and nontransforming growth factors are

present in medium conditioned by fetal rat calvariae. Proc Natl Acad Sci USA 82:7335–7339

Chambers TJ, Fuller K 1985 PTH stimulates bone resorption by inducing osteoblasts to expose bone mineral to osteoclastic contact. Calcif Tissue Int 37:162a

Deuel TF, Senior RM, Huang JS, Griffin GL 1982 Chemotaxis of monocytes and neutrophils to platelet-derived growth factor. J Clin Invest 69:1046–1049

Farley JR, Baylink DJ 1982 Purification of a skeletal growth factor from human bone. Biochemistry 21:3502–3507

Folkman J, Langer R, Linhardt RJ, Haudenschild C, Taylor S 1983 Angiogenesis inhibition and tumor regression caused by heparin or a heparin fragment in the presence of cortisone. Science (Wash DC) 221:719–725

Hanks CT, Kim JS, Edwards CA 1986 Growth control of cultured rat calvarium cells by platelet-derived growth factor. J Oral Pathol 15:476–483

Hauschka PV, Mavrakos AE, Iafrati MD, Doleman SE, Klagsbrun M 1986 Growth factors in bone matrix. Isolation of multiple types by affinity chromatography on heparin-Sepharose. J Biol Chem 261:12665–12674

Howard GA, Bottemiller BL, Turner RT, Rader RI, Baylink DJ 1981 Parathyroid hormone stimulates bone formation and resorption in organ culture: evidence for a coupling mechanism. Proc Natl Acad Sci USA 78:3204–3208

Jennings JC, Linkhart T, Mohan S, Lundy M, Baylink D 1986 Isolation of two beta transforming growth factors (beta TGF) from bovine bone. J Bone Miner Res 1 (Suppl 1):301

Lucas PA, Syftestad GT, Caplan AI 1986 Partial isolation and characterization of a chemotactic factor from adult bovine bone for mesenchymal cells. Bone (NY) 7:365–371

Mohan S, Linkhart TA, Jennings JC, Baylink DJ 1987 Identification and quantification of four distinct growth factors stored in human bone matrix. J Bone Miner Res 2 (Suppl 1):44

Olwin B, Hauschka SD 1986 Identification of the fibroblast growth factor receptor of Swiss 3T3 cells and mouse skeletal muscle myoblasts. Biochemistry 25:3487–3492

Pfeilschifter J, D'Souza S, Mundy GR 1986 Transforming growth factor beta is released from resorbing bone and stimulates osteoblast activity. J Bone Miner Res 1 (Suppl 1):294

Pfeilschifter J, Bonewald L, Mundy GR 1987 TGF-beta is released from bone with one or more binding proteins which regulate its activity. J Bone Miner Res 2 (Suppl 1):249

Rifas L, Shen V, Mitchell K, Peck WA 1984 Macrophage-derived growth factor for osteoblast-like cells and chondrocytes. Proc Natl Acad Sci USA 81:4558–4562

Rodan GA, Martin TJ 1981 Role of osteoblasts in hormonal control of bone resorption—a hypothesis. Calcif Tissue Int 33:349–352

Sampath TK, Muthukumaran N, Reddi AH 1987 Isolation of osteogenin, an extracellular matrix-associated, bone-inductive protein, by heparin affinity chromatography. Proc Natl Acad Sci USA 84:7109–7113

Seyedin SM, Segarini PR, Rosen DM, Thompson AY, Bentz H, Graycar J 1987 Cartilage-inducing factor-B is a unique protein structurally and functionally related to transforming growth factor-beta. J Biol Chem 262:1946–1949

Shimokado K, Raines EW, Madtes DK, Barrett TB, Benditt EP, Ross R 1985 A significant part of macrophage-derived growth factor consists of at least two forms of PDGF. Cell 43:277–286

Sullivan R, Klagsbrun M 1985 Purification of cartilage-derived growth factor by heparin affinity chromatography. J Biol Chem 260:2399–2403

Urist MR, Huo YK, Brownell AG et al 1984 Purification of bovine bone morphogenetic protein by hydroxyapatite chromatography. Proc Natl Acad Sci USA 81:371–375

DISCUSSION

Martin: You made the proposal that the matrix might be a store of growth factors, to be released during resorption. I have difficulty with this, not just because the factors might not survive proteolysis, but also because there is such a large store of factors there and your suggestion requires them to be released in controlled amounts that reach certain cells as they require them. I can't imagine how that could possibly be regulated sufficiently well.

Hauschka: I don't know what the release really involves. Suppose 1% of the total growth factor activity liberated during bone resorption actually survives osteoclastic degradation. One could envisage a retrograde diffusion of these mitogens into the perivascular space where osteoblasts and preosteoblasts are found (see Fig. 4). One percent of the total bone-derived growth factor (BDGF) activity amounts to 5–10 GFU/ml, which is enough to elicit a robust mitogenic response of osteoblasts *in vitro*. Enhanced proliferation of osteoblasts has been demonstrated histologically in the vicinity of the osteoclast. I am not trying to make a complicated mechanism out of this, but I am curious as to whether these released mitogens are important or not.

Caplan: You alluded to an experiment that many of us have tried, namely to design delivery vehicles for the re-introduction of purified molecules into *in vivo* test systems. What did you do with fibroblast growth factor (FGF)?

Hauschka: We did not have much of the acidic or basic FGFs to test for osteoinduction in our rat model system. We tried heparin fractionation of the whole guanidine extract from demineralized rat bone, followed by assay of fractions by ethanol reconstitution with inactive carrier bone particles; to date this has been unsuccessful. I think the purity of the growth factors, and the way in which one recombines them with the carrier particles is absolutely critical.

Caplan: To test whether a factor, or a group of factors, is effective, we shall need to put it back into an *in vivo* test system and obtain the cascade that we can see with demineralized bone powder or other molecules. We probably need a group of factors including chemotactic agents, mitogens and morphogens. What is the most appropriate *in vivo* system for testing bone-derived growth factors?

Hauschka: The reconstituted implant system used subcutaneously is suitable.

Caplan: But you are not going to affect bone cells at a subcutaneous site. You have a chance of affecting mesenchymal cells that would migrate to that site if there were a mesenchymal mitogen, for example.

Canalis: You are assuming that these factors act systemically and there is no proof of that. There is much evidence that basic FGF, for instance, is not a secreted protein.

Caplan: Dr Hauschka is not injecting the factors. He is delivering them locally as a solid substrate to a particular site. Is there a way to test the efficacy of individual factors *in vivo*, whether it's at a subcutaneous site, which is asking another question, or in a bone cell?

Hauschka: One has to dissect the process into its measurable parts. The reconstituted osteoinduction systems which we and others have used (Spampata et al 1987, Sampath et al 1987), and the demineralized bone implant that Dr Glowacki works with, both recruit some mesenchymal precursors and initiate an endochondral, bone-forming process. The question is: is one assaying purely that recruitment phenomenon, or is one assaying the differentiation of osteoblasts? Might the rate-limiting process involve other events such as neovascularization? We are exploring techniques for delivery of test substances at a time after initiation of the implant by direct injection into the implant. This two-stage approach is quite different from putting everything in at the beginning, because it's possible for the initial cells responding to the implant, polymorphonuclear leukocytes for instance, to destroy whatever factors have been put in.

Raisz: I hope that the neologism 'matricrine' is as short-lived as some of these growth factors These factors are going to be short-lived, and therefore time, as well as distance, becomes a problem. The reversal phase before bone formation starts in the remodelling cycle is about a week long. Substances released in bone during resorption are hardly likely to work at that distance in time.

Secondly, you are looking at two distinct questions here. One is the issue of making bone grow on old bone or on new sites adjacent to old bone, which would be for the treatment of osteoporosis at this stage. The other is healing lesions where there is no bone present. In the latter case, the induction model might be relevant because you could go through the cartilage process. You might be able to look at what the factor does when added in the induction model, not at the time of cartilage induction, but at the time of cartilage to bone transformation. Has anybody ever tried anything like that?

Termine: This is not so clear-cut as Dr Hauschka has suggested. He described an implant in which there is cartilage, bone and everything. In the first five or six days a broad spectrum of events occur.

Raisz: Is there a cartilage phase before the bone phase starts?

Caplan: No, that's not correct. There are lots of different separate steps going on simultaneously. If you inject the growth factor on Day 3 it may affect one stage, but not the other concurrent steps.

Glowacki: What they are both saying is not true! We have done many subcutaneous implants of demineralized bone powder into rats. When you use a narrow range of particles size, 75–250 µm, the process is very synchronous. By Day 5 no blood vessels have penetrated into the implant; they don't come in until about Day 7 or 8. There are already lots of chondroblasts being stimulated at this stage.

In response to the report by Caplan & Pechak (1987), I looked for the first appearance of alkaline phosphatase, because I thought that the chondrocytes would develop alkaline phosphatase when they became mineralized. However, on Day 5 there are many alkaline phosphatase-positive cells that don't look anything like chondrocytes or osteoblasts. What may be happening here first is that some progenitor, a mesenchymal-type cell, expresses the chondroblast phenotype in response to contact with the demineralized bone powder.

The way Dr Raisz asked his question is the way Professor Urist and I have been talking for years too. The chondrocyte then undergoes the normal endochondral sequence of events and you end up with bone. The presence of the alkaline phosphatase-positive cells at that early time may be suggesting that there is another process of osteoblast differentiation separate from chondrocyte differentiation. Once the chondroblasts are stimulated by the demineralized matrix, they may produce something in the matrix, or secrete some kind of factor that acts on a different progenitor cell and pushes it towards the preosteoblast or osteoblast phenotype.

Veis: In that model you could just as easily draw an arrow from demineralized bone powder to one chondrocyte progenitor and to another osteoblastic progenitor and have these things being produced at different rates. You have a huge mixed population of cells. I don't see the need to convert, in the same time frame, every cell which is interacting with your factor into a chondrocyte, chondroblast or osteoblast.

Glowacki: We thought of that possibility too and looked through lots of histological sections. We never see these alkaline phosphatase-positive cells at a distance from the chondrocytes. There's plenty of demineralized bone matrix in the subcutaneous site without induced chondrocytes nearby. The preosteoblasts are always located near cartilage.

Veis: The point is that osteogenesis is controlled by local interactions that occur in the matrix in a particular way when you do the experiment. Otherwise you would expect your chondrocytes to go uniformly to osteoblasts.

Glowacki: We do not know why every particle does not have cartilage around it. The point I'm trying to make is that you don't see the alkaline phosphatase-positive preosteoblasts at a distance from the chondroblast and its matrix.

Veis: That is where your concentration of factor is in the matrix you are implanting.

Canalis: It is important to remember that none of the sequenced factors is bone specific. The only three factors that we have found expressed in bone cells are IGF-1, TGF-β and β-2 microglobulin. We have not found mRNA for acidic or basic FGF; if mRNA for either is present, it is at a low concentration.

Krane: The osteoblast 'phenotype' has been arrived at by consensus. Alkaline phosphatase is not absent from other cells, but it is there at lower levels. We don't know what role this enzyme plays in production of bone matrix and its mineralization. The equation of an alkaline phosphatase stain with an osteoblast is too weak. When Dr Rodan transfected a non-expressing cell, the gene was expressed and this did not result in an osteoblast.

Glowacki: In this case, cells in this region, near the cartilage, will become *bona fide* osteoblasts in three days. As in Dr Caplan's model, the cartilage will be replaced by marrow and the peri-cartilage cells will become osteoblasts.

Krane: You are seeing them at one point in time, however, and you can't resection them.

Urist: If I follow Dr Glowacki's interpretation correctly, the geography of the tissue foretells its contribution to bone regeneration. The cartilage develops in the interior, in the avascular core, and the bone develops outside on the vascularized exterior surfaces.

Dr Hauschka, how did you recombine the extract with the matrix? I presented recombination experiments and showed induced bone formation ten years ago at a Gordon Conference where Dr Raisz remarked that this experiment was like going back into the swamp. In order to answer that objection, could you possibly tell us whether you have recombined, one at a time, the individual fractions that were isolated by heparin affinity chromatography?

Hauschka: The technique of reconstitution is very important. One starts with a guanidine-soluble extract which contains osteoinductive and mitogenic activity. If that material is recombined in the presence of 4 M guanidine with the inactive bone powder residue, and then dialysed to get rid of the guanidine before implantation, it causes no osteoinduction. If one co-precipitates with ethanol and uses ethanol to wash away all the guanidine, it works beautifully. Whatever the active components are, they are not irreversibly altered by ethanol. The precipitation with ethanol is critical.

Urist: What about the fractions?

Hauschka: We have not recovered enough osteoinductive activity in the heparin fractions to rigorously test them in the reconstitution assay.

Urist: I have done such experiments. What you say about the ethanol precipitation is correct. If you use absolute ethanol, it will not denature the BMP in the extract. However, when you get the composite that is active, it probably recombines with Gla protein in the matrix. These proteins have an enormous affinity for other non-collagenous proteins, not for collagen. People have tried to do this with reconstituted rat tail tendon soluble collagen and solubilized protein in culture, and it is not incorporated. There may still be some BMP and non-collagenous protein for which BMP has a high affinity left in the matrix. One such component is the 14 kDa material, matrix Gla protein (Price et al 1983).

Pierschbacher: I like your matricrine hypothesis but I would like to propose another alternative which might be called 'patricrine'. The matrix may determine how the cell responds to a growth factor rather than giving a growth factor signal itself. Then you wouldn't need a bone-specific growth factor. You would have a bone-type cell on a particular type of matrix. If that were the case, it should be possible to reconstruct it with the correct matrix and growth factors.

Caplan: Paul Lucas in my laboratory has taken purified type I collagen (Vitrogen) and combined it with water-soluble extracts from bone which contain no matrix Gla protein or insoluble proteins. Using a procedure that involves an ethanol step, he produced an implant material that has full biological activity in a muscle implant site. The importance of this is that it's a non-immunogenic purified collagen. All the bioactive materials used, whether partially purified or crude, were water soluble. When Vitrogen or lyophilized water-soluble material are used alone, no cartilage or bone is produced. The denaturing of the collagen with the ethanol probably delays the solubilization of the bioactive molecules and the delivery of this material. The delivery of bioactive proteins peaks around 2–4 days after the implantation. Either subcutanoeusly or intramuscularly, the initial events that occur are systemic events, the bringing together of circulating and systemic factors for this site.

The importance of Dr Hauschka's work is that we should now consider bone as a metabolic storehouse for bioactive factors, including mitogens, in the same way as we think about bone as a storehouse for calcium.

Veis: We obtain these growth factors by various denaturing procedures such as 4M guanidine. Then we talked about denaturing them. Is anything known about the secondary structure or reformation of secondary structure in any of these growth factors?

Hauschka: In our initial characterization of the individual BDGF activities isolated by heparin-Sepharose chromatography, we examined the effect of 4M guanidine treatment, using rapid microdialysis with suitable controls to remove the guanidine. Platelet-derived growth factor activity is totally unaffected by guanidine or boiling: it is very stable. aFGF and bFGF have peculiar sensitivity to guanidine. If dithiothreitol is present during the removal of guanidine, it actually helps the factors, especially aFGF, to recover their activity, but generally these mitogens are rather sensitive to guanidine.

Urist: When you dialyse against water to remove the guanidine, some proteins are obviously renatured with respect to their osteoinductive activity. When we isolate protein by electro-elution, the osteoinductive component is denatured. Fortunately, there are methods of renaturing denatured proteins, with acetone and other SDS solvents (Konigsberg & Henderson 1983).

References

Caplan AI, Pechak DG 1987 The cellular and molecular embryology of bone formation. In: Peck WA (ed) Bone and mineral research. Elsevier Science Publishers, Amsterdam vol 5:117–184

Konigsberg WH, Henderson L 1983 Removal of sodium dodecyl sulfate from proteins by ion pair extraction. In: Hirs CHW, Timasheff SN (eds) Methods in Enzymology. Academic Press, New York Vol 91:254–259

Price PA, Urist MR, Otawara Y 1983 Matrix gla protein, a new γ-carboxyglutamic acid containing protein which is associated with the organic matrix of bone. Biochem Biophys Res Commun 117:765–771

Sampath TK, Muthukumaran N, Reddi AH 1987 Isolation of osteogenin, an extracellular matrix-associated, bone-inductive protein, by heparin affinity chromatography. Proc Natl Acad Sci USA 84:7109–7113

Spampata RP, Werther JR, Hauschka PV 1987 Accelerated osteoinduction in a rat bone implant model system. J Bone Miner Res 2: Suppl (abstr)

Hormonal regulation of bone growth and remodelling

Lawrence G. Raisz

University of Connecticut Health Center, Farmington, Connecticut 06032, USA

Abstract. Many systemic and local hormones influence bone growth and remodelling. These include calcium regulating hormones, systemic growth regulators and local growth factors. Parathyroid hormone (PHT) is a potent stimulator of osteoclastic bone resorption and a direct inhibitor of osteoblastic collagen synthesis. However, intermittent low-dose PTH administration can increase bone formation *in vivo*. PTH may act indirectly via local factors. It has been shown to increase prostaglandin E_2 (PGE_2) and transforming growth factor β (TGF-β) release from bone. Both PGE_2 and TGF-β have complex effects on bone metabolism and are likely to be physiological regulators of bone remodelling. Oestradiol has been shown to inhibit bone resorption *in vivo* but not *in vitro*. While there is evidence for oestrogen receptors in cultured bone cells, the effect could still be indirect. Oestradiol can inhibit bone PGE_2 release in an *in vivo–in vitro* model in the rat. Glucocorticoids are potent inhibitors of bone formation and inhibit PGE_2 and interleukin 1 production both *in vivo* and *in vitro*. While many regulatory factors affect prostaglandin production in bone, the complex effects of PGE_2 on bone metabolism make it difficult to predict the ultimate response. The major effects of PGE_2 are stimulation of bone formation and resorption and an increase in bone turnover. However, opposite effects can occur at certain times and concentrations. Interactions among these factors could explain some physiological, pathological, and therapeutic responses in skeletal tissue.

1988 Cell and molecular biology of vertebrate hard tissues. Wiley, Chichester (Ciba Foundation Symposium 136) p 226–238

During the past 20 years there has been a remarkable change in our approach to the study of the regulation of bone metabolism. The idea that skeletal structure is determined by local forces is an old one, but the concept that there are local humoral factors which might be responsible for this regulation is relatively new. In addition, our understanding of systemic regulation of bone metabolism has been expanded by the finding of additional circulating factors besides parathyroid hormone (PTH). In 1960 PTH was the only calcium regulating hormone that was considered to act directly on bone. Since that time calcitonin and 1,25-dihydroxyvitamin D_3 ($1,25\text{-}(OH)_2D_3$) have been shown to act directly on bone, although they are probably less important as regulators than is PTH. In addition, direct effects of glucocorticoids and thyroid hormones and indirect effects of growth hormone and oestrogen have been identified.

The first local regulators to be defined were the prostaglandins, particularly PGE_2. However, the list has rapidly expanded and now includes osteoclast activating factors such as interleukin 1 and tumour necrosis factor, transforming growth factor beta (TGF-β), insulin-like growth factor 1 (IGF-1) and bone-derived growth factors, one of which has recently been found to be beta-2 microglobulin. The last two are also found in the circulation. Many other systemic and local factors have been shown to act on the skeleton. Rather than presenting a catalogue of these agents, I shall select four — PTH, oestradiol, cortisol, and PGE_2 — not only because there is interesting information about their actions on the skeleton, but also because they may play important roles in the pathogenesis and treatment of metabolic and local bone disease.

Parathyroid hormone

Since the demonstration, using parathyroid transplants and organ cultures, that parathyroid hormone could stimulate bone resorption directly, this response has been studied in great detail, although it still remains something of a mystery. PTH can increase the activity and number of osteoclasts in bone organ culture (Holtrop & Raisz 1979), but has little effect on isolated mature osteoclasts (McSheehy & Chambers 1987), even though these cells are still capable of resorbing bone. Both *in vivo* and *in vitro*, the major determinant of the magnitude of the resorptive response appears to be an increase in osteoclast number rather than changes in activity of individual osteoclasts, but both processes do occur. Since PTH has been shown to act directly on osteoblasts, the hypothesis that these cells mediate the increase in bone resorption has been widely accepted. While there is evidence that osteoblasts can produce substances which activate isolated osteoclasts, there is as yet no evidence that the increase in osteoclast number is mediated by an osteoblast product. In organ culture, this PTH-stimulated increase in the number of active osteoclasts does not depend on the multiplication of progenitor cells because it can occur in the presence of inhibitors of DNA synthesis (Lorenzo et al 1983). Thus PTH appears to stimulate the fusion of existing precursors into active osteoclasts. However, when cell division is taking place, there seems to be a selective incorporation of recently divided cells into the new osteoclasts which result from PTH stimulation.

Although PTH stimulates cyclic AMP production in osteoblasts, and agents which increase cyclic AMP production can stimulate bone resorption (Lorenzo et al 1986), it has been difficult to prove that this nucleotide is the major second messenger for the PTH response. Analogues which block PTH stimulation of adenylate cyclase do not inhibit bone resorption, and PTH analogues which have relatively little effect on cyclic AMP can stimulate bone resorption (Löwik et al 1985). It has been postulated that intracellular calcium provides an additional second messenger for stimulation of bone resorption, perhaps acting in concert with cyclic AMP. A dissociation between cyclic AMP activation and bone resorbing potency is also observed when comparing the amino terminal peptides of PTH itself and the PTH-related peptide (PTHrP) which is believed to be responsible for the humoral hypercalcemia of malignancy. Synthetic 1–34 PTHrP is more potent than PTH in stimulating cyclic AMP production in cells of the osteoblast lineage (Moseley et al 1987), but less potent in stimulating osteoclastic bone resorption in organ culture (Raisz L.G. & Simmons H.A., unpublished observations).

The effects of PTH on bone formation are variable, depending on time, dose and mode of administration. In high concentrations PTH inhibits osteoblastic collagen synthesis and can reduce alkaline phosphatase activity (Raise & Kream 1983). The osteoblasts assume a stellate shape and secrete collagenase and plasminogen activator which may be involved in initiating the bone resorptive process, perhaps by preparing the bone surface for the osteoclasts (Hamilton et al 1985). This catabolic phase of increased resorption and decreased formation can be sustained if the cells are continuously exposed to high concentrations of PTH. However, intermittent exposure to low concentrations of PTH results in an anabolic response which is quite reproducible *in vivo* (Gunness-Hey & Hock 1984) and has also been described *in vitro* (Howard et al 1981). The anabolic effect of PTH appears to represent an increase in the number as well as the activity of osteoblasts (Tam et al 1982). Both the extent of bone-forming surfaces and the rate of mineral and matrix apposition on these surfaces are increased. A number of hypotheses have been presented which attempt to explain the anabolic effect of PTH. Mediation by systemic hormones seems unlikely in view of the evidence for an *in vitro* effect, but has not been ruled out. PTH could increase bone formation by releasing a bone-derived growth factor from the matrix during the resorptive phase. There is some indirect evidence against this: effects on resorption and formation can be dissociated in terms of dose and time. Moreover, pretreatment with a diphosphonate or treatment with calcitonin during the early phase of intermittent low dose PTH administration does not decrease the anabolic response (Hummert et al 1987). PTH could produce its anabolic effect by stimulating the production of a local bone growth factor. PTH can stimulate the release of TGF-β from bone, but it is not clear whether this is derived from cells or matrix (Pfeilschifter & Mundy 1987). IGF-1 might be a good candidate for the anabolic mediator because it is produced in bone and because growth hormone deficiency, which presumably impairs local production of IGF-1, also impairs the anabolic response to PTH (J.M. Hock, unpublished observations). PTH can also stimulate the production of PGE_2 in bone (Raisz & Simmons 1985) and, as noted below, PGE_2 may be an important local stimulator of bone formation. However, the anabolic response to PTH in young growing rats is not abrogated by high doses of indomethacin (Gera et al 1987) and the pattern of increased bone formation in response to PTH is quite different from that described after a chronic administration of PGE_2 (Raisz & Martin 1984).

While the anabolic effect of PTH is not yet fully understood, its importance has been emphasized by recent data showing an increase in bone mass in patients with osteoporosis treated intermittently with low doses of human synthetic 1–34 PTH given in conjunction with small amounts of $1,25\text{-}(OH)_2D_3$ (Slovik et al 1986). On the basis of these results, it is possible that PTH, which is normally secreted at low concentrations and intermittently, is a physiological stimulator of skeletal growth as suggested by Parsons (1976). A catabolic effect of increased resorption and decreased formation would only occur when there was a marked and sustained increase in PTH concentration, because of diminished calcium intake or absorption, or in primary hyperparathyroidism. This catabolic effect could be enhanced in older individuals by an age-related impairment of osteoblast renewal or function. Thus, with calcium deficiency the metabolic needs of the body would be met at the expense of a reduction in bone mass.

Oestradiol

In contrast to the data concerning PTH effects on bone, there is remarkably little known about the effects of oestradiol and other sex hormones on bone metabolism in mammals. The most important source of information has been clinical studies on oestrogen withdrawal and replacement (Selby et al 1985). At the menopause there is an increase in bone resorption accompanied by a smaller increase in bone formation so that bone mass decreases. The rates of these changes vary considerably at different skeletal sites, but both axial and appendicular skeletal sites are involved. In postmenopausal oestrogen-deficient women, administration of oestradiol can produce a rapid, albeit small, decrease in serum calcium and phosphate and in urinary calcium and hydroxyproline excretion which are taken as evidence of decreased bone resorption. Oestradiol can also decrease serum calcium in patients with primary hyperparathyroidism which supports the early concept that oestrogen acts by opposing PTH-stimulated bone resorption (Selby & Peacock 1986). However, there is no evidence for a direct interaction between these two hormones, and the hormonal specificity of this response is not established. In hyperparathyroid patients, steroids which are predominantly progestins or androgens, as well as oestrogens, can reduce serum calcium and urinary calcium and hydroxyproline excretion. On the other hand, it may be that these other steroids act only after conversion to oestrogens.

Although oestrogen receptors have not been identified in freshly isolated bone or bone cells, cultured bone cells and osteosarcoma cells with an osteoblast phenotype have been shown to contain oestrogen receptors and to show metabolic changes when these receptors are activated (Komm et al 1987, Eriksen et al 1987). These changes, however, have not been related to the known effects of oestrogen on bone metabolism. A reasonable alternative hypothesis is that oestrogen acts indirectly to alter the production of one or more of the local regulators of bone metabolism. Evidence for this has been obtained in the female rat which shows a similar response to humans of increased bone turnover upon oestrogen withdrawal. Calvarial bones removed from oophorectomized rats produce substantially more PGE_2 than bones from sham-operated controls and pretreatment with oestrogen reduces PGE_2 production (Feyen & Raisz 1987). We were unable to show a direct effect of oestrogen on prostaglandin production in cultured bone from older rats, but this may have been due to the poor viability of these tissues. Because PGE_2 can stimulate both bone resorption and formation, an increase in its production could explain the changes in bone turnover observed with oestrogen withdrawal at the menopause.

Glucocorticoids

There is a well-established association between glucocorticoid excess and osteoporosis (Raisz 1980). The major causes for bone loss with glucocorticoid excess are decreased calcium absorption in the intestine, leading to secondary hyperparathyroidism, and increased bone resorption and a direct inhibitory effect on bone formation. The effects on bone formation have been studied extensively in organ culture. Paradoxically, treatment with low concentrations of glucocorticoids for short periods of time can actually increase bone matrix synthesis as measured by the incorporation of proline into collagen. This increase is not dose-related and can be reversed by high concentrations of glucocorticoids. Moreover, it is transient and is followed by progressive reduction in collagen synthesis as well as DNA synthesis and DNA content of bone. This decrease is dose-related. Studies of the time course suggest that the initial response is a decrease in DNA synthesis in bone which is probably also accompanied by a decrease in the differentiation of precursors into osteoblasts (Chyun et al 1984). As a result, the osteoblasts which were active at the initiation of glucocorticoid treatment are not renewed and as they complete their cycle of matrix production, bone formation rapidly decreases. The decrease in bone formation is accompanied by a marked decrease in PGE_2 production by bone. The glucocorticoid effect appears to be opposite to the effects of PGE_2 on osteoblast progenitor cell replication and precursor differentiation. Moreover, in cultures of bones treated with glucocorticoids, the inhibitory effects can be reversed by adding PGE_2 (Chyun & Raisz 1984). However, the inhibitory effects of glucocorticoids on bone formation are much greater than those of potent nonsteroidal inhibitors of prostaglandin synthesis, suggesting that there are additional pathways for the steroid effect.

Glucocorticoids have complex effects on bone resorption. They do not block bone resorption in cultures that have been stimulated by PTH, but can block resorption in unstimulated cultures and can block the resorptive response that occurs during escape from calcitonin inhibition. There is also some evidence that glucocorticoids can produce direct enhancement of bone resorptive activity in some *in vitro* systems (Teitelbaum et al 1981). However, this has not yet been studied in isolated osteoclast systems.

Prostaglandin E$_2$

The first evidence for a role of prostaglandins in bone metabolism was the observation that PGE$_2$ could increase cyclic AMP production in bone and stimulate resorption. Although there was a close parallel between the ability to increase cyclic AMP production, presumably in cells of the osteoblast lineage, and the resorptive response with different prostanoids, the association between these two responses is no more firmly established than it is for PTH (Raisz & Martin 1984). Prostaglandins are somewhat different from PTH in their action on bone resorption, showing a slower response in fetal rat long bone cultures and a greater dependence on cell replication than does PTH (J.A. Lorenzo, unpublished observations). However, the major differences between PTH and PGE$_2$ lie in their effects on bone formation. Although PGE$_2$ can inhibit collagen synthesis in bone selectively, in a manner similar to PTH, this only occurs at concentrations one or two orders of magnitude higher than those which stimulate bone resorption. With PTH the resorptive effects and the inhibition of collagen synthesis generally occur at the same concentrations (Raisz et al 1979). Lower concentrations of PGE$_2$, which still resorb bone, appear to enhance bone formation.

The effect on bone formation is relatively slow and again appears to involve replication and differentiation of osteoblast progenitors (Chyun & Raisz 1984). The effect on resorption is also biphasic. High concentrations of PGE$_2$ produce an initial inhibitory effect on motility and resorption in isolated osteoclast systems (Chambers & Ali 1983). This effect is relatively transient and is followed by a resorptive response presumably involving cell recruitment (Holtrop & Raisz 1979). While the transient initial inhibition has been observed in organ cultures as well, the resorptive response is certainly the predominant one in these systems (Conaway et al 1986).

Interest in the effects of PGE$_2$ on bone was greatly heightened when it was shown that many resorptive responses to bone growth factors as well as pathological responses to complement and toxins could be mediated by local production of PGE$_2$. This production is highly regulated. In addition to the hormones already discussed, IL-1 and TGF-β, as well as other growth factors, have been shown to stimulate prostaglandin production in bone. Prostaglandin production appears to be increased by mechanical forces (Somjen et al 1980), and there is evidence that prostaglandins may mediate both the rapid increase in resorption that occurs with immobilization and the pathological ectopic bone formation that occurs after injury or with implantation of demineralized bone matrix (Weintroub et al 1983). In the presence of so many other local factors, it is difficult to assess the relative importance of PGE_2 as a local regulator, but it is one of the few such factors which can directly increase both resorption and formation. The role for PGE_2 in pathological bone loss at sites of inflammation is supported by a number of indirect observations, but direct evidence has been difficult to obtain. A major limitation has been the difficulty in measuring local prostaglandin concentrations. Removal of tissues usually results in sufficient damage to produce an artifactual increase in phospholipid breakdown and prostanoid synthesis. Moreover, even if we could measure the total prostanoid concentrations in tissue accurately, we might not know what the local concentrations were. Presumably these local concentrations would determine both the magnitude and the site of the cellular response to such autocrine or paracrine factors.

Conclusions

This selection of four agents which act in quite different ways on skeletal tissue is obviously arbitrary but does provide us not only with examples of the multiple pathways involved, but also with some idea of numerous potential sites for interaction. The fact that there are many more factors which can act and interact in skeletal tissue underscores the complexity of the regulation of bone turnover. Nevertheless, a beginning has been made in identifying these factors and studying their actions. This should lead to a greater understanding of physiological regulation of bone growth and turnover and the pathological mechanism responsible for impairment of skeletal structure and function.

References

Chambers TJ, Ali NN 1983 Inhibition of osteoclastic motility by prostaglandins I_2, E_1, E_2, and 6-oxo-E_1. J Pathol 139:383–397

Chyun YS, Raisz LG 1984 Stimulation of bone formation by prostaglandin E_2. Prostaglandins 27:97–103

Chyun YS, Kream BE, Raisz LG 1984 Cortisol decreases bone formation by inhibiting periosteal cell proliferation. Endocrinology 114:477–480

Conaway HH, Diez LF, Raisz LG 1986 Effects of prostacyclin and prostaglandin E_2 on bone resorption in the presence and absence of parathyroid hormone. Calcif Tissue Int 38:130–134

Eriksen EF, Berg NJ, Graham ML, Mann KG, Spelsberg TC, Riggs BL 1987 Evidence of estrogen receptors in human bone cells. J Bone Miner Res 2 (Suppl 1):abstr 238

Feyen JHM, Raisz LG 1987 Prostaglandin production by calvariae from sham operated and oophorectomized rats: effects of 17β-estradiol in vivo. Endocrinology 131:819–821

Gera I, Hock JM, Gunness-Hey M, Fonseca J, Raisz LG 1987 Indomethacin does not inhibit the anabolic effect of parathyroid hormone on the long bones of rats. Calcif Tissue Int 40:201–211

Gunness-Hey M, Hock JM 1984 Increased trabecular bone mass in rats treated with human synthetic parathyroid hormone. Metab Bone Dis Relat Res 5:177–181

Hamilton JA, Lingelbach S, Partridge NC, Martin TJ 1985 Regulation of plasminogen activator production by bone-resorbing hormones in normal and malignant osteoblasts. Endocrinology 116:2186–2191

Holtrop ME, Raisz LG 1979 Comparison of the effects of 1,25-dihydroxy cholecalciferol, prostaglandin E2 and osteoclast-activating factor with parathyroid hormone on the ultrastructure of osteoclasts in cultured long bones of fetal rats. Calcif Tissue Int 29:201–205

Howard GA, Bottemiller BL, Turner RT, Rader JI, Baylink DJ 1981 Parathyroid hormone stimulates bone formation and resorption in organ culture: evidence for a coupling mechanism. Proc Natl Acad Sci USA 78:3204–3208

Hummert J, Hock JM, Fonseca J, Raisz LG 1987 Resorption is not essential for the stimulation of bone growth by hPTH 1–34 in rats in vivo. J Bone Miner Res 1 (Suppl 1):Abstr 83

Komm BS, Sheetz L, Baker M, Gallegos A, O'Malley BW, Haussler MR 1987 Bone related cells in culture express putative estrogen receptor mRNA and [125]I-17β-estradiol binding. J Bone Miner Res 2 (Suppl 1):abstr 237

Lorenzo JA, Raisz LG, Hock JM 1983 DNA synthesis is not necessary for osteoclastic responses to parathyroid hormone in cultured fetal rat long bones. J Clin Invest 72:1924–1929

Lorenzo JA, Sousa SL, Quinton J 1986 Forskolin has both stimulatory and inhibitory effects on bone resorption in fetal rat long bone cultures. J Bone Miner Res 1:313–318

Löwik CWGM, van Leeuwen JPTM, van der Meer JM, van Zeeland JK, Scheven BAA, Hermann-Erlee MPM 1985 A two-receptor model for the action of parathyroid hormone on osteoblasts: a role for intracellular free calcium and cAMP. Cell Calcium 6:311–326

McSheehy PMJ, Chambers TJ 1987 1,25-Dihydroxyvitamin D_3 stimulates rat osteoblastic cells to release a soluble factor that increases osteoclastic bone resorption. J Clin Invest 80:425–429

Moseley TM, Kubota M, Diefenbach-Jagger H et al 1987 Parathyroid hormone-related protein purified from a human lung cancer cell line. Proc Natl Acad Sci USA 84:5048–5052

Parsons JA 1976 Parathyroid physiology and the skeleton. In: Bourne GJ (ed) Biochemistry and physiology of bone. Academic Press, New York, vol 4:159–225

Pfeilschifter J, Mundy GR 1987 Modulation of type β transforming growth factor activity in bone cultures by osteotropic hormones. Proc Natl Acad Sci USA 84:2024–2028

Raisz LG 1980 Effect of corticosteroids on calcium metabolism. Prog Biochem Pharmacol 17:212–219

Raisz LG, Kream BE 1983 Regulation of bone formation. N Engl J Med 309:29–35, 83–89

Raisz LG, Martin TJ 1984 Prostaglandins in bone and mineral metabolism. In: Peck WA (ed) Bone and mineral research, Annual 2. Excerpta Medica, Amsterdam, p 286–310

Raisz LG, Simmons HA 1985 Effects of parathyroid hormone and cortisol on prosta-
    glandin production by neonatal rat calvaria in vitro. Endocr Res 11:59–74
Raisz LG, Lorenzo JA, Gworek S, Kream BE, Rosenblatt M 1979 Comparison of the
    effects of a potent synthetic analog of bovine parathyroid hormone with native
    bPTH-(1–84) and synthetic bPTH-(1–34) on bone resorption and collagen synthesis.
    Calcif Tissue Int 29:215–218
Selby PL, Peacock M 1986 Ethinyl estradiol and norethindrome in the treatment of
    primary hyperparathyroidism in postmenopausal women. N Engl J Med 314:1481–
    1485
Selby PL, Peacock M, Barkworth SA, Brown WB, Taylor GA 1985 Early effects of
    ethinyloestradiol and norethisterone treatment in postmenopausal women on bone
    resorption and calcium regulating hormones. Clin Sci (Lond) 69:265–271
Slovik DM Rosenthal DI, Dopelt SH et al 1986 Restoration of spinal bone in
    osteoporotic men by treatment with human parathyroid hormone (1–34) and 1,25-
    dihydroxyvitamin D. J Bone Miner Res 1:377–381
Somjen D, Binderman I, Beyer E, Harell A 1980 Bone remodeling induced by
    physical stress is prostaglandin E_2 mediated. Biochim Biophys Acta 627:91–100
Tam C, Heersche J, Murray T, Parsons JA 1982 Parathyroid hormone stimulates the
    bone apposition rate independently of its resorptive action: differential effects of
    intermittent and continuous administration. Endocrinology 110:506–512
Teitelbaum SL, Malone JD, Kahn AJ 1981 Glucocorticoid enhancement of bone
    resorption by rat peritoneal macrophages in vitro. Endocrinology 108:795–799
Weintroub S, Wahl LM, Feuerstein N, Winter CC, Reddi AH 1983 Changes in tissue
    concentration of prostaglandins during endochondral bone differentiation. Biochem
    Biophys Res Commun 117:746–750

DISCUSSION

Martin: You mentioned that the prostaglandin E (PGE) resorption dose-response has a time course which is long and shallow unlike the parathyroid hormone (PTH) dose-response. That was in the context of experiments in which you were trying to demonstrate inhibition. What happens if you mix PGE and PTH and look at the time course?

Raisz: We observed the short PTH time course and no further effect of PGE_2. We investigated whether PGE_2 could have its anti-resorptive effect against an early effect of PTH, but we did not see that. We have only been able to demonstrate an inhibitory effect of PGE_2 on the osteoclastic activity in our system in the absence of PTH.

Nijweide: PTH stimulates proliferation of bone cells in several cell systems, especially chick, but also mouse and human. Therefore, I am not surprised that you find a stimulation of bone formation. The rat is the only species where PTH does not seem to stimulate osteoblast proliferation in culture. However, it is possible that you isolate different differentiation stages from the different species; in the rat perhaps a differentiation stage that differs in its sensitivity to the hormone.

Raisz: We've tried to test the hypothesis that PTH is anabolic by stimulating the replication of preosteoblasts. It's a perfectly logical suggestion which is probably correct but we have been unable to demonstrate it experimentally. The studies in chicken which demonstrated an anabolic effect (Howard et al 1981) were at resorptive doses. They saw resorption as well as bone formation. They have not tested whether inhibition of resorption will block the formative response. The other problem is that the anabolic model that we've looked at *in vivo* is in rat and that's where we get this spectacular effect. Therefore we have to show the mechanism in the same animal before we can decide what it is.

Nijweide: The anabolic effect *in vivo* is measured over a much longer period than is normally used in *in vitro* cell systems.

Raisz: But we are able to observe a clear-cut anabolic effect at six days *in vivo*. Tam et al (1987) have shown it at three or four days. We see some evidence of an anabolic effect *in vitro*, but we haven't been able to reproduce a large effect.

Owen: Is it true that some workers have observed an anabolic effect of parathyroid hormone on trabecular bone mass but not on cortical bone mass?

Raisz: PTH has a large effect on trabecular bone mass in human patients with osteoporosis but no effect on cortical bone (Slovik et al 1986). I don't think the other studies gave information on cortical bone mass.

Hanaoka: I asked earlier whether preosteoclasts have PTH receptors. If preosteoclasts or mononuclear osteoclasts have such a receptor, does the scheme for the interaction between the osteoblast and the osteoclast need to be modified?

Raisz: The evidence that PTH has no effect on the osteoclast lineage is not conclusive. In PTH-treated bone in organ culture there is an increase in the number of osteoclasts. There are many other cell types present in these systems. The fact that you can't stimulate the mature osteoclast with PTH is not relevant because bone resorption is stimulated by parathyroid hormone mainly by increasing the number of osteoclasts on the bone surface, although there may also be some activation of existing osteoclasts. Because we can't identify and isolate an osteoclast precursor from all the other cell types present, there's no way in which we could show that PTH is not acting on this cell.

Martin: There are now ways that you can do that. At some stage of the osteoclast precursor, while it is still mononuclear, you can identify calcitonin (CT) receptors, for example. Now that we can prepare biologically active labelled PTH-related protein it will be possible to use double label experiments to find out whether there are PTH receptors on cells that have developed a CT receptor.

Krane: But that assumes that the calcitonin receptor is unique to the osteoclasts.

Martin: You are not putting kidney in with your bone organ culture, and I know of no evidence for calcitonin receptors on lymphocytes that are likely to be in bone. However, no one parameter will give the complete answer; another criterion is the development of tartrate-resistant acid phosphatase (TRAP). That also is insufficient in itself.

Rodan: A problem may arise if you find receptors for PTH but not for calcitonin.

Raisz: I can't even accept that. If PTH stimulates a preosteoclast to replicate and/or differentiate and fuse, the acquisition of calcitonin receptors may come at a later stage than, or simultaneous with, the PTH response. If you get both PTH and CT receptors on the same cell, you could argue that maybe that's some other cell, such as a residual lymphocyte.

Rodan: But some late passage UMR 106 cells do have calcitonin and para-thyroid hormone receptors.

Martin: But they are malignant cells which are presumably dedifferentiated in late passage to a primitive phenotype. That is a totally different question.

Raisz: Certainly the approach that Elizabeth Burger and Peter Nijweide have taken (Burger et al 1984) might help us isolate a cell which has calcitonin receptors, PTH or PTH-related peptide receptors, and is TRAP-positive.

Testa: The expression of receptors in haemopoietic cells when they are unexpected has been called 'lineage infidelity'. An example is lymphoid-type markers in myeloid leukaemic cells (Smith et al 1983). I wonder whether that would be a suitable term for these cells.

Nijweide: The acid phosphatase stain will not help you because TRAP positivity occurs late in differentiation. Only mononuclear osteoclasts are TRAP-positive, not the early stages.

Martin: The most important regulators are local and the ones that we can speculate least wildly about are the prostaglandins. Dr Raisz and Dr Chambers have shown that PGs can do everything: they can resorb bone; perhaps stimulate formation; and inhibit resorption. Certain cells, perhaps in the osteoblast lineage, produce prostaglandins which are able to act back, for example on lining cells to promote resorption, and also act on precursors to promote differentiation to cells capable of formation. As Dr Chambers said, PGs act as fine local regulators by migrating from their presumed site of production in lining cells to osteoclasts where they inhibit their activity. I can cope with that concept of a labile local regulator. We need to fit any stimulator of resorption into the pattern of acting through lining cells.

Dr Raisz referred to the PTH-related protein. When we consider the lining cells or osteoblast-like cells mediating the resorbing action of PTH, prostaglan-dins and other factors, the observation that PTH-related protein is a less potent bone resorber than PTH itself throws us into a state of confusion. For the hormones that acted through cyclic AMP, the relative effects of elevated cyclic AMP in osteoblast-like cells paralleled their effects on promoting bone resorp-tion. That was one of the arguments to support the idea that resorption needed the intervention of lining cells (Rodan & Martin 1987). However, this PTH-like protein is appreciably more effective than PTH itself in increasing cyclic AMP in osteoblasts and yet it is much less effective in promoting bone resorption. This is the first major deviation from the rule, and it raises the possibility that the hormones that act through cyclic AMP need something else in order to programme the osteoblast-like cell to activate the osteoclast. Hermann-Erlee et al (1983) suggested that calcium shifts in the osteoblast-like cell may be necessary. We must carefully compare PTH and PTH-related peptide for other biochemical effects on the osteoblast by itself.

Raisz: This issue of the cyclic AMP is also brought out by the finding that 3–34 PTH peptide can stimulate bone resorption (Löwik et al 1985) without an effect on cyclic AMP. The 3–34 analogue, which does block PTH-stimulated cyclic AMP production, albeit not completely, doesn't block the PTH response on resorption. That's true for five out of six research groups that have attempted that study.

Martin: I used to place another interpretation on these experiments, saying that it was probably just spare receptors. Dr M. Hermann-Erlee will be absolutely delighted if I have to eat my words

Rodan: Mike Rosenblatt and his colleagues (Horiuchi et al 1987) found that in rat the *in vivo* effects of the 1–34 sequence of the PTH-related peptide were very similar to those of 1–34 human PTH.

Raisz: In our hands, human 1–34 PTH *in vivo* in the rat is a very weak calcium mobilizer.

Martin: In those experiments it is very difficult to exclude a renal contribution; once you start any resorption you have then got to add the renal contribution, even if you have restricted calcium entry.

Raisz: I have made an arbitary selection in my paper of factors that might act locally on bone resorption. There is evidence that interleukin 1 is produced locally, and it could be an endogenous bone resorber from bone cells, not from macrophages. It would act by two pathways: a direct prostaglandin-independent bone-resorbing pathway; and a prostaglandin-dependent one. So Nature is clearly providing multiple routes for bone resorption.

References

Burger EH, van der Meer JWM, Nijweide PJ 1984 Osteoclast formation from mono-nuclear phagocytes: role of bone forming cells. J Cell Biol 99:1901–1906

Hermann-Erlee MPM, Nijweide PJ, van der Meer JM, Ooms MAC 1983 Action of bPTH and PTH fragments on embryonic bone in vitro dissociation of the cyclic AMP and bone resorbing response. Calcif Tissue Int 35:70–77

Horiuchi N, Caulfield MP, Fisher JE et al 1987 Similarity of synthetic peptide from human tumor to parathyroid hormone *in vivo* and *in vitro*. Science (Wash DC) 238:1566–1568

Howard GA, Bottemiller BL, Turner RT, Raider JI, Baylink DJ 1981 Parathyroid hormone stimulates bone formation and resorption in oxygen culture: evidence for a coupling mechanism. Proc Natl Acad Sci USA 78:3204–3208

Löwik CWGM, van Leeuwen JPTM, van der Meer JM, van Zeeland JK, Scheven BAA, Hermann-Erlee MPM 1985 A two-receptor model for the action of parathyroid hormone on osteoblasts: a role for intracellular free calcium and cAMP. Cell Calcium 6:311–326

Rodan GA, Martin TJ 1981 Role of osteoblasts in hormonal control of bone resorption—a hypothesis. Calcif Tissue Int 33:349–351

Slovik DM, Rosenthal DI, Dopett SH et al 1986 Restoration of spinal bone in osteoporotic men by treatment with human parathyroid hormone (1–34) and 1,25-dihydroxyvitamin D. J Bone Miner Res 1:377–381

Smith LJ, Curtis JE, Messner HA et al 1983 Lineage infidelity in acute leukemia. Blood 61:1138–1145

Tam C, Heersche J, Murray T, Parsons JA 1987 Parathyroid hormone stimulates the bone apposition rate independently of its resorptive action: differential effects of intermittent and continuous administration. Endocrinology 110:506–512

Cytokines

Stephen M. Krane, Mary B. Goldring and Steven R. Goldring

Department of Medicine, Harvard Medical School and the Medical Services (Arthritis Unit), Massachusetts General Hospital, Boston, Massachusetts 02114, USA

Abstract. During physiological and pathological skeletal remodelling, immune cells and stromal fibroblasts near active bone-forming and bone-resorbing surfaces might modulate the functions of skeletal tissue cells. Osteoblasts, osteo-clasts and their progenitor cells are the probable direct targets of these effector cells (e.g. lymphocytes and monocytes) which act through direct contact or the release of soluble ligands (e.g. interleukin 1 or tumour necrosis factor, lympho-toxins, transforming growth factors). These cytokines bind to specific cellular receptors, resulting in changes in the form and function of the target bone cells and variable activation of genes coding for extracellular matrix proteins and proteinases which are responsible for remodelling the matrix. The synthesis and release of eicosanoids such as prostaglandins (e.g. PGE_2) are frequent associ-ated events. PGE_2, in turn, affects several functions of the skeletal tissue cells as well as the lymphocytes and monocytes in their environment. The mesenchymal cells may also be induced to release ligands such as colony-stimulating factors, other cellular products or hormones resulting in a system of feedback and amplification loops. The cellular responses are thus subject to multiple controls not only determined by these ligands acting on their respective receptors but also by the pathways of signal transduction and how they, in turn, are influenced by interactions with molecules within the cells.

1988 Cell and molecular biology of vertebrate hard tissues. Wiley, Chichester (Ciba Foundation Symposium 136) p 239–256

It is generally accepted that cellular interactions play a critical regulatory role not only in physiological tissue remodelling, but also in inflammatory states and malignancy. In normal and pathological skeletal remodelling the function of osteoblasts and osteoclasts are influenced by each other as well as by other cells in their immediate environment, such as stromal fibroblasts, lympho-cytes, monocytes and mast cells, all of which are organized in a unique manner in the skeletal tissues. Elucidation of the mechanisms by which these cells communicate has been a major goal of cell biology. Possible mechanisms include: (1) induction of specialized membrane structures which result in cell-to-cell channels for direct communication between the interior of cells (gap junctions); (2) production and deposition of an extracellular matrix; (3) production and release of soluble mediators by some cells which influence the replication, form and function of other cells.

'Lymphokine' was the term used to describe soluble mediators derived from lymphocytes that could function to recruit other cells and/or modify their function, thereby modulating the immune response. By 1982 the term lymphokine became 'a household word for immunologists' although it was then stated that 'the often predicted cytokine explosion has not really occurred' (Pick 1982). The implication of this statement was that, although many biological activities had been ascribed to secreted cell products, a relatively limited number of these products could account for more than one function. It soon became apparent that cells other than lymphocytes could produce such soluble mediators. The term monokine was later used to describe mediators such as interleukin 1 (IL-1) which is characterized by its amino acid and cDNA sequences and genomic arrangement (Lomedico et al 1984, Auron et al 1984, Clark et al 1986, Furutani et al 1986). Subsequently, cells other than lymphocytes were found to secrete products identical to known lymphokines, leading to the increasing use of 'cytokine' to replace the more restrictive term. 'Cytokine' also has advantages over other terms such as 'growth factor'.

Source and properties of cytokines

Cytokines are thus soluble products produced by some cells. They can influence the growth, morphology and other specific functions of the same or other cells in their local environment. One cytokine may under some conditions stimulate growth and under other conditions inhibit growth. This is illustrated by the example of transforming growth factor β (TGF-β), a 25 kDa disulphide-linked polypeptide homodimer originally found in transformed fibroblasts and recently detected in osteoblasts (Sporn et al 1986, Massaqué 1987, Robey et al 1987). *In vitro*, TGF-β can strongly inhibit the proliferation of normal and certain tumour-derived epithelial cell lines but, in contrast, it induces the proliferation of some mesenchymal cells. The proliferation of other cell types may not be affected at all. In addition, the phenotypic expression of cells with a differentiating potential may be profoundly altered by TGF-β. For example, TGF-β can block adipogenesis, myogenesis and haemapoiesis while it can promote chondrogenesis and epithelial cell differentiation. Thus, even the name transforming growth factor is correct only in selected circumstances.

In skeletal tissue, mesenchymal cells such as fibroblasts and osteoblasts are often regarded as the major target cells, and immune cells such as monocytes and lymphocytes act as effector cells which regulate the function of these stromal cells. The distinction between 'effector' and 'target' is not at all clear, however, since the 'target' cells may also profoundly influence functions of the effector cells. The action of the soluble ligands, i.e. the cytokines, in these cellular interactions is mediated by binding to specific surface receptors

(Dower &'Urdal 1987, Nicola 1987). Initially the existence of many of these ligands was postulated on the basis of a biological response usually assayed on cells in culture. It was then a reasonable assumption that a single polypeptide product could be responsible for a specific observed effect. Over the past several years, however, the purification, cloning and sequencing of many of these polypeptides have provided definitive evidence for their existence and permitted their distinction from each other.

It is now apparent that multiple polypeptides, rather than single products, often subserve similar functions. For example, monocyte/macrophages can synthesize and secrete three very different polypeptides with relatively little or no aminoacid sequence similarity. These include: interleukin 1α (IL-1α), interleukin 1β (IL-1β) and tumour necrosis factor α (TNF-α). These peptides have many similar effects on mesenchymal target cells. Although there are significant similarities between IL-1α and IL-1β based on amino acid sequence and nucleotide sequence and organization of the respective genes (Lomedico et al 1984, Auron et al 1984, March et al 1985, Dinarello et al 1986, Furutani et al 1986, Clark et al 1986), there are no significant similarities between these polypeptides and TNF-α (Beutler & Cerami 1986). Whereas IL-1α and IL-1β probably act through the same receptor (Kilian et al 1986, Chin et al 1987, Dower & Urdal 1987), TNF-β acts through a different receptor (Aggarwal et al 1985). On the other hand, TNF-α is very similar to a gene product of T lymphocytes, called lymphotoxin or TNF-β (Gray et al 1984), which acts through the same receptor as TNF-α and produces similar effects. Furthermore, the phenotype of all monocyte/macrophages is not identical and indeed monocyte/macrophages under some circumstances produce IL-1 and under other circumstances TNF-α. Old (1987) has recently emphasized that TNF-α, along with TNF-β, IL-1α and β and other polypeptide mediators, forms part of a network of interactive signals that 'orchestrate inflammatory and immunological events that regulate the production, for example, of hematopoietic stem cells and their descendants'. For example, IL-1 or TNF, acting on stromal fibroblasts, can induce the production of colony-stimulating factors which alter erythropoietic or myelopoietic differentiation (Zucali et al 1986). In an analogous fashion, IL-1, TNF-α or TNF-β can influence the differentiation of osteoclasts by acting on osteoblasts or related cells (Thomson et al 1986, 1987).

Most cytokines produced by cells such as lymphocytes and monocytes, including platelet-derived growth factor (PDGF) (Ross et al 1986), mediate effects on target cells at concentrations in the pM range. These effects vary in a manner dependent upon the type of target cell, its stage of differentiation and the presence of other ligands. Major attention has been directed at defining the mechanisms by which cytokines alter cell replication. In some instances, e.g. PDGF or insulin-derived growth factor (somatomedin C), the precise phase of the cell cycle at which these ligands function can readily be demonstrated. On the other hand, IL-1 which also stimulates fibroblast proliferation (Rupp et al 1986) is neither a competency nor a progression factor and functions to alter replication only in the presence of other growth factors such as those present in serum (Bitterman et al 1982). As well as affecting growth, cytokines influence other important physiological or pathological events. These actions include the ability to alter cellular metabolism (e.g. glycolysis), movement, chemotaxis, transmembrane transfer of critical ions, production of enzymes such as the metalloproteinases that are capable of degrading the extracellular matrix, production of components of the extracellular matrix and the release of soluble factors that affect other cell types in the vicinity. Some of the highly purified, but very dissimilar recombinant preparations of these ligands may exert all of these effects, even on the same target cell, depending upon the conditions under which these responses are assayed.

Some biological examples of cytokine action

Thus, in this interactive network of target and effector cells, products released by effector cells may have profound and diverse effects on target mesenchymal cells. This type of target–effector system is illustrated by the interactions of cells from the inflammatory synovial lesion of rheumatoid arthritis with connective tissue cells present in synovium, articular cartilage and bone. We had observed that co-cultivation of synovial fibroblasts with monocyte/macrophages stimulated the synthesis and release of collagenase and prostaglandin E_2 (PGE_2) (Krane et al 1982). Medium conditioned by monocytes was capable of producing such stimulation, suggesting that cell contact was not essential for these effects. Articular chondrocytes can function similarly as target cells (McGuire-Goldring et al 1984). Although these biological activities were first demonstrated in crude monocyte-conditioned medium, subsequent evidence suggested that these activities could be attributed to IL-1 (Mizel et al 1981). IL-1α and IL-1β are among the components synthesized and released by monocytes that could be responsible for the action of the soluble products present in the crude conditioned medium. Following the discovery and characterization of TNF-α, however, it was soon found that TNF-α could induce collagenase and PGE_2 production by synovial fibroblasts (Dayer et al 1985). Indeed, in preliminary studies we have found that either TNF-α or TNF-β act on a number of these mesenchymal target cells in a similar fashion to IL-1, even though they do not all act through the same receptors as IL-1 (E.P. Amento, M. Hayes, S.M. Krane, unpublished).

Other profound and diverse effects on target mesenchymal cells result from the action of products released by mononuclear cells. These effects include changes in shape, replication and synthesis of matrix proteins. The effects on cell replication, as previously discussed, have now been shown to be properties of the most highly purified preparations of IL-1 as well as recombinant DNA preparations (Rupp et al 1986). Monocyte-conditioned medium or co-cultivation with monocytes can also be shown to sensitize target cells to exogenous PGE_2 under circumstances in which endogenous PGE_2 synthesis is inhibited by indomethacin (Dayer et al 1979, Goldring et al 1984). Recombinant preparations of IL-1α or IL-1β as well as several other polypeptide ligands, reproduce these effects of monocyte-cell conditioned medium. Moreover, IL-1 and TNF-α or TNF-β act in a synergistic fashion in the augmentation of hormone-induced cyclic AMP responses. We have found a similar synergism in the augmentation by cytokines of the synthesis of collagenase and the synthesis of PGE_2. We and others have not been able to show that IL-1α or β or TNF-α alone *directly* activate adenylate cyclase in intact cells. Potential *indirect* effects could still be exerted through alternate signal transduction pathways. We have preliminary data suggesting that, although IL-1 does not directly affect intracellular cyclic AMP levels, it may modulate the activity of the catalytic unit of adenylate cyclase, possibly through effects mediated by nucleotide regulatory proteins. Whether arachidonic acid metabolites induced by IL-1 could act directly as second messengers, as has been suggested by Piomelli et al (1987) in the marine mollusc *Aplysia*, is yet to be shown. In these cultured mammalian cell systems it is difficult to distinguish actions that are definitely intracellular from effects occurring initially at the cell membrane.

Potential mechanisms of cytokine action

It is thus not clear how the changes induced by cytokines such as IL-1, TNF, or even TGF-β are mediated. As discussed previously, the first step in the action of these mediators is binding to specific high affinity receptors present on the surface of different target cells. Variations in the number or affinity of these receptors for the specific ligands might explain differences in the responses of some target cells compared to others. For example, the presence of a large number of IL-1 receptors is one explanation for the apparent greater sensitivity to effects of IL-1 of rheumatoid synovial fibroblasts compared to most dermal fibroblasts. Several of the actions of IL-1 may be exerted through effects on the phosphoinositol pathway and subsequent activation of protein kinase C. The observations that IL-1-induced increases in collagenase synthesis can be mimicked by incubation of target cells with phorbol ester (Brinckerhoff et al 1979, Aggeler et al 1984) support this possibility. On the other hand, actions of IL-1 on synthesis of some extracellular matrix components can be dissociated from those of phorbol ester. Alternate mechanisms by which these ligands might potentially act could involve the cyclic AMP–protein kinase A system or steps of the phosphoinositol–protein kinase C activation pathway that have yet to be elucidated. Mechanisms such as inactivation of regulatory proteins by phosphorylation (Yoshimasa et al 1987) or alterations in the content of the regulatory proteins themselves could be involved. Whatever the signals that result from interaction of the cytokines with surface receptors, ultimately they must act at the level of the promoters/enhancers that regulate transcription of the genes which appear to be activated by these ligands. Transcription of several genes might be important in skeletal remodelling. These genes include those for the metalloproteinases which degrade the extracellular matrix and those for components of the extracellular matrix such as the collagens, fibronectin and other bone matrix glycoproteins.

When the secretion of collagenase by cells such as synovial fibroblasts or articular chondrocytes is increased by incubation with medium conditioned by monocyte/macrophages, these increases can be accounted for by increased synthesis and not simply by release of preformed proteins (Harris et al 1984). This stimulation of collagenase synthesis by soluble mononuclear cell products or by purified or recombinant preparations of IL-1α or IL-1β is readily shown by experiments in which increased incorporation of labelled amino acids is measured by immunocomplexing of collagenase protein with specific antibodies (McCroskery et al 1985, Stephenson et al 1987). In preliminary experiments we have found that TNF-α or TNF-β also induce increases in synthesis of collagenase by synovial cells (E.P. Amento, M. Hayes and S.M. Krane, unpublished). The labelled procollagenase released into culture medium under these conditions is present as a doublet of approximately 56 and 60 kDa proteins. As first shown by Nagase et al (1983), the high molecular mass protein is a glycosylated form of the procollagenase: it is retained on lectin-Sepharose columns and its labelling is inhibited by tunicamycin.

The cDNA for the human fibroblast collagenase has now been cloned and sequenced (Goldberg et al 1986, Whitham et al 1986). The sequence is essentially identical to that of the cDNA for a protein whose synthesis is induced by phorbol ester (Herrlich et al 1986). It was previously mentioned that phorbol ester or IL-1 can induce procollagenase synthesis by target cells. The procollagenase is similar at the protein and DNA sequence level to two other metalloproteinases which must be involved in the degradation of the extracellular matrix components. One of these enzymes degrades denatured but not native collagen and is called gelatinase and the other also attacks gelatin (but not native collagen) in addition to proteins such as cartilage proteoglycan core protein, type IV collagen, laminin and fibronectin (Chin et al 1985, Okada et al 1986). The latter enzyme has been termed stromelysin or metalloproteinase-3. Stromelysin can be induced in fibroblasts following infection with polyoma virus or Rous sarcoma virus or after transfection with either the middle T oncogene or the cellular oncogene H-*ras* (Matrisian et al 1986, Whitham et al 1986, Goldberg et al 1986).

Recently, the promoter regions of several phorbol ester-inducible genes such as those of collagenase and stromelysin have been found to share a conserved DNA sequence motif of nine base pairs (Angel et al 1987a, 1987b). These sequences are similar to those important in activating transcription of other unrelated genes, e.g. the proenkephalin gene (Comb et al 1986). These elements are recognized by a common cellular protein, the so-called transcription factor AP-1. It has been suggested that this factor is at the receiving end of a complex pathway responsible for transmitting the effects of phorbol esters from the plasma membrane to the transcriptional apparatus. The cytokines discussed previously may also act in this manner. Recombinant IL-1 not only increases the synthesis of collagenase but also increases the levels of collagenase mRNA in target synovial fibroblasts and chondrocytes (Stephenson et al 1987). These increases in the levels of collagenase mRNA parallel the rates of collagenase labelling and activity.

On the basis of analogies with the other systems described above, it is therefore likely that IL-1, presumably acting through receptors on the cell membrane, induces the production of factors which also recognize the inducible elements in the promoter regions of the collagenase (and other metalloproteinase) genes and activates transcription of these genes. Observations that genes activated by tumour promoters, oncogenic viruses and cellular oncogenes are also activated by cytokines produced in chronic inflammatory lesions, such as rheumatoid arthritis or periodontal disease or infiltrates in malignant tumors, should provide further clues useful in unravelling aspects of the pathogenesis of these lesions.

Interaction of cells with their extracellular matrix can have profound effects on their form and function, including modulation of cell attachment, spreading, differentiation and replication. Some of these effects may be exerted through specific receptors for these matrix proteins located on the surface of cells. The best studied of these cell surface receptors are those for fibronectin and other glycoproteins which contain similar cellular recognition sites. Another potential function of the extracellular matrix is to bind cytokines, e.g. granulocyte macrophage colony-stimulating factor (GM-CSF) (Gordon et al 1987). This binding could result in selective compartmentalization of cytokines and thus have an important function in the microenvironment of the marrow where haemapoietic differentiation is taking place.

Cytokines may also act to modulate levels of synthesis of extracellular matrix components. For example, interferon γ (IFN-γ), a product predominantly of T lymphocytes, profoundly inhibits collagen synthesis in several systems associated with decreases in the steady-state levels of the procollagen mRNAs (Jiminez et al 1984, Rosenbloom et al 1984, Amento et al 1985, Stephenson et al 1985, Goldring et al 1986). Depending upon the target cells, the levels of types I, II and III collagen mRNAs are all suppressed by IFN-γ. Based on preliminary experiments performed in collaboration with Drs Pellegrino Rossi and Benoit de Crumbrugghe and Dr Linda Sandell, it is likely that this inhibitory effect is exerted at the level of the transcription of these collagen genes, although the precise mechanism is yet to be elucidated. Other cytokines stimulate the synthesis of various collagens depending on the target cell. For example, crude mononuclear cell-conditioned medium, which contains IL-1, stimulates the synthesis of types I and III collagens and fibronectin by synovial fibroblastic cells. This stimulation is most apparent when the ambient levels of prostaglandins (which suppress collagen synthesis) are decreased by incubation with the cyclooxygenase inhibitor, indomethacin (Krane et al 1985). Furthermore, collagen synthesis by cultured synovial or articular chondrocytes is increased by IL-1 (either IL-1α or IL-1β), but again only if cyclooxygenase activity is blocked and ambient levels of PGE$_2$ are low (Goldring & Krane 1987). When recombinant IL-1 is incubated with these cells in the absence of indomethacin the synthesis of types I and III collagen is decreased. In dermal fibroblasts, in which the PGE$_2$ response to IL-1 is relatively small, IL-1 increases collagen synthesis and levels of procollagen mRNAs even in the absence of indomethacin (Kähäri et al 1987, Goldring & Krane 1987). The levels of types I and III procollagen mRNAs in these cells parallel those of the rates of collagen synthesis. In preliminary experiments from our laboratory, TNF-α and TNF-β act in a similar manner. Effects of IL-1 on type II collagen synthesis appear to be even more complex. For example, in cultures of costal chondrocytes, IL-1 suppresses type II collagen synthesis, as well as the levels of procollagen $\alpha1(II)$ mRNA. In contrast to the effects of IL-1 on synthesis of types I and III collagens observed in cultures of synovial fibroblasts and articular chondrocytes, the addition of indomethacin does not overcome the IL-1-induced inhibition of type II collagen synthesis and in fact potentiates it. Phorbol ester mimics the effects on IL-1 on types I and III collagen synthesis but is distinct from IL-1 with respect to effects on type II collagen synthesis.

It is not yet known whether these actions of IL-1 on collagen synthesis are exerted at transcriptional or post-transcriptional levels, although we favour the former possibility. In the case of TGF-β, however, effects on stimulating synthesis of type I collagen in mouse NIH 3T3 fibroblasts are accounted for by transcriptional activation. This has been demonstrated directly for the $\alpha2(I)$ procollagen gene. The transcriptional activation appears to be mediated by a specific sequence in the promoter which is a binding site for nuclear factor 1 (Oikarinen et al 1987) and requires the presence of a cell-specific transcriptional enhancer which is present in the first intron of the mouse $\alpha2(I)$ procollagen gene. It is tempting to speculate that the actions of other cytokines might also be exerted through the production, modification or alteration of binding of *trans*-acting factors which interact with similar or different critical elements of the target genes and thereby function as activators.

There are many other complexities in the systems that we have studied which result from the associated cytokine receptor-mediated synthesis of eicosanoids such as PGE_2. The PGE_2 released by cytokine-stimulated cells binds to specific PGE_2 receptors on the surface of the target cells or that of neighbouring cells to initiate yet another series of signal transduction events. The intracellular signals could operate in the same direction as that which results from the initial interaction of the cytokine with its receptor. For example, phorbol ester activates the proenkephalin gene and increases proenkephalin mRNA levels. Agents which increase cellular content of cyclic AMP *enhance* the expression of this gene in concert with phorbol ester. The DNA sequences required for the regulation by both phorbol ester and cyclic AMP map to the same region 5' to the mRNA cap site of the proenkephalin gene (Comb et al 1986). The short *cis*-acting element which serves as an inducible enhancer is similar in sequence to that of other phorbol ester-enhanced genes such as collagenase and stromelysin (Angel et al 1987a, 1987b). In other instances, the signals induced by cytokines could operate in the *opposite* direction from those produced by a second mediator such as PGE_2. For example, whereas IL-1 itself enhances types I and III collagen synthesis, PGE_2, whose levels are stimulated by IL-1, suppresses the synthesis of these collagens, which also may be regulated at the level of gene transcription. Thus, in such a complicated network, effects of ligands such as IL-1 and other cytokines not only are transduced by specific signals following binding of these ligands to their receptors but are also subject to modulation resulting from the simultaneous stimulation of synthesis of PGE_2 which in turn acts on its own receptors to induce different signals. Furthermore, the cytokines may directly or indirectly affect the responses to the very ligands (e.g. PGE_2) whose synthesis they induce. These modulations could well result from 'cross talk' between the various signal transduction pathways (Yoshimasa et al 1987, Gawler et al 1987).

Cytokines and the skeleton

Several of these cytokines have been identified in bone and characterized following extraction and purification. Some of them may be involved in mediating pathological bone resorption or formation and may eventually be shown to have a role in physiological skeletal tissue remodelling. For example, in *in vitro* systems IL-1 and TNF induce bone resorption (Gowen et al 1983, Gowen & Mundy 1986, Bertolini et al 1986) and they elevate serum calcium concentrations when infused in animals. Colony-stimulating factors can induce the development of osteoclasts in bone *in vitro* (Lorenzo et al 1987). On the other hand INF-γ inhibits bone resorption (Gowen & Mundy 1986). TGF-β stimulates bone cell growth and collagen synthesis (Centrella et al 1987) and the activity of TGF-β in bone cultures is increased by IL-1 and hormones such as parathyroid hormone and 1,25-$(OH)_2$ vitamin D_3 (Pfeil-schifter & Mundy 1987). IL-1 may, under certain circumstances, increase the synthesis of collagen and other components of the extracellular matrix in bone and under other circumstances inhibit synthesis (Canalis 1986). These effects are probably mediated by IL-1 binding to cell surface receptors and then activating specific genes as described.

Even though these factors may act on target cells that are present in skeletal tissues and can be demonstrated in extracts of bone and cartilage or found in fluids draining these tissues, their precise role in physiological and pathological remodelling is unknown. It will be necessary to design rigorous experiments to establish the importance of any of these putative mediators in any particular cellular response *in vivo*. Identification of mutations in the genes coding for these mediators that would result in the absence of the gene product or a non-functional gene product, followed by correction of this defect by adding back the factor would be a definitive way of establishing the role of a specific cytokine. Methods might also be developed which would enable deletion of an individual bone cell type or a specific cellular product that would then provide the proof of functional importance. Characterization of the specific cellular defect(s) in animal or human disorders of the skeleton, such as osteogenesis imperfecta, and correction of the bone disease by selective cellular reconstitution by transplantation is possibly a promising approach. Another possibility which might be adapted to the study of cytokines arises from experiments in which acinar cells in the pancreas of transgenic mice were specifically ablated (Palmiter et al 1987). A chimeric gene was constructed which contained the pancreas-specific elastase I promoter/enhancer fused to a gene for the toxic diphtheria toxin A polypeptide. This was then micro-injected into the mouse germline permitting selective targeting of toxin gene expression. Using similar experimental designs, we may eventually be able to determine how the cytokines regulate the function and proliferation of cells in tissues such as the skeleton. It may then be possible for us to understand how each factor acts in a complex cellular network filled with amplification and inhibition loops.

Acknowledgements

Original work discussed here was supported by USPHS grants AR-03564, AR-34390 and AR-07258 and a grant from MedChem Products, Inc. We thank Michele Angelo for preparation of the manuscript.

References

Aggarwal BB, Eessalu TE, Hass PE 1985 Characterization of receptors for human tumour necrosis factor and their regulation by γ-interferon. Nature (Lond) 318:665–667
Aggeler J, Frisch SM, Werb Z 1984 Collagenase is a major gene product of induced rabbit synovial fibroblasts. J Cell Biol 98:1656–1661
Amento EP, Bhan AK, McCullagh KG, Krane SM 1985 Influences of gamma inter-

feron on synovial fibroblast-like cells. Ia. Induction and inhibition of collagen synthesis. J Clin Invest 76:837–848

Angel P, Baumann I, Stein B, Delius H, Rahmsdorf HJ, Herrlich P 1987a 12-O-tetradecanoyl-phorbol-13-acetate induction of the human collagenase gene is mediated by an inducible enhancer element located in the 5′-flanking region. Mol Cell Biol 7:2256–2266

Angel P, Imagawa M, Chiu R et al 1987b Phorbol ester-inducible genes contain a common *cis* element recognized by a TPA-modulated *trans*-acting factor. Cell 49:729–739

Auron PE, Webb AC, Rosenwasser LJ et al 1984 Nucleotide sequence of human monocyte interleukin 1 precursor cDNA. Proc Natl Acad Sci USA 81:7907–7911

Bertolini DR, Nedwin GE, Bringman TS, Smith DD, Mundy GR 1986 Stimulation of bone resorption and inhibition of bone formation *in vitro* by human tumour necrosis factors. Nature (Lond) 319:516–518

Beutler B, Cerami A 1986 Cachectin and tumour necrosis factor as two sides of the same biological coin. Nature (Lond) 320:584–588

Bitterman PB, Rennard SI, Hunninghake GW, Crystal RG 1982 Human alveolar macrophage growth factor for fibroblasts. Regulation and partial characterization. J Clin Invest 70:806–822

Brinckerhoff CE, McMillan RM, Fahey JV, Harris ED Jr 1979 Collagenase production by synovial fibroblasts treated with phorbol myristate acetate. Arthritis Rheum 22:1109–1115

Canalis E 1986 Interleukin-1 has independent effects on deoxyribonucleic acid and collagen synthesis in cultures of rat calvariae. Endocrinology 118:74–81

Centrella M, McCarthy TL, Canalis E 1987 Transforming growth factor β is a bifunctional regulator of replication and collagen synthesis in osteoblast-enriched cell cultures from fetal rat bone. J Biol Chem 262:2869–2874

Chin J, Cameron PM, Rupp E, Schmidt JA 1987 Identification of a high-affinity receptor for native human interleukin 1β and interleukin 1α on normal human lung fibroblasts. J Exp Med 165:70–86

Chin JR, Murphy G, Webb Z 1985 Stromelysin, a connective tissue-degrading metalloendopeptidase secreted by stimulated rabbit synovial fibroblasts in parallel with collagenase. Biosynthesis, isolation, characterization, and substrates. J Biol Chem 260:12367–12376

Clark BD, Collins KL, Gandy MS, Webb AC, Auron PE 1986 Genomic sequence for human prointerleukin 1 beta: possible evolution from a reverse transcribed prointerleukin 1 alpha gene. Nucl Acids Res 14:7897–7914

•Comb M, Birnberg NC, Seasholtz A, Herbert E, Goodman HM 1986 A cyclic AMP-and phorbol ester-inducible DNA element. Nature (Lond) 232:353–356

Dayer J-M, Goldring SR, Robinson DR, Krane SM 1979 Effects of human mononuclear cell factor on cultured rheumatoid synovial cells. Interactions of prostaglandin E_2 and cyclic adenosine 3′,5′-monophosphate. Biochim Biophys Acta 586:87–105

Dayer J-M, Beutler B, Cerami A 1985 Cachectin/tumor necrosis factor stimulates collagenase and prostaglandin E_2 production by human synovial cells and dermal fibroblasts. J Exp Med 162:2163–2168

Dinarello CA, Cannon JG, Mier JW et al 1986 Multiple biological activities of human recombinant interleukin 1. J Clin Invest 77:1734–1739

Dower SK, Urdal DL 1987 The interleukin-1 receptor. Immunol Today 8:46–51

Furutani Y, Notake M, Fukui T et al 1986 Complete nucleotide sequence of the gene for human interleukin 1 alpha. Nucl Acids Res 14:3167–3179

Gawler D, Milligan G, Spiegel AM, Unson CG, Houslay MD 1987 Abolition of the expression of inhibitory guanine nucleotide regulatory protein G_i activity in diabetes. Nature (Lond) 327:229–232

Goldberg GI, Wilhelm SM, Kronberger A, Bauer EA, Grant GA, Eisen AZ 1986 Human fibroblast collagenase. Complete primary structure and homology to an oncogene transformation-induced rat protein. J Biol Chem 261:6600–6605

Goldring MB, Krane SM 1987 Modulation by recombinant interleukin 1 of synthesis of types I and III collagens and associated procollagen mRNA levels in cultured human cells. J Biol Chem 262:16724–16729

Goldring MB, Sandell LJ, Stephenson ML, Krane SM 1986 Immune interferon suppresses levels of procollagen mRNA and type II collagen synthesis in cultured human articular and costal chondrocytes. J Biol Chem 261:9049–9056

Goldring SR, Dayer J-M, Krane SM 1984 Rheumatoid synovial cell hormone responses modulated by cell-cell interactions. Inflammation 8:107–121

Gordon MY, Riley GP, Watt SM, Greaves MF 1987 Compartmentalization of a haematopoietic growth factor (GM-CSF) by glycosaminoglycans in the bone marrow microenvironment. Nature (Lond) 326:403–405

Gowen M, Mundy GR 1986 Actions of recombinant interleukin 1, interleukin 2, and interferon-γ on bone resorption in vitro. J Immunol 136:2478–2482

Gowen M, Wood DD, Ihrie EJ, McGuire MKB, Russell RGG 1983 An interleukin 1 like factor stimulates bone resorption in vitro. Nature (Lond) 306:378–380

Gray PW, Aggarwal BB, Benton CV et al 1984 Cloning and expression of cDNA for human lymphotoxin, a lymphokine with tumour necrosis activity. Nature (Lond) 312:721–724

Harris ED Jr, Welgus HG, Krane SM 1984 Regulation of the mammalian collagenases. Collagen Relat Res 4:493–512

Herrlich P, Angel P, Rahmsdorf HJ et al 1986 The mammalian genetic stress response. Adv Enzyme Regul 25:485–504

Jimenez SA, Freundlich B, Rosenbloom J 1984 Selective inhibition of human diploid fibroblast collagen synthesis by interferons. J Clin Invest 74:1112–1116

Kähäri V-M, Heino J, Vuorio E 1987 Interleukin-1 increases collagen production and mRNA levels in cultured skin fibroblasts. Biochim Biophys Acta 929:142–147

Kilian PL, Kaffka KL, Stern AS et al 1986 Interleukin 1α and interleukin 1β bind to the same receptor on T cells. J Immunol 136:4509–4514

Krane SM, Goldring SR, Dayer J-M 1982 Interactions among lymphocytes, monocytes, and other synovial cells in the rheumatoid synovium. In: Pick E, Mizel S (eds) Lymphokines. Academic Press, New York & London, vol 7:75–136

Krane SM, Dayer J-M, Simon LS, Byrne MS 1985 Mononuclear cell-conditioned medium containing mononuclear cell factor (MCF), homologous with interleukin 1, stimulates collagen and fibronectin synthesis by adherent rheumatoid synovial cells: effects of prostaglandin E_2 and indomethacin. Collagen Relat Res 5:99–117

Lomedico PT, Gubler U, Hellman CP et al 1984 Cloning and expression of murine interleukin-1 cDNA in *Escherichia coli*. Nature (Lond) 312:458–462

Lorenzo JA, Sousa SL, Fonseca JM, Hock JM, Medlock ES 1987 Colony-stimulating factors regulate the development of multinucleated osteoclasts from recently replicated cells in vitro. J Clin Invest 80:160–164

March CJ, Mosley B, Larsen A et al 1985 Cloning, sequence and expression of two distinct human interleukin-1 complementary DNAs. Nature (Lond) 315:641–647

Massaqué J 1987 The TGF-β family of growth and differentiation factors. Cell 49:437–438

Matrisian LM, Leroy P, Ruhlmann C, Gesnel M-C, Breathnach R 1986 Isolation of the oncogene and epidermal growth factor-induced transin gene: complex control in rat fibroblasts. Mol Cell Biol 6:1679–1686

McCroskery PA, Arai S, Amento EP, Krane SM 1985 Stimulation of procollagenase synthesis in human rheumatoid synovial fibroblasts by mononuclear cell factor/ interleukin 1. FEBS (Fed Eur Biochem Soc) Lett 191:7–12

McGuire-Goldring MB, Meats JE, Wood DD, Ihrie EJ, Ebsworth NM, Russell RGG 1984 In vitro activation of human chondrocytes and synoviocytes by a human interleukin-1-like factor. Arthritis Rheum 27:654–662

Mizel SB, Dayer J-M, Krane SM, Mergenhagen SE 1981 Stimulation of rheumatoid synovial cell collagenase and prostaglandin production by partially purified lymphocyte-activating factor (interleukin 1). Proc Natl Acad Sci USA 78:2474–2477

Nagase H, Brinckerhoff CE, Vater CA, Harris ED Jr 1983 Biosynthesis and secretion of procollagense by rabbit synovial fibroblasts. Inhibition of procollagenase secretion by monensin and evidence for glycosylation of procollagenase. Biochem J 214:281–288

Nicola NA 1987 Why do hemopoietic growth factor receptors interact with each other? Immunol Today 8:134–140

Oikarinen J, Hatamochi A, de Crombrugghe B 1987 Separate binding sites for nuclear factor 1 and a CCAAT DNA binding factor in the mouse α_2(I) collagen promoter. J Biol Chem 262:1064–1070

Okada Y, Nagase H, Harris ED Jr 1986 A metalloproteinase from human rheumatoid synovial fibroblasts that digests connective tissue matrix components. J Biol Chem 261:14245–14255

Old LJ 1987 Polypeptide mediator network. Nature (Lond) 326:330–331

Palmiter RD, Behringer RR, Quaife CJ, Maxwell F, Maxwell IH, Brinster RL 1987 Cell lineage ablation in transgenic mice by cell-specific expression of a toxin gene. Cell 50:435–443

Pfeilschifter J, Mundy GR 1987 Modulation of type β transforming growth factor activity in bone cultures by osteotropic hormones. Proc Natl Acad Sci USA 84:2024–2028

Pick E 1982 Preface. In: Pick E, Mizel S (eds) Lymphokines. Academic Press, New York & London, vol 7:xi–xiv

Piomelli D, Volterra A, Dale N et al 1987 Lipoxygenase metabolites of arachidonic acid as second messengers for presynaptic inhibition of Aplysia sensory cells. Nature (Lond) 328:38–43

Robey PG, Young MF, Flanders KC et al 1987 Osteoblasts synthesize and respond to transforming growth factor-type β (TGF-β) in vitro. J Cell Biol 105:457–463

Rosenbloom J, Feldman G, Freundlich B, Jimenez SA 1984 Transcriptional control of human diploid fibroblast collagen synthesis by γ-interferon. Biochem Biophys Res Commun 123:365–372

Ross R, Raines EW, Bowen-Pope DF 1986 The biology of platelet-derived growth factor. Cell 46:155–169

Rupp EA, Cameron PM, Ranawat CS, Schmidt JA, Bayne EK 1986 Specific bioactivities of monocyte-derived interleukin 1α and interleukin 1β are similar to each other on cultured murine thymocytes and on cultured human connective tissue cells. J Clin Invest 78:836–839

Sporn MB, Roberts AB, Wakefield LM, Assoian RK 1986 Transforming growth factor-β: biological function and chemical structure. Science (Wash DC) 233:532–534

Stephenson ML, Krane SM, Amento EP, McCroskery PA, Byrne M 1985 Immune interferon inhibits collagen synthesis by rheumatoid synovial cells associated with decreased levels of the procollagen mRNAs. FEBS (Fed Eur Biochem Soc) Lett 180:43–50

Stephenson ML, Goldring MB, Birkhead JR, Krane SM, Rahmsdorf HJ, Angel P 1987 Stimulation of procollagenase synthesis parallels increases in cellular procollagenase mRNA in human articular chondrocytes exposed to recombinant interleukin 1β or phorbol ester. Biochem Biophys Res Commun 144:583–590

Thomson BM, Saklatvala J, Chambers TJ 1986 Osteoblasts mediate interleukin 1 stimulation of bone resorption by rat osteoclasts. J Exp Med 164:104–112

Thomson BM, Mundy GR, Chambers TJ 1987 Tumor necrosis factors α and β induce osteoblastic cells to stimulate osteoclastic bone resorption. J Immunol 138:775–779

Whitham SE, Murphy G, Angel P et al 1986 Comparison of human stromelysin and collagenase by cloning and sequence analysis. Biochem J 240:913–916

Yoshimasa T, Sibley DR, Bouvier M, Lefkowitz RJ, Caron MG 1987 Cross-talk between cellular signalling pathways suggested by phorbol-ester-induced adenylate cyclase phosphorylation. Nature (Lond) 327:67–70

Zucali JR, Dinarello CA, Oblon DJ, Gross MA, Anderson L, Weiner RS 1986 Interleukin 1 stimulates fibroblasts to produce granulocyte-macrophage colony-stimulating activity and prostaglandin E_2. J Clin Invest 77:1857–1863

DISCUSSION

Raisz: Is there any down-regulation of the interleukin 1 (IL-1) response when you go for longer periods of time? You went up to 21 days and it looked as though it was reaching a plateau.

Krane: We could still find the IL-1 response in cells that were incubated with IL-1 continuously for three weeks. The indomethacin effect goes away because all the prostaglandin receptors are down-regulated.

Raisz: Is that because the prostaglandin receptors or the prostaglandin production pathways are down-regulated?

Krane: It's the prostaglandin receptors. We still see large amounts of prostaglandin present.

Caplan: The conclusion is very complicated. You are careful to state that you don't know if the same cell is producing effector molecules and then responding to them. In chondrocyte long-term cultures, a phenotypic transition can occur as the chondrocytes become more fibroblastic. Therefore, the cell shape changes and the conformation of the membrane receptors and the sensitivity to cytokines might also be expected to change. The relationship between type I and type II collagen may be an indicator of the number of cells in the culture which maintain a vigorous chondrogenic phenotype.

The point is well taken about the multiple signals in the differential response and I think your model of two genes responding to the same signal in a completely different way is absolutely correct. It is going to be difficult to manipulate this in a therapeutic or diagnostic sense.

Krane: Yes, that's true, but even when the cells are making practically no type I collagen in these cultures, a short term incubation with IL-1 turns off synthesis of type II collagen and induces the expression of type I collagen.

Caplan: Are we changing the cell's phenotype into a down-regulated chondrocyte which is going to produce more type I collagen?

Rodan: Do you see a common mechanism for the change in production of type I collagen in chondrocytes in continuous culture and the changes in IL-1 and PGE_2?

Krane: Yes, and this made us wonder whether there could be transformation of chondrocytes into non-chondrogenic cells. The stimulus for this might be, for example, a ligand introduced from the vascular system during development. This is partly analogous to Dana Giulian's suggestion that in the central nervous system the cells that make IL-1 are the ameboid microglia (Giulian et al 1986 a,b). There is correlation between the appearance of astroglial proliferation and the production of IL-1 and other mediators in the developing brain.

Caplan: We must distinguish between a developmental transition and a physiological change, because the inherent mechanisms are different. In a physiological change, alterations in the shape of the chondrocyte may cause a change in the balance of molecules synthesized, but if the stimulus is removed the cell will revert to its original phenotypic state. In a developmental change, a mesenchymal cell differentiates into a chondrocyte which causes a change in the spectrum of molecules that are synthesized; this change is not reversible when the stimulus is removed. These different processes will not be regulated in the same way.

Prockop: I don't know why it is necessary to introduce this complex terminology. You seem just to be talking about reversible or non-reversible changes.

Raisz: Is that what Dr Caplan is talking about? Many phenotypes that appear to be permanent are reversed under disease conditions.

Caplan: I think that's what you have to be careful about.

Martin: Dr Krane, you blocked prostaglandin production in target cells and the addition of IL-1 restored the synthesis of PGE_2. Can you restore the production with phosphodiesterase inhibitor or forskolin?

Krane: We have never tried to do that. It is possible that this effect is not specific and we could produce it with some other ligand that could also induce cyclic AMP in the cell.

Martin: Does γ interferon (IFN-γ) block the synergism or individual action of tumour necrosis factor α (TNF-α) and IL-1?

Krane: Yes, IFN-γ blocks the collagen stimulation that we induce with IL-1 or TNF. We don't see synergism with IFN-γ.

Termine: You probably do have a permanent change in phenotype because the type IX and type III collagen genes also changed.

Glowacki: Are the chondrocyte cultures done at low or high density?

Krane: These are mostly confluent cultures. For some reason or other, these human cells stay like looking like chondrocytes for weeks in culture on plastic. This differs from other species. The cells are polygonal, even though they are mostly confluent. When they are incubated with IL-1, for example, then you do see a shape change: they become more fibroblastic.

Glowacki: What about proliferation of the flattened cells? Do they undergo division?

Krane: Yes, they divide. We have not seen major effects of the ligands on replication under these circumstances, but under other circumstances we do see them. The effects vary according to the cells we use. In synovial fibroblasts, IL-1 suppresses replication, but you can reverse this with indomethacin. In other cells, such as dermal fibroblasts or human bone cells, we mostly see an increase in replication; but this is not a direct growth effect of IL-1 because IL-1 requires other growth factors which are provided by the serum. IL-1 is neither a competency nor progression factor (Bittermann et al 1982). That is, IL-1 doesn't either move the cells into G1 from G0 or move them along.

Prockop: What is the time course of your effect?

Krane: We are trying to measure the kinetics. In the experiments so far we see the stimulation by IL-1 at the earliest time points because the prostaglandins haven't been stimulated yet. As soon as the prostaglandins appear you see the inhibitory effects on type I collagen production. We haven't enough data yet on the type II. We see these effects within three or four hours.

Osdoby: Have you tried growing these cells on different substrates and seeing if you get the same kind of response?

Krane: No, but this is obviously the thing to do. We tried to keep chondrocytes from adhering because that's the best way that others have found to maintain their phenotype: either you grow them in suspension culture on agarose, or try to grow them on some other matrix.

Hauschka: Was the monocyte-conditioned medium in all your experiments essentially identical in its effects to recombinant IL-1, or were there some differences?

Krane: There have been differences. Every time we make monocyte-conditioned medium we use a different source. Therefore, we have a different mixture which may contain, for example, different amounts of TNF and IL-1, and different amounts of IFN-γ. Because these mixtures are not absolutely free of lymphocytes and other ligands, we observe plus/minus effects. That's where much of the confusion arises. Perhaps the crude mixtures reflect real life and the pure ligands are artificial!

Hauschka: Have similar experiments been done with osteoblasts to see whether there are any changes in collagenase activity?

Krane: We've not tried any human osteoblasts. We have tried so-called bone cells. They have some features similar to those of synovial fibroblasts but we have not yet looked at the collagenase.

Fleisch: The complexity of the situation with so many interrelated factors and cells involved makes endocrinology look like Kindergarten! What approach would you take in the future to unravel the interrelationships? Would you take the *in vitro* approach or would you go, as you suggested, to *in vivo* systems using genetic engineering technology?

Krane: To make antibodies and to get them to the site is far too complex to attempt. The drug effects observed so far are almost impossible to interpret. I think that the best approach is going to be the genetic engineering approach *in vivo*. We have been doing some experiments with R. Jaenisch that are designed to tell us whether the collagenase is important in bone resorption. We have made a mutation that codes for substitution of isoleucine for proline at the collagenase cleavage site and we rescued the *Mov 13/Mov 13* homozygous cells with this gene.

Our preliminary results indicate that in those cultured cells the normal α2 chain is rescued in a similar way to that in the report by Schnieke et al (1987). We found that a triple helical type I collagen is secreted but it is not cleaved by rheumatoid synovial collagenase. We plan to inject this clone into the germline to produce a type I collagen *in vivo* that can't be cleaved by collagenase.

Prockop: The transgenic mice experiments are very exciting, but I would suggest that we take a simpler approach. The use of diphtheria toxin to wipe out cells will have very widespread effects. I would like us to investigate more straightforward cases, such as over-production of these factors. This can be easily done with growth hormone (Palmiter et al 1983).

Krane: That is one succesful approach. We are faced with enormous complexity; we don't yet know which cytokine is important *in vivo*.

Raisz: Where does IFN-γ work against IL-1? Is it a generalized effect so that none of the IL-1 responses occur?

Krane: IFN-γ does not block all IL-1 responses. For example, it does not block the prostaglandin induction. While turning on some genes, it also turns off a large number of genes. Mary Goldring has done preliminary experiments with Pellegrino Rossi and Benoit de Crombrugghe and it appears that the increment in collagen gene expression induced by transforming growth factor β is blocked by IFN-γ. Presumably, if TGF-β acts by inducing nuclear factor 1 which binds to a site in the promoter of the α1(I) or α2(I) chain (Rossi et al 1988) then IFN-γ must interfere either with the production of nuclear factor 1 or its binding at the nuclear factor 1 site on the promoter.

References

Bitterman PB, Renard SI, Hunninglake GW, Crystal RG 1982 Human alveolar macrophage growth-factor for fibroblasts—regulation and partial characterization. J Clin Invest 70:806–822

Giulian D, Allen RL, Baker TJ, Tomazawa Y 1986a Brain peptides and glial growth. I. Glia-promoting factors as regulators of gliogenesis in the developing and injured central nervous system. J Cell Biol 102:803–811

Giulian D, Baker TJ, Shih L-CN, Lachman LB 1986b Interleukin 1 of the central nervous system is produced by ameboid microglia. J Exp Med 164:594–604

Palmiter RD, Norstedt G, Gelinas RE, Hammer RE, Brinster RL 1983 Metallothio-
nein human GH fusion genes stimulate growth of mice. Science 222:809–814

Rossi P, Karsenty G, Roberts AB, Roche NS, Sporn MB, de Crombrugghe B 1988 A
nuclear factor 1 binding site mediates the transcriptional activation of a type I
collagen promoter by transforming growth factor-β. Cell 52:405–414

Schnieke A, Dziadek M, Bateman J et al 1987 Introduction of the human pro-alpha 1(I)
collagen gene into pro-alpha 1(I)-deficient *mov*-13 mouse cells leads to formation of
functional mouse human hybrid type I collagen. Proc Natl Acad Sci USA
84:764–768

Haemopoietic growth factors: their relevance in osteoclast formation and function

N.G. Testa, T.D. Allen, G. Molineux, B.I. Lord and D. Onions*

*Paterson Institute for Cancer Research, Christie Hospital & Holt Radium Institute, Withington, Manchester M20 9BX, UK and *Veterinary School, The University, Bearsden Road, Bearsden, Glasgow G61 1QH, UK*

Abstract. The major recent advance in our knowledge of the haemopoietic system has been the purification and characterization of a family of haemopoietic growth factors, and their availability in recombinant form. In the bone marrow the sequences of differentiation and proliferation leading to the production of mature cells that these factors regulate may be determined by the relative availability of the factors in microenvironmental domains. The observation that growth factor-producing cells and haemopoietic progenitor cells are not evenly distributed in the bone marrow leads us to expect that the overall effect of growth factors (and other regulatory molecules) on the production and function of macrophages and osteoclasts may differ when *in vivo* or *in vitro* assays are used as end-points and, in the latter case, when whole marrow or purified cell populations are tested. The availability of an *in vitro* assay in which osteoclast-like cells are generated will allow these concepts to be tested.

1988 Cell and molecular biology of vertebrate hard tissues. Wiley, Chichester (Ciba Foundation Symposium 136) p 257–274

Colony-stimulating factors

The colony-stimulating factors (CSFs) which regulate the survival, proliferation and differentiation of haemopoietic cells have been purified and, recently, molecularly cloned. This subject has been reviewed (Metcalf 1984, Nicola 1987). All CSFs stimulate undifferentiated haemopoietic cells which, if cultured in permissive *in vitro* conditions, will proliferate and produce a colony of differentiated progeny. The predominant lineage of the colony cells will be determined by the type of CSF used: granulocyte-CSF (G-CSF) stimulates the formation of colonies composed of neutrophilic granulocytes; macrophage-CSF (M-CSF, also called CSF-1) induces colonies which contain monocytes and macrophages; GM-CSF stimulates colonies which contain neutrophils, eosinophils, monocytes and macrophages; and interleukin 3 (IL-3, also known as multi-CSF) will induce the production of all those cell

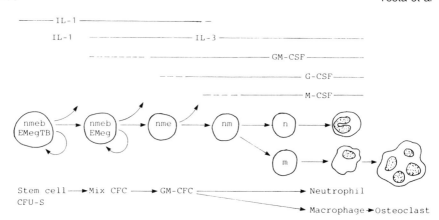

FIG. 1.   Range of action of growth factors showing the overlap in the target populations of haemopoietic cells at progressive stages of differentiation along the neutrophilic and macrophage lineages. The potential for differentiation at each stage is indicated within the cell diagrams. T, T cell; B, B cell; E, erythroid; Meg, megakaryocyte; e, eosinophil; n, neutrophil; m, macrophage; b, basophil; CFU-S, colony-forming unit-spleen (multipotential cells in mice). At least GM-CSF and M-CSF stimulate the formation of osteoclast-like cells *in vitro*.

types and, in addition, megakaryocytes and mast cells. Furthermore, if either of the latter two factors is used in conjunction with erythropoietin, erythroid cells will also develop.

There are, however, some concentration-dependent differences in the cell lineages observed in the colonies stimulated by the different CSFs: although G-CSF and M-CSF are generally lineage specific, at high concentrations they will stimulate the formation of macrophages and granulocytes respectively. GM-CSF and IL-3 will induce eosinophilic, megakaryocytic and erythroid cell production at higher concentrations than those needed for the neutrophilic and macrophage lineages (reviewed by Nicola 1987). These data imply, firstly, that there is some degree (which may be considerable) of overlap in the progenitor cell populations which respond to the different CSFs (Fig. 1) and secondly, that there may be receptors for more than one CSF in the same cell. It is known that CSFs do not compete directly for the same binding sites, and that there are specific receptors for each CSF (Table 1). Furthermore, there is a hierarchical organization in the cell lineages which have receptors for the different CSFs. Within a lineage, the numbers of receptors is related to the state of maturation of the cells (Table 1). Thus IL-3 receptors are found in several myeloid lineages in fairly low numbers per cell (but a small proportion of bone marrow cells have several thousands per cell). Receptors for GM-CSF are also found in the lineages which are also positive for IL-3 receptors. Interestingly, the numbers of receptors per cell may decrease with cell maturation (reviewed by Nicola 1987). Receptors for G-CSF are found in all

TABLE 1 Receptors for colony-stimulating factors in murine cells

Receptor for	Relative molecular mass	Found in	No per cell	Changes in number	References
IL-3	60 000 75 000	Neutrophilic, eosinophilic, monocytic lineages	<1000[a]	Decrease with cell maturation	Nicola & Metcalf 1986, Park et al 1986
GM-CSF	51 000	As above	<1000	As above	Walker & Burgess 1985, Nicola & Metcalf 1986, Nicola 1987
G-CSF	150 000	Neutrophilic lineage Monocyte precursors	<1000 <100	Increase with cell maturation	Nicola & Paterson 1986
M-CSF	165 000	Monocytes, macrophages	~50 000	Increase with cell maturation	Morgan & Stanley 1984, Rettenmier et al 1986

[a] <1% of bone marrow cells had several thousand receptors per cell.

TABLE 2   Direct (D) and indirect (I) effects of cytokines on the neutrophilic and macrophage lineages

Cytokine	Stem cells	Neutrophilic lineage	Macrophage lineage
IL-1	D	I	I
IL-3	D	D	D
GM-CSF	D	D	D
G-CSF	I	D	Small effect
M-CSF	I?	Small effect	D
TNF	I	D,I	I,D

the cells within the neutrophilic lineage but, in contrast to the observations described above, the numbers of receptors per cell increase with cell maturation in murine (but not in human) cells. Very small numbers are also found in promonocytes and monocytes. Receptors for M-CSF have been found in all the cells of the monocyte-macrophage lineage, in numbers which increase markedly with cell maturation (Table 1).

It is therefore apparent that: (a) cells of the neutrophilic and monocyte-macrophage lineages express receptors for the four types of CSFs; (b) eosinophils express only IL-3 and GM-CSF receptors; (c) receptors for G-CSF are restricted (with the exception of some mononuclear cells) to the neutrophilic lineage; and (d) receptors for M-CSF are restricted to the monocyte-macrophage lineage. The latter two types of CSF (M and G) are those known to induce the expression of several specific functions of mature macrophages and neutrophils respectively (Sklar et al 1985, Platzer et al 1985).

The CSFs do not compete directly with each other for binding sites. However, some can change the response to other CSFs by modulation of their receptors in bone marrow cells: IL-3 may transmodulate GM-CSF and M-CSF receptors, and GM-CSF the receptors for M-CSF. In contrast, G-CSF and M-CSF will primarily down-modulate their own receptors. Here again we see a hierarchical organization which is reminiscent of the relative specificity for different cell lineages shown by the different CSFs, and which may be relevant in determining their differentiation- or proliferation-inducing activities, with IL-3 and GM-CSF having a more important role in inducing proliferation and G-CSF and M-CSF in the induction of differentiation (Nicola 1987). These concepts highlight the complex regulation which maintains a stable output of mature cells in a normal steady state, and which is able to show a prompt, flexible and adequate response to haemopoietic injury.

It is important to differentiate between direct and indirect effects of growth factors on target cell populations. For example, G-CSF stimulates the formation of mixed colonies which contain several myeloid lineages through indirect effects on regulatory populations in the bone marrow, but, as described above, stimulates only the formation of neutrophils when acting on purified populations of colony-forming cells (Souza et al 1986). Other factors, which do not induce colony formation *per se*, may act indirectly in any of several ways. For example, interleukin 1 (IL-1) can potentiate the response to CSFs by inducing stem cells to become sensitive to their action (Moore et al 1987a). There is a complex network of interacting cytokines. These cytokines include the CSFs, IL-1, IL-2 and tumour necrosis factor (TNF) that regulate haemopoiesis either directly, by interacting with target cells, or indirectly, by modulating the production of other cytokines (Fig. 2 and Table 2). Some cytokines may do both: TNF inhibits the effect of G-CSF on neutrophil production, but stimulates the production of GM-CSF (Moore et al 1987b).

Structure of the haemopoietic tissue

The regulatory networks illustrated in Fig. 2 emphasize the importance of cell interactions in haemopoiesis. The growth factors produced within the bone marrow may exert their effect through cell–cell contact (Dexter et al 1984) or may be bound to specific components of the extracellular matrix (Gordon et al 1987). Those two characteristics may define regulatory domains. In this context, the realization that the bone marrow has a well-defined tissue structure is important for understanding its regulation. The data summarized in Table 3 indicate that stem cells, haemopoietic progenitors and at least some types of regulatory cells are not homogeneously distributed in the femoral

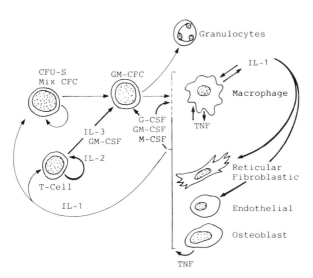

FIG. 2.   Regulatory networks in haemopoiesis acting on the neutrophilic and macrophage lineages. Mix-CFC and GM-CFC, haemopoietic progenitors; CSF, colony-stimulating factors; CFU-S, colony-forming unit-spleen (multipotential cells assayed *in vivo* in mice); M, macrophage; G, granulocyte; GM, granulocyte-macrophage; IL, interleukin; TNF, tumour necrosis factor.

TABLE 3   Incidence of haemopoietic and regulatory cells in the mouse femoral cavity

Cell type	Peak incidence (μm from bone)	Lowest incidence
Early stem cells	400–450	0–100
Late stem cells	0–100	400–450
Multipotential progenitor cells (Mix-CFC)	50	
Granulocyte-macrophage progenitors (GM-CFC)	150–200	300–450 and 0–50
Early erythroid progenitors (BFU-E)	200	0–100
Late erythroid progenitors (CFU-E)	400–450	0–100
Macrophages producing stem cell inhibitory factor	250–450	0–150
Macrophages producing stem cell stimulatory factor	0–100	250–450

Data taken from Lord & Testa 1988. The radius of the femoral cavity is approximately 450 μm.

marrow in the mouse. The collected data are compatible with the concept that differentiation from stem cells takes place preferentially in areas close to the bone surface, where the late stem cells concentrate and where they are found to be in active cell cycle (reviewed by Lord & Testa 1988). The concentration of early progenitor cells is higher in the vicinity of this area but as they mature and differentiate they are found progressively closer to the centre of the bone. This gives an orderly spatial and temporal sequence of maturation and differentiation. Data obtained from other murine bones and from human ribs support the concept that a defined tissue structure is likely to be a general feature of active bone marrow.

The study of osteoclasts in vitro

The use of culture techniques has allowed much progress in the understanding of the ontogeny and function of osteoclasts (reviewed by Mundy & Roodman 1987). These techniques can be subdivided in three groups.

(a) The isolation of fresh osteoclasts from the bone surface has been useful in the study of the phenotypic and functional characteristics of the harvested cells. The major potential problem is the viability of the cells kept *in vitro*, which may restrict their usefulness to short-term experiments. However, a large amount of information, such as that on the response of osteoclasts to osteotropic hormones (Chambers & Magnus 1982), has been produced by these techniques.

(b) Bone organ culture systems have been used successfully for identifying

TABLE 4   **Percentage of cells positive for Fc receptors in bone marrow cultures which generate osteoclast-like cells**

Days of culture	Spherical monocytes	Flattened monocytes	Osteoclast-like cells
24	82	53	8
51	94	46	17
82	87	61	22

Feline bone marrow cultures were established as described by Testa et al (1981).

factors that stimulate or inhibit bone resorption, but are not useful for distinguishing direct from indirect effects. The latter, as described for the CSFs, are relevant for the unravelling of regulatory networks.

(c) Cell culture techniques using cell suspensions of bone marrow tissue allow the study of the ontogeny, formation and function of multinucleated osteoclast-like cells. Our report (Testa et al 1981) established that such cells could be generated for several weeks in cultures of feline bone marrow. The cells had ultrastructural features characteristic of osteoclasts, including a clear zone and a rudimentary ruffled border. Similar cells have been further characterized in cultures of feline bone marrow by other workers (Ibbotson et al 1984) as well as in cultures of baboon, human and rabbit bone marrow (Roodman et al 1985, MacDonald et al 1987, Fuller & Chambers 1987), and have fulfilled several criteria which support their characterization as osteoclasts, including response to osteotropic hormones and calcium resorption from bone.

These cultures can be used to follow the development and differentiation of osteoclast precursors. The results in Table 4 show that mononuclear cells which maintain a more spherical shape even when attached to the culture surface have Fc receptors. When these cells flatten (a step we have always observed to be necessary before cell fusion can occur; Allen et al 1981) the percentage of cells with Fc receptors decreases. Finally, only a few multinucleated cells manifest these receptors. This is compatible with the concept that we can distinguish the immediate precursor of the multinucleated cell. These precursor cells are still phagocytic: studies in which glutaraldehyde-fixed red cells were added to cultures for 60 minutes showed that phagocytic cells had fused to the periphery of multinucleated cells, but the cytoplasm of the recently fused mononuclear cells could still be recognized by the localization of the phagocytosed red cells. This recent fusion may explain the small percentage of multinucleated cells still showing Fc receptors (Table 4).

When fetal rat calvaria labelled with ^{45}Ca was added to feline bone marrow cultures, release of ^{45}Ca to the culture medium indicated the presence of bone-resorbing cells (Fig. 3). Calcium resorption has also been demonstrated by Ibbotson et al (1984).

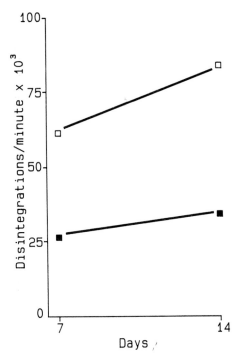

FIG. 3.   Feline bone marrow cultures were maintained for 10 days, as described by Testa et al (1981). Fetal rat calvariae, labelled *in vitro* by injecting an 18-day pregnant rat with 0.2 mCi $^{45}CaCl_2$ one day previously, were dissected and maintained in culture medium overnight and then were added to the cultures (□) or to control flasks containing culture medium only (■). The amount of ^{45}Ca released into a 2 ml aliquot of the medium was measured. These cultures were not stimulated exogenously to increase Ca release.

   The multinucleated cells are remarkably mobile in cultures. Some show a flattened roughly circular shape but, using time-lapse video, one can see waves of cytoplasmic movement which result in sliding motion. This phenomenon is similar to that observed by Gaillard (1955) in osteoclasts on the bone surface. Motile osteoclasts may show lobulated and convoluted cytoplasmic edges (Allen et al 1981), which change shape and size very actively (Fig. 4). These motile cells became quiescent after the addition of calcitonin (50 ng per ml) to the culture (Fig. 5), and all cytoplasmic movement ceases. The changes observed under time-lapse video are strikingly similar to those described by Chambers & Magnus (1982) when calcitonin was added to harvested osteoclasts kept *in vitro*. A comparison of several parameters found in fresh osteoclasts and in osteoclast-like cells generated in culture is shown in Table 5. These results validate the classification of the latter as osteoclasts and encourages the extension of studies designed to investigate the role of

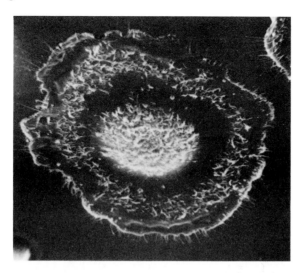

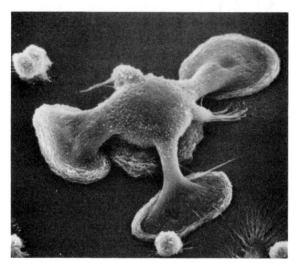

FIG. 4.   Scanning electron micrographs of osteoclast-like cells in a six-week tissue culture from cat bone marrow. (a) Typical large cell (150 μm in diameter) showing firm peripheral attachment and surface ruffles some distance from the cell periphery, and over the central region. This size of cell would be likely to contain 30–50 nuclei. (b)   Another large cell of similar size to that in (a), showing typical cytoplasmic extensions associated with very active movement when viewed in time-lapse video. Such extensions may be 50 μm in diameter, and may be extended and resorbed over distances of 200 to 300 μm at speeds of 10–15 μm/min. Active peripheral ruffling is apparent at the edges of the cytoplasmic extensions, and ventral membrane ruffling is seen in each extension. Reduced activity after exposure of this type of cell to calcitonin leads to the assumption of an appearance similar to that in (a).

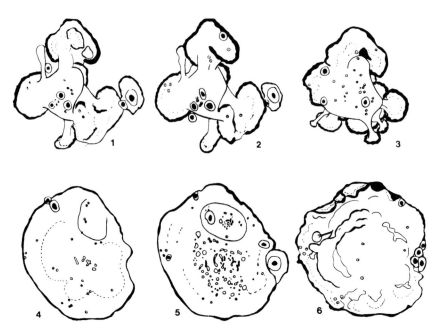

FIG. 5. Series of tracings from time-lapse video record of osteoclast before, during and after the addition of calcitonin to a culture. Frame 1. (Time zero). Active osteoclast, showing many cytoplasmic extensions, and active ventral membrane ruffling (indicated by dotted lines). Frame 2. (Time 6 min). Addition of calcitonin (CT). Frame 3. (Time 4 h 15 min). Beginning of withdrawal of cell process, leading to the shape shown in Frame 4. Frame 4. (Time 6 h 25 min). Assumption of typical 'resting' osteoclast outline; considerably reduced ventral membrane ruffling, which subsequently stops completely. Frame 5. (Time 8 h). Resumption of ventral membrane ruffling. Frame 6. (Time 15 h). Active ventral membrane ruffling and reformation of cytoplasmic extensions resumes.

CSFs and other cytokines in the regulation of osteoclast formation and function. GM-CSF and M-CSF enhance the formation of these cells in culture (MacDonald et al 1986), and the addition of 1,25-dihydroxyvitamin D_3 enhanced their effect, which was the result of increased proliferation of osteoclast precursors, and of increased cell fusion. An opposite effect, decreased proliferation and inhibition of cell fusion, is seen after treatment with steroids (Suda et al 1983). Whether M-CSF, which induces several mature cell functions in macrophages, is able to stimulate bone resorption awaits investigation. The effect of other cytokines on bone resorption are described elsewhere (Krane, this volume) but it would be of interest to compare their effects *in vivo*, or on isolated osteoclast-like cells in *in vitro* systems where the effects of controlled additions of characterized cell populations may be investigated. In particular, control of osteoclast generation and control of cell

TABLE 5 Characteristics of freshly harvested osteoclasts and osteoclast-like cells generated in culture

Parameter	Osteoclasts harvested from bone[a]	Osteoclast-like cells originated in bone marrow culture[b]	References
Acid phosphatase, Tartrate-resistant acid phosphatase	+	+	Testa et al 1981, Ibbotson et al 1984
Fc receptors	–	–	Table 3 (this paper)
Phagocytosis	–	–	Suda & Testa, unpublished[c]
Calcium release from labelled bone	+	+	Fig. 3, Ibbotson et al 1984
Calcitonin response	+[b]	+	Fig. 5, MacDonald et al 1987
Resorption lacunae	+	+	Fuller & Chambers 1987

[a] From the review by Mundy & Roodman 1987.   [b] From Chambers & Magnus 1982.   [c] Phagocytosis was assayed by incubating bone marrow cultures with glutaraldehyde-fixed red cells for 60 min.

functions are now amenable to study, and the direct versus indirect actions of regulators can begin to be unravelled.

Acknowledgements

N.G. Testa, T.D. Allen, G. Molineux and B.I. Lord are supported by the Cancer Research Campaign of Great Britain. N.G. Testa is grateful to Professor T.M. Dexter for suggesting her participation in this symposium.

References

Allen TD, Testa NG, Suda T et al 1981 The production of putative osteoclasts in tissue culture. Ultrastructure, formation and behaviour. In: Johari O et al (eds) Scanning electron microscopy. Scanning Electron Microscopy Inc, AMF O'Hare, Chicago, vol III:347–355

Chambers TJ, Magnus CJ 1982 Calcitonin alters behaviour of isolated osteoclasts. J Pathol 136:27–40

Dexter TM, Spooncer E, Simmons P, Allen TD 1984 Long-term marrow cultures: an overview of techniques and experience. In: Wright DG, Greenberger JS (eds) Long-term bone marrow culture. A R Liss, New York, p 57–96

Fuller K, Chambers TJ 1987 Generation of osteoclasts in culture of rabbit bone marrow and spleen cells. J Cell Physiol 132:441–452

Gaillard PJ 1955 Parathyroid gland and bone in vitro. Dev Biol 1:152–181

Gordon MY, Riley TP, Watt SM, Greaves MF 1987 Compartmentalization of haematopoietic growth factor (GM-CSF) by glycosaminoglycans in the bone marrow environment. Nature (Lond) 326:403–405

Ibbotson KJ, Roodman GD, McManus LM, Mundy GR 1984 Identification and characterization of osteoclast-like cells and their progenitors in cultures of feline marrow mononuclear cells. J Cell Biol 99:471–480

Krane SM, Goldring MB, Goldring SR 1988 Cytokines. In: Cell and molecular biology of vertebrate hard tissues. Wiley, Chichester (Ciba Found Symp 136) p 239–256

Lord BI, Testa NG 1988 The hemopoietic system, structure and regulation. In: Testa NG, Gale RP (eds) Hematopoiesis: long-term effects of chemotherapy and radiation. Marcel Dekker, New York, p 1–26

MacDonald BR, Mundy GR, Clark S et al 1986 Effects of human recombinant CSF-GM and highly purified cells with osteoclast characteristics in long-term bone marrow cultures. J Bone Miner Res 1:227–233

MacDonald BE, Takahashi N, McManus LM, Holahan S, Mundy GR, Roodman GD 1987 Formation of multinucleated cells that respond to osteotropic hormones in long term human bone marrow cultures. Endocrinology 120:2326–2333

Metcalf D 1984 The hemopoietic colony stimulating factors. Elsevier, Amsterdam

Moore MAS, Waven DJ, Souza LM 1987a In vivo and in vitro action of G-CSF and IL-1 in myelosuppressed mice. J Cell Biochem, in press

Moore MAS, Welte K, Gabrilove J, Souza LM 1987b Biological activities of human granulocyte colony stimulating factor (rh G-CSF) and tumor necrosis factor: in vivo and in vitro analysis. Haemat Blood Transf 31:210–220

Morgan CJ, Stanley ER 1984 Chemical crosslinking of the mononuclear phagocyte specific growth factor CSF-1 to its receptor at the cell surface. Biochem Biophys Res Commun 119:35–41

Mundy GR, Roodman GD 1987 Osteoclast ontogeny and function. In: Peck WA (ed)

Bone and mineral research. Elsevier, Amsterdam, vol 5:209–280

Nicola NA 1987 Why do hemopoietic growth factor receptors interact with each other? Immunol Today 8:134–140

Nicola NA, Metcalf D 1986 Binding of iodinated multipotential colony-stimulating factor (interleukin-3) to murine bone marrow cells. J Cell Physiol 128:180–188

Nicola NA, Peterson L 1986 Identification of distinct receptors for two hemopoietic growth factors (granulocyte colony-stimulating factor and multipotential colony-stimulating factor) by chemical cross linking. J Biol Chem 261:12384–12389

Park LS, Friend D, Gillis S, Urdal DL 1986 Characterization of the cell surface receptor for a multi-lineage colony stimulating factor (CSF-2). J Biol Chem 261:205–210

Platzer E, Welte K, Gabrilove J 1985 Biological activities of a human pluripotent hemopoietic colony stimulating factor on normal and leukemic cells. J Exp Med 162:1788–1801

Rettenmier CW, Sacca R, Furman WL et al 1986 Expression of the human c-fms proto-oncogene product (colony-stimulating factor-1 receptor) on peripheral blood mononuclear cells and choriocarcinoma cell lines. J Clin Invest 77:1740–1746

Roodman GD, Ibbotson KJ, MacDonald BR, Kuehl TJ, Mundy GR 1985 1,25-Dihydroxyvitamin D_3 causes formation of multinucleated cells with several osteoclast characteristic in cultures of primate marrow. Proc Natl Acad Sci USA 82:8213–8217

Sklar MD, Tereba A, Chen BD, Walker WS 1985 Transformation of mouse bone marrow cells by transfection with a human oncogene related to c-myc is associated with the endogenous production of macrophage colony stimulating factor. J Cell Physiol 125:403–408

Souza LM, Boone TC, Gabrilove J et al 1986 Recombinant human granulocyte colony-stimulating factor: effects on normal and leukemic myeloid cells. Science (Wash DC) 232:61–65

Suda T, Testa NG, Allen TD, Onions D, Jarret O 1983 Effect of hydrocortisone on osteoclasts generated in cat bone marrow cultures. Calcif Tissue Int 35:82–86

Testa NG, Allen TD, Lajtha LG, Onions D, Jarret O 1981 Generation of osteoclasts in vitro. J Cell Sci 44:127–137

Walker F, Burgess AW 1985 Specific binding of radioiodinated granulocyte-macrophage colony-stimulating factor to hemopoietic cells. EMBO (Eur Mol Biol Organ) J 4:933–939

DISCUSSION

Chambers: I am not happy about your identification of multinucleate cells as osteoclasts. Multinuclearity is a very good guide for osteoclastic phenotype only if you take cells directly from bone—there's nothing else present that is multinucleate. However, when you grow cells in culture macrophages very commonly become multinucleate. Moreover, there are some problems with the data that you invoked to distinguish your cells from multinucleate macrophages. While tartrate-resistant acid phosphatase (TRAP) is quite useful, in fresh tissue, to distinguish osteoclasts from macrophages, it's been shown by several groups that TRAP develops in macrophages in culture. It is also known that macrophages gradually lose the Fc receptors and phagocytic capacity in culture. The calcitonin (CT) responsiveness you showed is difficult to interpret from drawings. What's missing is bone resorption. When you first described these cells five years ago, Alan Boyd found that they were not resorptive: that is very strong evidence against their being osteoclasts.

Testa: I am not postulating that each and every one of the cells is an osteoclast. What I am saying is that this is a system in which osteoclasts are present. As you yourself have shown in a similar system, the cells do resorb bone (Chambers, this volume). Your own data show that there are osteoclasts present in the system. Like all phenotypic characterizations, it is wrong to put too much stress on any single parameter that you study. If you study 20 different parameters and every one fits, then the data becomes more convincing.

Dr M. Horton (personal communication) is using the monoclonal antibodies specific for osteoclasts that you referred to. They do react with some of the multinucleated cells in both his and our cultures. In our system, the mononuclear cells lose neither the Fc receptors in culture, even over a period of several weeks (Table 4), nor the phagocytic capacity. In all the marrow cultures, the cells we produce are normal cells, but may not be quite as normal as the harvested cells. This is to be expected. In erythroid cells you see megaloblastic changes, when you see cells which are undoubtedly erythroid but with very little haemoglobin. The differences are a matter of degree. I quite accept that the multinucleated cells may not be quite happy as osteoclasts. These cells haven't seen bone in their lives! But when taken together the results suggest that this system is useful for study of progeny, and, with any luck, function.

Martin: Direct demonstration of calcitonin receptors would be more convincing than any data you have presented.

Testa: Absolutely! We have already demonstrated that they respond to calcitonin (Fig. 5).

Raisz: In this discussion between you and Professor Chambers, it seems that what needs to be done is to quantitate the number of these polykaryons which cause resorption pits from your system. Has that been done?

Chambers: I think that the point about these cultures in general is that they contain macrophages, and, if you are lucky, osteoclasts. Macrophages can be mononuclear or multinucleate, and osteoclasts can be mononuclear or multinucleate. So why count osteoclasts as multinucleate cells? You might just as well count osteoclasts as the mononuclear cells. You need another criterion: excavation pits as evidence of resorption. Alan Boyd did those studies, but saw no resorption.

Testa: As I said, I am not using the multinuclearity as a criterion. It is what started me off in trying to investigate what these cells were. It bothered us that these cells did not resorb dentine.

Chambers: Osteoclasts resorb dentine very happily.

Testa: Dentine is not a physiological substratum for osteoclasts. Besides, these were unstimulated cultures. Perhaps you need to stimulate osteoclasts in culture.

Chambers: You don't need to stimulate osteoclasts for them to resorb mineralized tissue *in vitro*.

Testa: Perhaps not harvested osteoclasts, but certainly you stimulate those generated in culture.

Nijweide: I am always worried when I see a marrow culture in which most of the nuclei are supposed to be osteoclasts. In marrow and bone the number of osteoclast nuclei that are formed *in vivo* by differentiation is very small compared, for example, to the number of macrophage nuclei. We shall not be able to come to a definite conclusion until we are sure that every cell that looks like an osteoclast really can resorb bone.

Testa: These are very difficult experiments. Dr Raisz asked whether I had related the number of cells to the number of resorption pits. I am not even sure that one osteoclast will produce one resorption pit. I don't know whether all the cells will be functional at the same time, or whether there are refractory stages for the cells. It's very nice to have osteoclast resorption pits in cultures, but to quantify them is much more difficult.

Baron: After many years of working on osteoclasts I still don't know whether the giant cells formed from macrophages differ from osteoclasts. We should beware of blindly following dogmas. How strong is the evidence that giant cells formed from macrophages are different from osteoclasts? How strong is the evidence that all osteoclasts *in vitro* can make a resorption pit on a slice of dead bone? Do we know if osteoclasts reach a particular number of nuclei and then stop resorbing? I still question whether macrophage fusion leads to osteoclasts or not. It would be a mistake to consider this a closed issue.

Testa: Macrophages are the most protei-form cells in the body and what they do will depend largely on the circumstances in which they find themselves. I'm not sure which differences in phenotype between giant cells and osteoclasts are critical. I agree with Dr Baron that this is an open question.

Chambers: I do not think that it is so open. We find that macrophages and macrophage giant cells don't resorb bone. I have never seen an excavation beneath multinucleate macrophages. Osteoclasts resorb bone within hours, whereas cells derived from monocytes or from monocytic cell lines don't resorb bone after prolonged periods of incubation. If we grow bone marrow cultures without vitamin D, we see the same cells that Dr Testa has described—macrophages and multinucleate cells. If there's no vitamin D, as in her cultures, we never see excavations. Osteoclasts don't need vitamin D to resorb bone; they do it spontaneously. That shows me that multinuclearity is no guide to osteoclastic phenotype.

Testa: I think we are getting into a circular argument: you have just stated that osteoclasts generated in culture do need stimulation to resorb bone.

Freshly harvested osteoclasts do not. Similarly, we cannot use the argument that antibodies which have been postulated to be specific for osteoclasts react with cells in culture in a non-specific way, or that the loss of Fc receptors, or of phagocytic ability are non-specific, 'untrue' phenotype characteristics when the argument suits us. I think the generation of osteoclasts and the induction of a functional state in them may well be separate events. There are other examples of induction of macrophage functions in culture, independently of the generation of the cells themselves.

Osdoby: Dr Krane (this volume) stated that different cells behave differently in different environments and in response to different cells. If we don't listen to that we shall be conducting this argument for the next 20 years! Cells can be expressing different proteins in different functional states. We need to come to some consensus on phenotypic criteria and the conditions that these criteria are assayed in. This is important if cells respond to different environments in different ways depending on the assay conditions and the individuals conducting the assay.

Urist: What happened to the ultrastructural criterion in which the centriole was found to be in the very centre of an osteoclast (Matthews et al 1967), whereas in a foreign body giant cells the centriole is anywhere in the cell?

Chambers: In the foreign body giant cell that's correct. However, in the Langerhans giant cell, as seen in many pathological conditions, which is a more organized version of the same thing, the centrosomes are central again. It is a matter of the degree of specialization of the macrophages. I think you can talk about different functional states and different phenotypic markers in the osteoclast. However, ultimately osteoclasts are cells that resorb bone, and that is the only reliable criterion.

Baron: We all agree on that. However, the data can be interpreted in several different ways. Perhaps the cells that you call giant multinucleate cells in your cultures have, at that stage, gone beyond whatever they could do. Perhaps they needed 1,25-$(OH)_2D_3$ the first day but not the second. There is accumulating troubling evidence that this is more complex than was thought. Some other cells under the right conditions might do the job of an osteoclast.

Chambers: I've seen no evidence that macrophages can be induced to osteoclastic differentiation.

Fleisch: To obtain a macrophage lineage do you need first IL-1 and then M-CSF or will M-CSF alone be sufficient?

Testa: M-CSF will be sufficient in most instances. If you purify haemopoietic cell populations so as to get rid of the late committed progenitor cells and you purify very immature progenitor cells, then M-CSF is not sufficient. But you can induce the cells with IL-1 to become responsive to M-CSF. Exposure to IL-1 alone won't result in generation of macrophages. What IL-1 does is to allow or induce differentiation up to the stage where the cells become responsive to M-CSF (Dexter et al 1988).

Fleisch: Will IL-1 without IL-3 do that?

Testa: IL-1 plus M-CSF. You don't need IL-3 in this system, although IL-3 can generate macrophages as well, without the action of other CSF.

Osdoby: Have you done any experiments where you have removed the marrow intact and analysed the change in the progenitor cells? Is the bone itself acting as a gradient for regeneration of the progenitor, because you have an increase in progenitors close to the bone with all the growth factors involved and everything else?

Testa: The experiments done to establish the concentration of different cell populations in different regions of the femoral marrow are done by extracting cylinders of haemopoietic tissue of different diameters from the femoral cavity, and determining, for each bone, the location of the cells extracted, and the location of the cells left behind, and then determining the numbers of stem, progenitor or stromal cells in both populations. Their concentrations are plotted as a function of their distance from the bone surface. The maximum concentrations of committed progenitor cells are fairly close to the bone, but we do not know if the bone itself will influence this distribution. By grinding the bone, we can recover very few stem cells and they are at the mature end of the stem cell compartment, close to the stage of differentiation of progenitor cells.

Raisz: The hydrocortisone results are very surprising. 10^{-9} M hydrocortisone is a sub-physiological concentration. What was the concentration of hydrocortisone in the cultures that had none added?

Testa: The control had 10% fetal calf serum.

Raisz: 10% fetal calf serum has 10^{-8} M hydrocortisone, yet you add 10^{-9} and get an effect?

Testa: We observed a very small effect. If you assume that 10^{-9} M concentration of hydrocortisone in 10% serum after our addition of hydrocortisone we have 2×10^{-9} M. The effect is clear at higher concentrations.

Nijweide: What do you think the driving forces for the cells are? You stated that the stem cells in the resting phase are in the centre of the marrow. They go to the periphery to proliferate and then the more differentiated forms return to the centre.

Testa: I wish I knew! We have discussed in this symposium the importance of the extracellular matrix. There might be a change in the extracellular matrix at different sites of the bone cavity. Growth factors may be produced or presented to their target cells differently in different environmental domains. It appears to be a very ordered sequence of events.

Williams: In their papers Dr Krane (this volume) and Dr Testa discuss important concepts which will be lost if we focus solely on the definition of osteoclasts. We are discovering certain types of message molecules which flow between cell systems. If we consider how they act, we must be aware of a number of very interesting possibilities. Nearly everything described in these papers was a positive response and all the messages were proteins or peptides.

We should know if the cells express new enzymes on the surfaces for destroying these message molecules while they respond to them. If they do, then the regulation of the cell population along the pathway is just a true morphogenic field. We could get therefore into the whole theory of morphogenic fields very beautifully in this system, which we need to do. However, we can't do it unless we have some destructive as well as some positive feedback. For example, if you consider the nerve message peptides, it is very clear that what is needed is not just a synthesis of small peptides like enkephalin, but proteases and peptidases to destroy them.

Testa: That's right. I do not know about proteases, but there are ways to control the system. Tumour necrosis factor (TNF) can act in at least two ways. It will displace the dose response to make the cells less responsive to GM-CSF. That may be a direct effect because it has been observed by using purified target cell populations. On the other hand, TNF may increase the production of M-CSF. That is a way of modulating the balance between the production of granulocytes and macrophages when more macrophage production is needed.

Williams: We need to know the life-time of the growth factor in the system. For example, some epidermal growth factors have a short life-time whereas that of insulin is long. These growth factors are chosen to be modulated by their life-time because they control different cell populations in different ways. They should be positional peptidases.

Testa: CSFs are glycoproteins which appear to be fairly stable in serum and under other conditions in culture. I do not think there are data on their biological half-life *in situ* in bone marrow.

References

Chambers T 1988 The regulation of osteoclastic development and function. In: Cell and molecular biology of vertebrate hard tissues. Wiley, Chichester (Ciba Found Symp 136) p 92–107

Dexter TM, Ponting ILO, Roberts RA, Spooncer E, Heyworth C, Gallagher JT 1988 Growth and differentiation of haemopoietic stem cells. J Gen Physiol, in press

Krane SM, Goldring MB, Goldring SR 1988 Cytokines. In: Cell and molecular biology of vertebrate hard tissues. Wiley, Chichester (Ciba Found Symp 136) p 239–256

Matthews JL, Martin JH, Race GJ, Collins EJ 1967 Giant cell centrioles. Science (Wash DC) 155:1423–1424

Final general discussion

Clinical implications

Glimcher: We shall now consider the clinical implications and applications of the basic science information presented during the symposium. I suggest that we focus first on local responses of bone to injury and disease. Orthopaedic surgeons are mostly concerned with the *local* repair and healing of bone rather than more general afflictions of the skeleton. The required repair may be of a simple, single, healing fracture, or it may necessitate the knitting together of more than two segments of bone. Sometimes bone formation must be promoted to fill in local losses of bone mass, such as occur in cysts of various sizes, in the resorption of bone tissue due to tumours, or in infections.

Another problem is how to promote the repair of dead bone *in situ*. Although this disease (osteonecrosis) is initially limited to bone, its clinical manifestations are seen in the adjacent joint which, as I shall describe, is eventually entirely destroyed in almost all cases in adults. In children, the adjacent joint is only rarely destroyed, but deformities in the size and shape of the femoral head, for example, and late changes in the joint are common.

Orthopaedic surgeons also need to induce new bone formation so as to fuse adjacent bones in continuity. Examples are the fusion of a hip (two bones) and the fusion of adjacent vertebrae in long segments of the spine, such as occurs in scoliosis. They also face the opposite problem: how to deal with the formation of too much bone. For instance, after major hip surgery, a certain number of patients develop *de novo* formation of bone in the soft tissues adjacent to the hip joint. This is not simply calcification (the deposition of apatite crystals in the soft tissues) but true bone formation—that is, ectopic ossification. This can cause many clinical problems.

Similarly, many patients with severance of their spinal cord develop a rapid and massive *de novo* bone formation around their hips. In these instances, no surgery has occurred in the vicinity so that unintentional scattering of potential bone cells in the wound cannot be responsible. For reasons that are not wholly understood, transection of the spinal cord gives rise to the proliferation of mesenchymal stem cells in the soft tissues, mostly about the hip, but also rarely about other joints. In the early stages, at least, these cells differentiate almost exclusively to osteoblasts, resulting in the formation of bone. The pathogenesis of this remarkable example of induced bone formation in areas and tissues not normally ossified is unknown. At this symposium we have not yet discussed the role of the nervous system in bone formation and resorption, its effect on stem cell and progenitor cell proliferation, and on differentiation of these cells to functioning osteoblasts, osteoclasts and their regulation.

As an aside, for Arnold Caplan's sake (see Caplan, this volume), we should note that ectopic ossification after hip surgery or spinal cord transection arises directly from the production and differentiation of mesenchymal stem cells to osteoblasts without the formation of cartilage. As I shall describe, this phenomenon of bone formation without preceding cartilage formation also occurs in the repair of dead coarse cancellous (trabecular) bone.

Bone is one of the few tissues that is capable of repairing itself by synthesis of the original kind of tissue. Its repair differs greatly from that of skin and other connective tissues where a successful result is the formation of a fibrous scar. In bone, repair by fibrous tissue or even by cartilage is a biological and clinical failure.

Incidentally, many fractures occur in the spine and hips of elderly postmenopausal women who have lost bone mass as a result of osteoporosis. Clearly, a better understanding of abnormalities and balance of cell functions in this disease, leading to preventive therapy, will be more useful than better clinical management of the local fractures, even if the latter is based on newer knowledge of bone induction, and cell differentiation, function and regulation.

Our understanding of heterotopic *calcification*, as opposed to heterotopic ossification, and therefore our ability to treat or prevent such conditions as the deposition of mineral crystals in shoulder tendons or bursae, articular cartilage or in non-skeletal sites such as the large arteries and coronary arteries, will depend almost entirely on further definition of the underlying mechanism. This will evolve from basic biological and physical/chemical studies. In this regard, Dr Veis's suggestion that certain macromolecules, such as the phosphoproteins, may have specific domains which facilitate calcification and others which inhibit calcification is intriguing, because it might be possible to alter the individual regions of such macromolecules and thus control tissue calcification with precision.

Local bone repair is described by concepts of bone cells already presented here: the origin and nature of the stem cells; the progenitors and the phenotypic osteoblasts and osteoclasts; the stimuli which incite or inhibit these processes; their regulation; cell–cell and cell–matrix interactions; and their specific phenotypic expressions. Many growth stimulators and osteoinductive factors have been identified, and they have an enormous, immediate and long-range potential for clinical use in some cases where bone healing is either slow or absent. We need to relate the effect of these various agents on each of the cells involved with the eventual production of phenotypic osteoblasts and osteoclasts in different locations so that we can tailor the repair to the problem at hand.

We sometimes need to stop the repair process in one type of bone tissue and to increase bone formation and decrease bone resorption in another type of bone tissue. Additional knowledge about the specific factors that stimulate the formation of osteoblasts or osteoclasts should help us to alter our therapy according to the type of bone tissue involved. The biological process of bone healing is not as simple as it is sometimes made out to be and does not consist only of the production of more and more osteoblasts, making more and more bone, which is then remodelled.

The nature of the cells responding to an injury or to a repair is highly dependent on the microscopic organization of bone. This can easily be demonstrated by examining the cellular response and subsequent tissue changes in the simplest case of bone healing, the repair of a segment of dead bone with the bone in continuity—where there is no fracture, no motion of the fracture fragments, no infection, and no massive absence or loss of bone mass to complicate the biological response (Glimcher & Kenzora 1979a,b,c, Kato & Glimcher 1974). The segment of bone involved may be a portion of compact, cortical bone of the diaphysis of a long bone, or a segment of coarse cancellous (trabecular or spongy) bone in the metaphysis or epiphyseal region of long or flat bone.

In osteonecrosis of both compact and cancellous bone, all the cells in the bone (osteoblasts, osteocytes, osteoclasts, stem cells, progenitor cells) and their progeny, including those in the marrow, are dead and cannot contribute to the healing process. The eventual goal of repair is the replacement of dead with living bone. The nature of the cells recruited to repair compact cortical bone from the diaphysis of a long bone and the sequence of events which results from the action of these cells is the opposite of that observed with coarse cancellous bone.

The first sign of repair of compact cortical bone is the appearance of osteoclasts on the surfaces of the long bone over the dead portions, and later deep within the cortex, especially in the central Haversian and other vascular canals. The osteoclasts resorb bone first on the surface of the bone and then within the substance of the compact bone by forming cutting cones. Very few osteoblasts are seen early in the process; almost all the new bone cells involved are osteoclasts. Thus the initial consequence of this response is a decrease in bone mass. New osteoblasts appear later, distal to the cone-cutting osteoclasts, but they synthesize new living bone more slowly than the dead bone is resorbed by the osteoclasts. This discrepancy in the rates of bone formation and resorption persists for a long period, so that bone mass continues to decline during the repair.

The healing of compact cortical bone is an extremely slow process: hence the slow repair of fractures in which large segments of a long bone are deprived of blood supply with subsequent death of the bone cells. Large long bone grafts, with or without accompanying articular ends of the bone, likewise require much time before the compact cortical bone is replaced. A 5 mm piece of the cortex of a long bone in a monkey was killed by freezing with liquid nitrogen. After one year, approximately 20% of the dead bone had not yet been replaced. By the end of three years, essentially all the dead bone appeared to have been replaced by living bone. One can imagine how long it might take a large adult human bone transplant to be repaired and incorporated as living bone!

Although the process is slow and bone formation lags behind bone resorption so that bone mass is decreased significantly until the last stages of repair, eventually the original mass of dead bone is replaced by new living bone. Clinically, depending on the size of the dead segment, the decreased bone mass in the long bone can seriously decrease the strength of the bone so that early assumption of too much mechanical load can result in fracture. Clearly, new biologically based therapy must provide more diffusely distributed vascular penetration of the compact bone, and the formation first of osteoclasts to resorb bone. Then should follow, in the correct sequence, the proliferation of osteoblast stem cells and progenitors and their differentiation to osteoblasts that would replace dead bone with an equal amount of new bone.

In contrast to what happens in compact cortical bone, after the death of the cells in coarse cancellous bone (trabecular or spongy bone), stem and progenitor cells in the adjacent living bone marrow proliferate. These cells are 'mobile' and, with new capillaries, they proceed to the dead bone and invade the marrow spaces between the dead trabeculae. There is little or no differentiation of the proliferating stem and progenitor cells and certainly no phenotypic osteoblasts are formed. New bone is synthesized in the living bone in and from which the proliferating cells are derived and from which they advance into the dead bone. That is, differentiation of these proliferating cells to osteoblasts occurs only when they are in the marrow spaces of the dead trabecular (cancellous) tissue. This differentiation occurs only on the surface of the dead bony trabeculae and not in the bulk of the marrow space. Bone is formed directly, and *not* by an intermediate cartilage phase.

Not until the surface of a dead trabeculum is almost completely covered by new living bone does one usually see the appearance of osteoclasts and cutting cones burrowing into the dead bone of the central core of the trabecular 'sandwich'. New bone is then formed in the cavities of the dead bone carved out by the osteoclasts in the central portion of the trabeculum. Thus the original dead bone in the centre of the trabeculum is replaced by new living bone. Eventually, at least in small experimental animals, all the dead coarse cancellous bone of the distal and of the bone adjacent to the articular cartilage is replaced by new living bone. Contrast the *decrease* in the mass of bone during the early and mid stages of the repair of compact cortical bone of the diaphysis with *the increase* in the mass of bone during the healing of coarse cancellous bone.

If the process stopped at this stage, things would be fine. Unfortunately, the signals to some stem cells to stop proliferating, dividing and differentiating to osteoclasts and to stop resorbing bone either fail to be sent or go unheeded, with the result that the repair process continues to and includes the subchondral, compact bony plate. This elicits a completely different cellular response from that observed in the compact bone of the diaphysis of a long bone. At first, like the response seen in compact diaphyseal bone, most of the cells are osteoclasts which proceed to resorb the bone. However, there is less vigorous proliferation of osteoblasts, resulting in less synthesis of new bone and an increasing loss of subchondral bone mass.

Often, almost all the subchondral bony plate is resorbed. Moreover, the repair process does not stop at the bone–cartilage junction, but continues into the deep and even mid portions of the articular cartilage, destroying it locally and inciting a typical, degenerative process of the articular cartilage. These processes seriously weaken the cartilage–bone junction and eventually lead to subchondral fracture, collapse of the articular end of the bone with distortion, and complete destruction of the cartilage on both sides of the joint with loss of joint function.

Therapy based on cell biology faces several problems. (1) How could one provide for a significant increase in the rate at which the tissue invades the spaces between the trabecular bone, and the rate at which new osteoblasts are formed by differentiation from the progenitor cells? Which of the mitogens and/or osteoinducers could be used locally, or what drug systemically? (2) How does one turn off (regulate) the repair process once it reaches the subchondral plate? How could one turn off the tendency of the repair process to go unimpeded into the articular cartilage? There appears to be a serious lack of understanding of the cell biology of osteonecrosis and misdirected rationale in some of the current surgical therapies at the stage when the disease affects only the coarse cancellous bone of the epiphyseal end of a long bone. The goal of many procedures is to *stimulate* the full healing response, regardless of the fact that it is the continuation of the repair *per se* that results in the eventual destruction of the joint. If we can't tailor and control the cell response, at least we must try to constrain or *inhibit* the repair so as to prevent subchondral resorption and cartilage invasion, not stimulate it (Glimcher & Kenzora 1979a,b,c).

Even in the relatively simple circumstances of osteonecrosis one can see the variations and complexities in the repair which result from the different type and sequence of bone cells that are recruited to repair the dead tissue. These differences depend on the microscopic architecture of the bone tissue. The repair response is marked in one case by an increase in the proliferation of progenitor cells which eventually differentiate to osteoblasts and, in another case, by stimulation of the proliferation of progenitor cells with their eventual differentiation to osteoclasts. Similar careful analyses of the cell biology must be taken note of in the repair of small and large cysts, bone tissue defects, and so on. To make matters more complex, the healing of bone includes not only the replacement or repair of bone but the replacement of bone tissue in its original and proper microscopic organization and overall orientation.

To sum up, we need a greater understanding of the complex system of cell–cell and cell–matrix systems of communication and other factors which regulate cell proliferation and differentiation to specific cell types, so that we can tailor the repair and healing process for different types of bony tissue, in order to achieve the regulation of bone cell populations and formation to fit the specific task of repair at hand.

Glowacki: I would like to summarize our experience using demineralized bone implants at the Children's Hospital and Brigham and Women's Hospital in Boston. An example of the challenges we face is an identical twin whose craniofacial development was compromised *in utero*. About 20 years ago Dr Paul Tessier in France developed heroic surgical techniques for correcting such deformities. The procedure involved massive harvesting of ribs, iliac crests, and tibial segments of bone and transplanting those into the craniofacial region to correct the skeletal deformity.

There are several problems with the technique of autografting for craniofacial reconstruction. A harvesting operation is involved which may take up to 14 or 16 hours. Much of the operative time is devoted to harvesting the donor bone. There is also a great deal of blood loss from children, in particular. Complications arise from the harvesting of ribs; pneumothorax is encountered in 10% of cases in which ribs are harvested. There is always an inadequate quantity of bone and the shape of the bone is never what is needed for the reconstruction.

The most serious biological issue, however, is unpredictable resorption. Craniofacial and oral surgeons find this a very severe problem; it is not as severe for orthopaedic applications of bone grafts. In the craniofacial region, 70% of the operations are repeat operations. Children that have had reconstructions lose bone over three, four or five years and revert to a similar deformity which requires more operations.

About 15 years ago we started to apply the principles that Dr Urist (1965) had pioneered and that were extended by Reddi & Huggins (1972) to craniofacial bone reconstruction. This takes advantage of the mysterious but remarkable quality of demineralized bone powder to induce production of cartilage, even in fibroblastic or subcutaneous space, by fibroblastic or mesenchymal-type cells. Once new blood vessels penetrate, the cartilage is replaced by bone and marrow. We began with craniofacial and maxillofacial deformities in 1976. We have now used demineralized bone in more than 300 procedures.

One craniofacial patient was a boy who was born with a cloverleaf skull deformity (Glowacki et al 1981). During the first four months of his life neurosurgical operations were repeatedly done in an attempt to remove the prematurely fused sutures to keep the bones apart so that it would grow into the normal shape. However, the deformity persisted.

When the patient was seven years old his calvaria was excised from just above the orbits, 19cm back. His brain also had the cloverleaf shape because its shape is determined by the cranial vault around it. His skin was closed and for two weeks he wore a football helmet in the hospital, during which time his brain resumed the normal shape. In that time we processed his own bone and made demineralized bone chips and powder out of it. The demineralized bone chips were laid out over the dura to maximize contact with target cells. Four months after the operation his cranium was solid. About 1½ years later a computer-assisted tomogram (CAT) scan showed thick bone formed at the level of the operation. There were still some gaps and some remodelling continuing at this stage. Two years later the skull had a normal rounded configuration.

This is an example of the 'low tech' approach that we are using right now, and sets the stage for the 'high tech' approach of Professor Urist and for speculation as to what we may be using in the future.

Urist: I presented the experimental basis (Urist 1965) and some preliminary clinical trials (Urist 1968) of osteoinductive preparations of hydrochloric acid-demineralized bone matrix at a Ciba Foundation Symposium on Hard Tissue Growth, Repair and Remineralization (Urist 1973) and I reviewed the evidence for and against the transfer of a bone morphogenetic protein (BMP) in the process of induced bone formation in rats.

In 1965, the observation of bone formation induced by implants of HCl-demineralized bone matrix called attention to the fact that reports of implants of bone demineralized in EDTA by Irving & Handelsman (1963), Ray & Sabet (1963) and many others had failed to mention induced bone formation. Why should implants of bone demineralized with EDTA at relatively mild, nearly physiological conditions of pH 7.2 to 7.4 not induce bone formation as well as matrix demineralized in HCl, citric acid and other solutions of low pH? Furthermore, why was HCl-demineralized bone matrix-induced bone formation not reported before 1965? Had not HCl (muriatic acid)-demineralized bone been implanted to repair bone defects by Senn (1889) in the USA? Why was the work of Senn and his followers in the USA rejected by Bier (1892) and many surgeons in Europe at the turn of the century? Why were implants of HCl-demineralized bone untested in animals for over 70 years, until the work by our research group in the 1960's?

It may have been that the right matrix (HCl-demineralized) was implanted in a bone defect of less than the critical size which regenerated chiefly by proliferation of preexisting osteoprogenitor cells in the host bed and obscured the bone deposits developed from the matrix-induced morphogenesis. Another answer might be that the wrong matrix (degraded by EDTA-activated proteases) was implanted in the right host bed (muscle) to reveal induced bone formation by differentiation of perivascular connective tissue cells in response to BMP. Or the right matrix (HCl-demineralized bone) in the right host bed (muscle), was implanted in the wrong species (dog, monkey, or human) in which there is a long-lag phase and associated cofactors of unknown nature, possibly suppressing the osteoinductive response. Reports of scanty deposits of new bone in muscle injected with epiphyseal bone extracts with acid alcohol (osteogenin) or on surfaces of alcohol-devitalized old bone implanted in soft parts in rabbits (La Croix 1951) confused the issue. 'Osteogenin', prepared from alcohol/acid extracts to a 22 kDa protein in guanadine extract of bone matrix in rats, shows the osteoinductive response only when the putative 22 kDa protein is recombined with bone matrix. These species and matrix discrepancies are the basis of an interesting controversy that is fueled by reports from three other research groups working on osteoinductive proteins with comparable molecular mass but three different names.

The EDTA paradox is analysed in my earlier paper (Urist 1973). Briefly stated, data obtained from detailed experiments on EDTA-demineralized bone demonstrated that at neutral pH endogenous proteases were activated. These enzymes (BMPases) degraded BMP selectively and abolished the osteoinductive activity. That raised the question of whether it was a BMP or the entire bone matrix that was responsible for induced formation. Dr Hauschka's matricrine concept (Hauschka, this volume) reviews this discussion.

Evidence for the involvement of a discrete BMP is detailed in a recent review which included a report on the purification of a 19 kDa BMP (Urist et al 1987). A basic assumption is that while the whole matrix is Nature's delivery system, a discrete low molecular mass protein or protein aggregate is transferred by diffusion (or haptotaxis) to receptors in perivascular mesenchymal-type cells of muscle or marrow stroma before induced bone formation occurs. In adult rat muscle, the target consists of cells which, in the lifetime of the individual, underexpressed by BMP, would never have entered into a bone morphogenetic pathway of development.

For bioassay, many research groups recombine BMP (or its equivalent) with bone matrix residues deactivated by extraction with guanidine HCl. One bioassay consists of connective tissue outgrowths of rat muscle *in vitro* onto a substrate of either cellulose acetate (Takahashi & Urist 1986) or deactivated matrix with BMP in the culture medium. Another consists of BMP implanted in the mouse quadricepts muscle pouch either as a lyophilized 19 kDa BMP or as an aggregate of BMP, MGP, and other associated calcium-binding low molecular mass bone matrix non-collagenous proteins (BMP/NCP). Our research group is gradually replacing bioassay with BMP radioimmunoassay (Urist 1984, Urist et al 1985 a,b).

Human BMP/NCP has been implanted in a series of 32 patients (Johnson et al 1988). It was prepared from bone of human organ donors as described by Urist et al (1983). The patients had ununited fractures and large segmental defects of long bones which failed to respond to multiple non-surgical and surgical treatments and therefore represented difficult problems. The working hypothesis assumes that the regeneration process would be augmented by means of implants of BMP/NCP.

Augmentation was obtained by implantation of a composite of autogeneic cancellous bone and bone marrow and hBMP/NCP into large bone defects. hBMP/NCP alone was implanted for defects of one to two cm³. Intractable nonunion and large segmental defects were located in the femoral, tibial, or humeral diaphysis. hBMP/NCP was implanted only after debridgement, skin grafts, and stabilization (by either internal or external fixation). Autogeneic cancellous bone plus the hBMP/NCP in the delivery system of polylactic–polyglycollic acid (PLA–PGA) copolymer was inplanted in 20 of the 32 cases.

The 32 cases included 3 humeral, 14 femoral, and 9 tibial nonunions; six had total circumferential segmental tibial defects measuring 3, 4, 5, 8, 13, and 17cm. All of the patients with tibial nonunions had previously been advised to consider such conventional treatment as amputation. The average preoperative duration of symptomatic nonunion or segmental defect was 25.8 months (range 4.1–68.3 months). The preoperative intervals of total disability were: tibial segmental defects, 23.6 months; tibial nonunion, 27.1 months; femoral nonunion, 22.9 months; humeral nonunion, 12.8 months. There were a total of 104 previous surgical operations in 28 patients (average 3.7 procedures/patient); only four patients had been treated initially by plaster cast immobilization.

Doses of 50 to 100mg of lyophilized aggregates of hBMP and water-insoluble non-collagenous proteins (hBMP/iNCP) were implanted either in ultra thin capsules, or a strip of PLA–PGA copolymer. The implants were placed either as an onlay strip across, or as capsules inside, the fracture gap. Continuity of the bone was restored by the one operation in 26/32 cases. The average time to union was: humeral nonunions, 3.4 months; tibial nonunions 3.6 months; femoral nonunions, 4.7 months; and the segmental tibial defects, 5.7 months.

Overall, at the time of this report, the follow-up averages 25.2 months (range 6 to 54.7 months). Five of the six segmental tibial defects had bone continuity restored *per primam*; the sixth after a second implantation of BMP/NCP. Six operations (four tibial nonunions, one femoral nonunion, and one tibial segmental defect) failed but the fracture united with a second operation for reimplantation of hBMP/NCP; two of the four tibial nonunions had recurrent infection. Against advice, one patient removed his plaster cast and incurred a refracture. One patient with femoral nonunion had sufficient stabilization but healing followed refixation with bone plate fixation and reimplantation of BMP/NCP. One tibial segmental defect of 4 cm in another noncompliant patient was healed but incurred a fatigue fracture at the proximal bone end; the defect united after reimplantation of hBMP/iNCP and reimmobilization in a cast. The course of wound healing was remarkably uncomplicated in all pa-

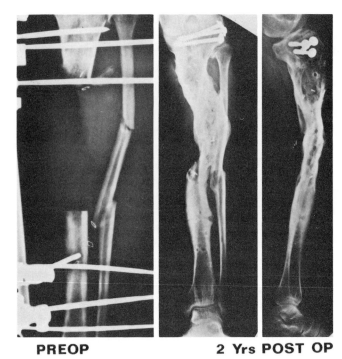

PREOP   **2 Yrs POST OP**

FIG. 1. (Urist). Roentgenograms of tibia of 22-year-old man showing a 13 cm defect, bone (left), and after (right), implantation of autogeneic iliac cancellous bone and bone marrow with an onlay strip of PLA-PGA measuring 15 × 1 × 0.2 cm delivering 50 mg of BMP/NCP. The strip is marked with four vascular clips, two at each end. Note complete regeneration of the shaft of the tibia (right); two years after the operation.

tients. There was no unusual swelling, hyeremia, or evidence of immune reaction.

A typical segmental defect of the tibia in a 22-year-old man before and two years after implantation with 50mg of BMP/NCP in an onlay strip of PLA–PGA with autologous iliac cancellous bone for the marrow stroma target cells is shown in Fig. 1. hBMP/NCP augmentation appeared to favourably influence bone regeneration in the majority of these very difficult conditions. We are doing further clinical investigations with a new protocol, aiming to distinguish the effects of hBMP/NCP from the effects of new or improved methods of external skeletal fixation and autogeneic cancellous bone alone. Further investigations must solve how to measure the amount of new bone formation produced from preexisting host bone cells and to determine the amount of new bone developed from bone marrow perivascular stroma target cells that are recruited to enter an osteogenetic pathway of development under the influence of hBMP. The second phase protocol calls for a consecutive double blind, randomized series of matched cases.

Glimcher: Another set of questions arises from Dr Prockop's paper on osteogenesis imperfecta (OI) (Prockop et al, this volume). The major clinical manifestations of OI are repeated fractures of bone which are often caused by an apparently insignificant amount of force. As a result, the disease is often referred to as 'brittle bones'. Darwin Prockop has reviewed the remarkable data which he and others have obtained about the mutations in type I procollagen genes in OI. It is certainly an example of the potential of molecular biology to unravel some complex bone diseases.

We should first ask if the bone really is brittle: is it less ductile or malleable than normal bone? Second, are the deficiencies in the structural properties of bone as a whole due to changes in the material properties and organization of bone substance, of bone tissue, or of the organization or orientation of the bone tissue elements? How are these changes brought about, if at all, by the mutations of type I procollagen gene? Many years ago my colleagues and I tried to determine if OI bone material was mechanically brittle. It was very difficult: we never had enough bone and could never shape the bone plugs properly for the testing machine. But even under the worst circumstances it seemed clear that the OI bone substance was not more brittle than normal bone, or if it was, only minimally so. Why then do patients break their bones so easily?

It may be that the deficiency in the structural properties of bone in OI occurs because there is less bone substance or mass in a particular bone. This could result from failure to make enough collagen fibrils for incorporation into the tissue or proteolysis of collagen that is more rapid than synthesis, or both. In growing bone this would have two major effects. If we simplify the situation and consider that the growth of a long bone occurs by appositional periosteal new bone formation and endosteal resorption, then it follows that if the rate of appositional periosteal new bone deposition were decreased while endosteal resorption continued at the same rate or was increased, the outside diameter of the long bone would be smaller than normal and the cortex would be thinner. These anatomical changes would tend to significantly decrease the strength of the bone, especially in bending. Indeed, in more severe non-lethal cases of OI many of the long bones have smaller outside diameters and thinner cortices than bones of normal children of the same age. However, this is not true of all cases. Many children with the clinical manifestations of OI have relatively normal diameter long bones with normal cortical thickness. Clearly size and cortical thickness alone cannot account for the structural weakness of the bone and the frequent fractures in all patients.

What about bone tissue organization? This is a more complicated issue. Most histological studies, including our own, demonstrated that some of the compact cortical bone of long bone diaphyses in children with active OI contains a variable but significant quantity of coarse cancellous bone tissue. Moreover, at least part of the bone substance in most patients (but not all) is woven bone and not lamellar bone. This occurs when essentially all the cortical bone of healthy children of the same age consists of compact lamellar bone. There are many regions of long bones in patients with OI that consist of lamellar bone, even at a very young age, which indicates that the cells of most patients are capable of organizing bone tissue and substance into normal compact lamellar bone. Therefore, the presence of coarse cancellous woven bone in portions of the cortices of long bone indicates, at the least, that the maturation and developmental processes of part of the bone tissue and bone substance in these patients are delayed. Such defects in the organization of bone tissue at the microscopic level might also contribute to a deficiency in the structural properties of bone, in addition to the fact that the presence of coarse or fine cancellous bone in the diaphyses would decrease the mass of bone.

What about the material properties of bone substance? This depends on the chemistry and structure of the inorganic crystals, the chemistry and mechanical properties of collagen, and the relationship, both physical and spatial, between the two components, including the nature and strength of any chemical bonds between them. The mechanical properties of bone substance will also depend on the orientation and organization of the collagen fibrils and the mineral phase particles. I know of no in-depth study of the inorganic crystals in OI. Until recently, studies of bone crystals were limited to X-ray diffraction of the apatite crystal lattice form, crystallinity, and the basic chemistry of the crystals. New techniques for studying synthetic and biological crystals, such as ^{31}P NMR spectroscopy (Aue et al 1984, Roufosse et al 1984) and Fourier-Transform infrared spectroscopy with deconvolution of the spectra (Rey et al, submitted), should enable us to examine the short-range order in the crystals and possibly collagen–mineral interactions. The detailed properties of the inorganic crystals and their interaction with the collagen and non-collagenous proteins, should provide important information on the actual defect in OI bone which may cause brittleness.

So far, there is no evidence that there are gross abnormalities in the collagen fibrils, the major structural protein in bone: the fibrils generate normal wide and low-angle diffraction patterns, and appear normal in size and axial length periodicity by electron microscopy. The reducible and non-reducible cross-links do not seem to be fewer than in normal collagen fibrils. Again, these characteristics are not the only ones contributing to the structural properties of collagen or to bone substance. The structural properties of bone substance do not seem to be affected in OI as evidenced by indirect data, although no direct mechanical measurements have been made.

The recent work on the non-collagenous phosphoproteins of bone and Dr Veis' work (this volume) in defining the domains within the phosphoproteins in dentine which may have specific functions such as facilitating the nucleation of calcium phosphate crystals, and linking them to the organic matrix and to cells, may make it possible to uncover the defect which so radically changes the structural properties of bone in osteogenesis imperfecta and other bone diseases. It is intriguing that although so many gene mutations and possible deletions have been uncovered in the type I procollagen gene, the relationship between these molecular defects and the structural mechanical properties of bone and the frequent fractures remains unknown.

Prockop: Dr Glimcher has made the major points about osteogenesis imperfecta. We now know of several mutations. None of them are the same in unrelated people. The mutations in typeI procollagen genes are all different. Therefore each patient presents a different situation to study. In addition to the mutations already defined, there is likely to be a category of mutations in which the amount of collagen synthesized and the stability of the protein are normal, but the abnormality is loss of a binding site for a phosphoprotein, or a site for cross-linking the monomers in fibrils. We don't have any examples of mutations in this category because we lack the necessary techniques to detect them.

The most complicated part of the story is that within a given family you see tremendous heterogeneity in how the disease manifests itself in different individuals with the same mutation and the same changes in the procollagen molecule. The collagen molecule is like a brick which is used to build structures such as bone and other tissues, but the way that brick is used may differ in different individuals in ways we can't fully sort out. Some people who clearly have a defective collagen molecule seem to have bones that are normal in size and shape and that withstand normal stresses.

In addition to mutations in the structural genes for type I procollagen, there are certainly defects in enzymes required to synthesize procollagen, as Dr Krane pointed out. I should have mentioned in my paper that the first genetic disease of collagen was a deficiency of lysine hydroxylase found in a patient with Ehlers–Danlos syndrome by Dr Krane, Dr Glimcher and their associates (Pinnell et al 1972). I confined my remarks to patients with osteogenesis imperfecta, where most mutations are in type I procollagen genes.

Krane: How does the single base mutation lead to a single amino acid substitution that becomes translated into this extraordinarily disturbed morphological picture in the bone?

Glimcher: That's a good point. Even more difficult to see, how do you get patients who don't have abnormal-looking bone and are still breaking it?

Hanaoka: The bone in osteogenesis imperfecta is more fragile in childhood. As the children grow up the bone becomes less fragile. Is there a biomolecular explanation?

Glimcher: From the morphological point of view, if you biopsy the patients with age, there appears to be in most patients more and more of the lamellar compact bone replacing the coarse cancellous bone.

Termine: That is usually observed in type I osteogenesis imperfecta. There are many cases, however, where that does not occur. Most of the recessive OIs, according to the data we heard in Pavia, don't link to either collagen gene. At that meeting, Dr Stoss looked at dominant and recessive OI inheritance patterns. He saw very similar morphological characteristics in all cases. It remains a puzzle as to how the OI defects, whether they come from a dominant or recessive inheritance pattern, yield closely identical clinical phenotypes.

Martin: Understanding the interaction between the cells of the osteoblast series and the osteoclast is an important step in the approach to new ways to influence bone resorption and formation. The Paget's disease paradigm is very useful. There is a large increase in both resorption and formation. The latter is described as compensatory but the actual activity of the osteoblast in the local areas in Paget's disease is much greater than you would expect from the increase in resorption.

A much less substantial osteoblastic response occurs in hyperparthyroidism. Therefore something extra happens to the osteoblast in Paget's disease. A small number of the patients develop malignancies, which raises the possibility of a transforming process in the osteoblast. We need to find out what causes that.

A few major questions present themselves. What are the factors produced by cells of the osteoblast series to activate the osteoclasts? It may be some transmitter already known to act in other organs. Because it is proving difficult to isolate these factors, it might help to determine how the osteoclast is activated, just as we know that calcitonin inhibits osteoclast activity by increasing cyclic AMP production. There are only a few obvious general mechanisms and it seems likely that one of these is involved in the early activation of the osteoclast.

We know that calcitonin inhibits osteoclastic activity and generation by acting on precursors and preventing precursor maturation to osteoclasts. We should aim to purify and clone the calcitonin receptor. That opens the way to modelling with the aim of finding low molecular mass drugs, which would be easier to use than calcitonin, to act through the calcitonin receptor both in Paget's disease and possibly also in osteoporosis.

Finally, in terms of inactivation of the osteoclast, I would like to know how γ interferon acts directly on the osteoclast. That too might help us to understand these interactions. It is highly unlikely to be through cyclic AMP and it should not be all that difficult to find out.

Future perspectives

Fleisch: We should take advantage of the knowledge developed in the haemopoietic field on the development from stem cells to mature blood cells and we should investigate whether a cascade of factors is necessary in bone differentiation from stem cells to bone-forming cells. This means looking at the regulation, proliferation, differentiation, survival and function of the various cells involved. That will present us with the same problems as those that face people studying the osteoclast: we have insufficient markers of the cells at the various stages of differentiation. The monoclonal antibodies should help us there. We shall also need homogeneous cell populations to work with, preferably at various stages of differentiation. This should enable us to investigate the various relevant factors, such as cytokines and matrix components, and might help us to understand better the mechanisms of osteoporosis in patients where bone is formed in inadequate quantities, leading to treatment for this disease.

Nijweide: I agree with Dr Fleisch. We need to know more about the differentiation of osteogenic cells and of osteoclastic cells. We now know that osteoclasts originate from haemopoietic stem cells and we are able to identify the end cell but we need to elucidate what happens in between. Until we do this, we shall not understand regulation and homeostasis of bone.

I would also like to pursue the role of osteocytes. We shall soon be able to isolate them in chick and that may provide opportunities to study their properties and functions.

Ruch: It should be possible to devise some experimental approach to the question of why osteoclasts have to be polynucleated cells.

As far as odontogenesis is concerned, it will be important: (1) to determine whether all the neural crest-derived dental papilla cells are potential odontoblasts; (2) to evaluate the possible cell kinetic-dependent typological hierarchy of both dental papilla and epithelial cells; and (3) to elucidate the molecular mechanisms involved in commitment and terminal differentiation.

Glowacki: Will we find evidence to support or reject the hypothesis that the osteoclast is a macrophage polykaryon which, as a consequence of cellular interactions with the bone matrix, develops specialized features that enable the cell to digest the bone in a regulated manner? Can we look at multinucleated giant cells and say 'You are what you eat'? I think this information may influence our pharmacological rationale for many skeletal diseases.

Owen: In the haemopoietic system we have examples of irreversible commitment of cells to differentiation along a particular lineage and of well-defined lineage-specific regulators such as erythropoietin and M-CSF. I would like to know whether there are committed progenitors for osteoblastic cells and whether lineage specific factors for these and other lines of the stromal system exist.

Chambers: In the future we intend to use the assays that are now available for osteoclast function and detection (Chambers, this volume) to identify the mechanisms of osteoclastic regulation, the lineage relationships of osteoclasts, and the mechanisms by which osteoclastic differentiation is regulated.

Termine: Over the next few years the definition of the lineage of the osteoblast phenotype should be achieved at a molecular level. Then elegant experiments could be done along the lines that we have heard here. The big task for this group, and the field as a whole, is to develop *in vitro* systems of bone resorption and bone formation that more accurately reflect what goes on in the adult or growing human patient; presently, our techniques are basically bone modelling ones, based on embryos or neonates.

Slavkin: From the perspective of a developmental biologist, the most promising and exciting opportunities in the tooth and bone field are being previewed in work with invertebrates, particularly Drosophila. Researchers are beginning to illustrate that there are regulatory genes which determine segmentation patterns, and which identify head, thorax and limb position. These homeotic regulatory genes may be archetypal for molecular determination of skeletal forms.

It appears that methods are available to begin to pursue these questions in mammalian models. I suspect two to three years from now regulatory genes will complement the current emphasis on structural genes and promoters of structural genes.

Hauschka: It is important to focus on the actions of osteoblasts and osteoclasts at their interface with the bone matrix. We should investigate what determines the fluxes of biosynthetic components into and out of the matrix. For example, when osteoblasts produce osteocalcin, is it all secreted vectorally into osteoid, or does a fraction escape apically to appear in blood plasma? What processes control the flux of matrix components out of the matrix for potential action on neighbouring cells? This would include osteoclastic release of matrix components such as growth factors. Finally, we should examine the proteinase activities of osteoblasts and osteoclasts, because it appears that proteolytic fragments of matrix components could act as messengers between the two cell types.

Osdoby: There is a somewhat polar orientation of osteoblasts and osteoclasts and bone matrix, but there is also a vascular component and vascular cells interdigitated in this bone remodelling unit. It's quite possible that the vascular cells in the bone environment may have specific modulating activities, both on osteoblasts and osteoclasts, and may act as an intermediate in communications between these cells. This is a neglected area that needs further investigation.

Baron: Because I do not think that mature osteoblasts can talk to osteoclasts, I believe we should find out what the differences are between mature osteoblasts and lining cells, these latter being the only ones that could 'talk' to osteoclasts. Also, it is essential to identify, characterize and define culture conditions for osteoclast precursors because we need to generate these cells *in vitro* for future studies. Finally, osteoclasts do talk to osteoblasts and it is important to know how they do that. Therefore, we also need to understand how the osteoclast functions.

Williams: First, it seems to me that the cell system is poorly defined at the molecular level. Without that definition one can't go on to discuss spatial and temporal relationships relative to the cellular systems as they develop.

Second, from the side of bone material science, I should like to see more emphasis on developing substitute implant materials that react in a favourable way with the normal bone environment. Glasses are being produced as such implants and I should like to see a great effort put in here. Possibly interaction with the matrix would shorten the period of time for the development of the repair system.

Testa: One of the things which emerged from this meeting is who talks to whom and about what! Concerning the relationship between haemopoietic and bone tissue, Dr Owen's experiments have shown that within the haemopoietic tissue there are cells that are relevant to bone formation. Bone marrow is able to produce the factors that haemopoietic cells need for their own development. The question arises as to whether the host of factors in bone have something to say to haemopoietic cells; that is not being addressed at the moment. Conversely, the roles of haemopoietic growth factors in the generation and function of osteoclasts also require extensive exploration.

Weiner: Lack of information in one particular area, the basic organization of the major constituents of bone, is a barrier to progress in several areas. For example, Dr Glimcher discussed bone which fractures for unknown reasons. Dr Prockop raised the question about how a one-point mutation which changes the collagen 'building brick' in bone can affect mineralization. We don't yet have the knowledge to address these questions, but we do have the tools and we should be working at it. The functions of non-collagenous proteins, for example, are only going to be solved when we can map their locations into the system. That requires an understanding of the molecular organization of bone. We have not addressed this topic here, but it is essential to all these related areas.

Vies: Is there some commonality in strategy in the way mineralized tissues are formed and in the set of interactions that cause the specific deposition of mineral in the right place? What factors influence those events? We must address this, not just in bone or dentine, but in other mineralized tissue systems as well. We may discover some simpler systems and be able to describe the macromolecular events more cleanly, and from there we could progress to more particular cases.

Another exciting question about stimulation of bone formation is whether or not all mineralized tissues, not specifically bone, contain the same factors.

Butler: There remain non-collagenous matrix proteins on which we have little or no information. I shall be considering whether these materials are embedded in the mineralized matrix and are simply present in fossilized state, with no future implications, or whether they are destined for release, as BGP might be, for future activities. Do they have activity against cells, as does osteopontin, or were they involved in nucleation and mineral growth?

Pierschbacher: I should like to reiterate the point that matrix molecules and their receptors will determine how cells respond to various factors and various environments. In thymus, for example, it might be a reasonable goal to construct and study a transgenic organ rather than a transgenic mouse. This would avoid the problem that the molecules that are involved in a particular organ could also be involved at some other place.

Glimcher: A great area for development is the study of what stimulates and regulates cells. As we have seen, there are very serious clinical problems. In particular, the coarse cancellous and compact bone examples showed that different cells are recruited for different jobs in healing; we don't know why. This area could be explored so that we really can tailor some of the phenomena that we are trying to treat.

Our knowledge of the matrix is a mess because there are so many molecules. Even for the molecules that we isolate and work on, we don't know at what time they come out of the cell, where they go in the matrix and what relationship they have with one another and to the collagen fibril. If we did, we could do what Professor Williams would like and look at the three-dimensional conformation to investigate their function, for example in mineralization. We must address this problem, but I don't know exactly how.

I would agree with Dr Weiner's suggestion about transgenic mice or other methods where you specifically eliminate one of these molecules at a time to see physiologically or histochemically what happens.

Gazit: We need to make more progress in studying the role of bone marrow in the aetiology of osteoporosis and we require greater understanding of the local regulatory mechanisms in osteogenesis in general, and in the expression of the osteogenic potential of bone marrow, in particular.

The osteogenic potential of bone marrow has been well established in experimental models, and the decreased number and/or activity of marrow stromal cells may play an important role in the aetiology of osteoporosis. The osteogenic potential of human bone marrow has been demonstrated in diffusion chambers. In addition, the marrow stroma provides the microenvironment that is essential for the haemopoietic system and thus may play an important role in the clinical success of marrow transplantation.

Healing bone marrow produces an osteogenic growth factor activity. This activity may be part of a local regulatory mechanism which is involved in osteogenesis and expression of the osteogenic potential of bone marrow. I hope that detailed characterization of the growth factors comprising the local regulatory mechanisms will add an important aspect to our knowledge of the aetiology and pathogenesis of osteoporosis.

Caplan: I would like to see our knowledge base refined so that we have predictive information which can be used in repair or regeneration of skeletal tissues. Examples might be the use of markers in serum to define disease states such as arthritis or to give insight into a repair process such as at a bone-break site. This will involve complex molecular pathways and highly specific probes.

Prockop: I would like to see a better definition of how the synthesis of connective tissues is regulated at the molecular level. We are making some progress in understanding how the collagen molecule is deposited and assembled appropriately in the extracellular matrix. Also, we are learning how the other macromolecules are assembled and incorporated. The aspect of the story that I find most perplexing is how connective tissues re-model themselves. We see cells making an external matrix that is insoluble. After a time, the cells somehow know that the matrix that they made earlier has outlived its usefulness. So they make it disappear and synthesize it anew. It is an amazing process, and one for which we still lack a conceptual model.

On a more practical level, I would like us to examine osteoporosis with the same hypothesis that has proved succesful in osteogenesis imperfecta—that brittleness of bone can be caused by mutations in the structural gene for type I procollagen. We are still limited by technology in exploring this because of the difficulty of finding mutations as small as a single base change in large genes. However, I am optimistic that the technology will improve soon.

Hanaoka: In normal bone tissue the haemopoietic stem cells are usually observed in the secondary spongiosa and are surrounded by bone marrow myeloid cells, but the immature perivascular cells in the top of the primary spongiosa that seem to be the precursor of the chondroblasts are not surrounded by myeloid cells. No haemopoiesis happens there, only the disruption of the calcified cartilage. I believe the natural phenomenon should be economical and logical, so it is hard for me to connect haemopoietic stem cells in the secondary spongiosa with immature perivascular cells on the top of the primary spongiosa. In addition, I don't think there is any definite evidence that the osteoclast outside the bone collar in the early stage of endochondral ossification of the long bone is derived from circulating haemopoietic stem cells.

As a clinician, I have been using allografts in many cases of bone tumours. The main sources for allografts were healthy parts of amputated limbs from tumour patients. However, we are now doing more limb-saving surgery, so bone sources are more limited. In addition, we need to screen the sources with the Wasserman's test and the hepatitis B test, and we now face the problem of AIDS as well. Therefore we are gradually reducing the use of allografts and adopting hydroxyapatite treatment instead.

Krane: We have not yet identified the critical features of the phenotypes of the cells that we are dealing with. I was hoping that Dr Rodan's experiment transfecting the gene for alkaline phosphatase into cells that didn't have it would provide the answer, but it did not.

I am impressed by the extraordinary experiments by Young et al (1987) in which gap junctions have been reconstituted. They isolated the 27 kDa protein, put it into liposomes and recreated the electrical physiological characteristics of the gap junction. If we could perform similar experiments we might identify the critical functions of cells that make them behave in the ways we expect.

Receptors are being put into cells that shouldn't have them, reproducing the expected biological responses.

Urist: Over the past 50 years research on osteoporosis has had a limited theme. Work on bone densitometry has been done better than it's ever been done before, but the basic information about osteoporosis that has come out of conferences does not seem to have changed. Patients with osteoporosis are studied from the point of view of treatment rather than from the point of aetiology. We would like to know what causes osteoporosis.

The patients are easily recognisable and represent a quarter of white Caucasian females in the USA. Even if you argue that osteoporosis has a multifactorial form, stereotypical treatment has been used for 50 years: calcium in Britain, and oestrogen in the USA. Many observers are insecure about the evidence for sex hormone treatment as preventative medicine. Exercise has always been considered a good thing but it's difficult to know exactly what effect it has on the natural course of the disease.

A better approach seems in order. One that deserves attention was suggested by a report at a Ciba Foundation Symposium on Autoimmunity and autoimmune disease: it was shown that diabetes in the aged can be an autoimmune disease (Feldman 1987). An old theory of osteoporosis holds that the disorder is a form of ageing characterized by organ specific, accelerated, uncontrolled resorption of old bone without a corresponding replacement with new bone. A modern theory of ageing is that it is an autoimmune phenomenon (Walford 1974). The concept of osteoporosis as an autoimmune phenomenon should be investigated so far as the course of the disease shows various signs of inheritance, initiation, and dysregulation (Feldman 1987, Urist et al 1985a,b). A plausible, but unproven, theory views osteoporosis as a manifestation of BMP autoimmune disease (Urist et al 1985a,b). Interestingly, a recent report demonstrates higher amounts of circulating interleukin1 in osteoporotic women *and men* than in age-matched controls (Pacifici et al 1987).

Raisz: Osteoporosis probably encompasses a series of specific pathogenetic mechanisms, all of which are unknown. We have in myeloma a mimic of osteoporosis that is clinically identical, except that it's more rapid. As far as we know, the bone loss is produced by tumour necrosis factor β or lymphotoxin. We have prostaglandins, interleukins, a whole series of resorbing substances which might be pathogenetic, and growth factors, deficiency of which might be important in osteoporosis. We can use the localization methodologies of molecular biology and immunocytochemistry. My hope would be that by the next of these conferences we would know where prostaglandins and interleukins are made in bone and what changes in these factors occur in disease.

References

Aue WP, Roufosse AH, Glimcher MJ, Griffin RG 1984 Solid state ^{31}P nuclear magnetic resonance studies of synthetic solid phases of calcium phosphate: potential models of bone mineral. Biochem 23:6110–6114

Bier A 1892 Osteoplastische necrotomie nebst bemerkungen ber die an der Kieler chirurgischen. Kliniek Ausgefuhrten methoden der necrotomie. Arch Klin Chir 43:121–135

Chambers TJ 1988 The regulation of osteoclastic development and function. In: Cell and molecular biology of vertebrate hard tissues. Wiley, Chichester (Ciba Found Symp 136) p 92–107

Feldman M 1987 Regulation of HLA class II expression and its role in autoimmune disease. In: Autoimmunity and Autoimmune disease. Wiley, Chichester (Ciba Found Symp 129) p 88–108

Glimcher MJ, Kenzora JE 1979a The biology of osteonecrosis of the human femoral head and its clinical implications. I. Tissue biology. Clin Orthop 138:284–309

Glimcher MJ, Kenzora JE 1979b The biology of osteonecrosis of the human femoral head and its clinical implications. II. The pathological changes in the femoral head as an organ and in the hip joint. Clin Orthop 139:283–312

Glimcher MJ, Kenzora JE 1979c The biology of osteonecrosis of the human femoral head and its clinical implications. III. Discussion of the etiology and genesis of the pathological sequelae; comments on treatment. Clin Orthop 140:271–312

Glowacki J, Kaban LB, Murray JE, Folkman J, Mulliken JB 1981 Application of the biological principle of induced osteogenesis for craniofacial defects. Lancet 1:959–963

Hauschka PV, Chen TL, Mavrakos AE 1988 Polypeptide growth factors in bone matrix. In: Cell and molecular biology of vertebrate hard tissues. Wiley, Chichester (Ciba Found Symp 136) p 207–225

Irving JT, Handelsman CS 1963 Bone destruction by multinucleated cells. In: Sognnaes RF (ed) Mechanisms of hard tissue destruction. Am Assn Adv Sci, Washington DC vol 1:515–530

Johnson EE, Urist MR, Finerman GAM 1988 Repair of segmental defects of the tibia with cancellous bone grafts augmented with human bone morphogenetic protein (hBMP): a preliminary report. Clin Orthop, in press

Kato F, Glimcher MJ 1974 The relationship between cellular modulation and architectural structures in bone healing. J Jpn Orthop Assoc 48:395–401

La Croix P 1951 The Organization of bones. Blakiston Co, Phildephia

Pacifici R, Rifas L, Teitelbaum S et al 1987 Spontaneous release of interleukin 1 from human blood monocytes reflects bone formation in idiopathic osteoporosis. Proc Natl Acad Sci USA 84:4616–4620

Pinnell SR, Krane SM, Kenzora JE, Glimcher MJ 1972 Heritable disorder of connective-tissue—hydroxylysine-deficient collagen disease. N Engl J Med 286:1013

Prockop DJ, Kadler KE, Hojima Y et al 1988 Expression of type I procollagen genes. In: Cell and molecular biology of vertebrate hard tissues. Wiley, Chichester (Ciba Found Symp 136) p 142–160

Ray RD, Sabet TY 1963 Bone grafts: cellular survival versus induction. J Bone Jt Surg 45A:357–350

Reddi AH, Huggins CB 1972 Biochemical sequences in the transformation of normal fibroblasts in adolescent rats. Proc Natl Acad Sci USA 69:1601–1605

Rey C, Collins B, Goehl G, Dickson IR, Glimcher MJ 1988 The carbonate environment in bone mineral. A resolution enhanced Fourier-transform infrared spectroscopy study. Submitted

Roufosse AH, Aue WP, Roberts JE, Glimcher MJ, Griffin RG 1984 An investigation of the mineral phases of bone by solid state ^{31}P magic angle sample spinning nuclear magnetic resonance. Biochem 23:6115–6120

Senn N 1889 On healing of aseptic bone cavities by implantation of antiseptic decalcified bone. Am J Med Sci 98:219

Takahashi S, Urist MR 1986 Differentiation of cartilage on three different substrates under the influence of an aggregate of morphogenetic protein and other bone tissue non-collagenous proteins (BMP/NCP). Clin Orthop 207:227–238

Urist MR 1965 Bone: formation by autoinduction. Science (Wash DC) 150:893–899

Urist MR 1968 Surface-decalcified allogeneic bone (SDAB) implants. Clin Orthop 56:37–50

Urist MR 1973 Enzymes in bone morphogenesis: endogenous enzymic degradation of the morphogenetic property in bone in solutions buffered by ethylenediaminetetraacetic acid (EDTA). In: Hard tissue growth, repair and remineralization. Associated Scientific Publishers, Amsterdam (Ciba Found Symp 11) p 143–160

Urist MR 1984 Radioimmunoassay of bone morphogenetic protein in serum: a tissue specific parameter of bone metabolism. Proc Soc Exp Biol Med 176:472–475

Urist MR, Nilsson OS, Hudak RT et al 1985a Immunologic evidence of a bone morphogenetic protein in the *milieu interior*. Ann Biol Chem 43:755–766

Urist MR, Hudak RT, Huo YK, Rasmussen JK 1985b Osteoporosis: a bone morphogenetic protein autoimmune disease. In: Dixon A, Sarnet BG (eds) Second Intern Conf on Bone Growth. Alan R Liss, New York p 77–96

Urist MR, Sato K, Brownell A et al 1983 Human bone morphogenetic protein (hBMP). Proc Soc Exp Biol Med 173 (2):194–199

Urist MR, Chang JJ, Lietze A, Hou YK, Brownell AG, DeLange RJ 1987 Methods of preparation and bioassay of bone morphogenetic protein and other bone tissue non-collagenous proteins (BMP/NCP). Methods Enzymol 146:294–312

Veis A 1988 Phosphoproteins from teeth and bone. In: Cell and molecular biology. Wiley, Chichester (Ciba Found Symp 136) p 161–177

Walford RL 1974 Immunological theory of aging: current status. Fed Proc 33:2020–2026

Young JD-E, Cohn ZA, Gilula NB 1987 Functional assembly of gap junction conductance in lipid bilayers. Demonstration that the major 27-KD protein forms the junctional channel. Cell 48:733–743

Chairman's summary

G.A. Rodan

Department of Bone Biology and Osteoporosis Research, Merck Sharpe & Dohme Research Laboratories, Division Merck & Co Inc., West Point, Pennsylvania 19486, USA

The preceding discussion of the future perspectives for cell and molecular biology of vertebrate hard tissues has highlighted many areas in which there is much experimental work required before we can fully understand the mechanisms of bone and tooth formation and resorption. The papers that were presented on bone formation encouraged us to review traditional approaches and consider whether, for instance, cartilage is necessary for bone formation. Other unresolved issues include: the origin and function of osteoclasts; the identity of osteoprogenitor cells; and the basis and functional significance of osteoblast diversity. In many areas our need for molecular markers and additional assays for cell types and functions has been apparent. The detailed characterization of local regulatory factors and their contribution to tissue formation and destruction under physiological and pathological conditions has only started.

Despite the advances already achieved in the identification and characterization of a number of matrix constituents, our understanding of matrix construction and of how genetic differences in the macromolecular structure of the matrix determine bone strength and predisposition to disease, such as osteoporosis, is still limited. The most striking lesson from this symposium, which was highlighted by heated discussions, was the realization that this is a young field of investigation and that we are in need of large amounts of additional data. It is my hope that we have gained a clearer understanding of all that has been achieved in this area so far and an enthusiasm and inspiration for the work that is still to be done.

In closing, I would reiterate the question I posed at the beginning of the symposium: to what extent can we start to apply our incomplete knowledge of the cellular and molecular basis for the structure and function of bones and teeth to medical problems?

297

Index of contributors

Non-participating co-authors are indicated by asterisks. Entries in bold type indicate papers; other entries refer to discussion contributions

Indexes compiled by John Rivers

*Allen, T.D., **257**
*Argraves, S., **131**
Baron, R., 74, 103, 126, 128, 129, 138, 157, 271, 272, 290
Butler, W.T., 38, 73, 75, 173, 176, 203, 204, 205, 291
Canalis, E., 21, 37, 75, 85, 101, 102, 159, 193, 194, 195, 220, 222
Caplan, A.I., **3**, 16, 17, 18, 19, 20, 21, 40, 58, 72, 73, 75, 87, 128, 136, 138, 158, 220, 221, 224, 252, 253, 292
Chambers, T.J., **92**, 100, 101, 102, 103, 128, 269, 270, 271, 272, 289
*Chen, T.L. **207**
*Constantinou, C.D., **142**
*Dedhar, S., **131**
*Dombrowski, K.E., **142**
*Evans, J., **22**
*Fincham, A., **22**
Fleisch, H., 54, 55, 56, 73, 85, 89, 158, 200, 254, 272, 273, 288
*Friedenstein, A.J., **42**, 54
Gazit, D., 58, 292
Glimcher, M.J., 17, 18, 36, 89, 102, 175, 191, 195, 196, 275, 285, 287, 292
Glowacki, J., 19, 20, 100, 123, 221, 222, 223, 253, 254, 280, 289
*Goldring, M., **239**
*Goldring, S., **239**
Hanaoka, H., 18, 57, 76, 101, 104, 127, 129, 235, 287, 293
Hauschka, P.V. 36, 55, 85, 102, 127, 192, 199, 200, **207**, 220, 221, 223, 224, 254, 290
*Heath, J.K., **78**
*Hojima, Y., **142**
*Kadler, K., **142**

Krane, S.M., 16, 59, 74, 100, 126, 127, 128, 140, 158, 176, 195, 199, 222, 223, 235, **239**, 252, 253, 254, 255, 287, 293
*Krukowski, M., **108**
*Kuivaniemi, H., **142**
*Lau, E.C., **22**
*Lord, B.I., **251**
*Luo, W., **22**
*MacDougall, M., **22**
Martin, T.J., 55, 74, 88, 100, 101, 122, 139, 172, 204, 220, 234, 235, 236, 237, 253, 270, 288
*Molineux, G., **257**
*Mavrakos, A.E., **207**
*Nakamura, M., **22**
Nijweide, P.J., 17, 61, 72, 73, 74, 75, 76, 89, 90, 101, 122, 200, 234, 235, 236, 271, 273, 289
*Noda, M., **78**
*Oliver, P., **22**
*Olthof, A.A., **61**
*Onions, D., **257**
Osdoby, P., 21, 72, 74, **108**, 122, 123, 126, 128, 139, 254, 272, 273, 290
*Oursler, M.J., **108**
Owen, M.E., 20, **42**, 53, 54, 55, 56, 57, 58, 59, 235, 289
Pierschbacher, M.D., 40, 123, **131**, 136, 137, 138, 139, 140, 223, 291
Prockop, D.J., 56, 57, 58, 127, 139, **142**, 156, 157, 158, 159, 193, 253, 254, 255, 287, 292
Raisz, L.G., 16, 17, 53, 73, 74, 85, 87, 100, 122, 123, 126, 140, 193, 204, 205, 221, **226**, 234, 235, 236, 237, 252, 253, 255, 270, 273, 294
Rodan, G.A., **1**, 19, 37, 54, 57, **78**, 85, 86, 87, 88, 89, 90, 101, 136, 139,

157, 173, 193, 200, 205, 236, 237, 253, **296**
*Rodan, S.B., **78**
Ruch, J.V., 35, 36, 37, 157, 289
*Ruoslahti, E., **131**
*Salino-Hugg, T., **108**
Slavkin, H.C., 20, **22**, 35, 36, 37, 38, 39, 40, 86, 136, 174, 196, 205, 290
*Snead, M.L., **22**
*Suzuki, S., **131**
Termine, J.D., 21, 35, 40, 75, 102, 127, 139, 156, 158, **178**, 192, 193, 194, 195, 196, 197, 199, 200, 205, 221, 253, 287, 289
Testa, N.G., 18, 53, 56, 100, 129, 137, 236, **257**, 270, 271, 272, 273, 274, 291

*Tromp, G., **142**
Urist, M.R., 39, 59, 75, 76, 90, 129, 192, 193, 199, 200, 223, 224, 272, 281, 294
*van der Plas, A., **61**
Veis, A., 38, 39, 86, 127, 159, **161**, 173, 174, 175, 176, 177, 192, 194, 195, 200, 205, 222, 224, 291
*Vogel, B., **142**
Weiner, S., 37, 40, 87, 127, 174, 194, 197, 201, 291
Williams, R.J.P., 57, 58, 88, 125, 126, 137, 138, 176, 177, 197, 198, 199, 200, 201, 273, 274, 290
*Yoon, K., **78**
*Zeichner-David, M., **22**

Subject index

Acid phosphatases, 126
Adipocytes, 50, 53, 58
Adipogenic precursor cells, 50
Alcohol, bone resorption and, 95
Alkaline phosphatase, 19–21, 44–46,
    50, 74, 79, 80–82, 85, 86, 88, 180,
    181, 213, 221, 222, 228, 293
Ameloblasts, 23, 26, 27, 28, 29, 36
    autocrine regulation of, 32, 33, 35
    differentiation, 33
Amelogenesis imperfecta, 33
Amelogenins, 24, 26, 27, 29, 36, 37
    DNA sequence, 40
    polyclonal antibodies against, 28
    See also Enamel proteins, Enamelins
29C Antigen, 117, 120
121F Antigen, 111, 113, 114, 115, 117,
    119, 120, 122, 126
Antisense RNA, 193
2ar *See Osteopontin*
Arg-Gly-Asp sequence *See RGD
    sequence*
Atrial natriuretic peptide receptors, 55

Bone γ-carboxyglutamic acid-containing
    protein (BGP) *See Osteocalcin*
Bone cells
    extrinsic influences on, 4, 5, 6
    intrinsic properties, 6
    regulation, 292
Bone-derived growth factor
    (BDGF), 209–219, 220, 227
Bone development, 3–21, 76, 108
Bone environment, osteoclast terminal
    differentiation and, 119, 120
Bone extracellular matrix *See
    Extracellular matrix*
Bone, fetal, 178, 185
Bone formation, 5, 6, 83, 89, 161
    cartilage and, 16–21, 276, 278
    ectopic, 13, 14, 15, 275
    extracellular matrix in, BDGFs
        controlling, 216–218
    in vitro systems for, 290

lipid environment and, 53
long bone, 6–13, 15, 19, 20, 21
    osteogenic and vascular elements
        in, 15, 19, 20, 21, 208, 290
    resorption and, coupling in, 216
    transmembrane, 90
Bone fusion, 275
Bone Gla-proteins *See Osteocalcin*
Bone growth and remodelling,
    hormonal regulation, 226–238
Bone implants, 280–284, 291, 293
Bone, local healing and repair, 14, 15,
    275, 276–280, 292
Bone maintenance and
    replacement, 4, 5, 6
Bone marrow, *See Marrow*
Bone mineralization, 161, 166, 176,
    178, 179, 291
    oxygen and, 57, 58
    procollagen gene mutation
        and, 152, 153
Bone, molecular organization, 291
Bone morphogenic protein
    (BMP), 6, 208, 281, 282, 283,
    284
Bone, non-collagenous proteins
    of, 166, 178–202
Bone phosphoprotein *See
    Osteopontin*
Bone remodelling, 108, 161, 226–
    238
Bone resorption, 83, 208
    bone formation and, coupling
        in, 216
    in vitro systems for, 290
    osteoclastic, 92, 94–96, 98, 102, 127,
        270–272
Bone sialoproteins, 180, 181, 182, 184,
    191, 208
    type I *See Osteopontin*
    type II, 182, 184, 191, 205
Bone, tissue, as, 86
    haemopoietic tissue and, 291
Bone, woven, 87

Brittle bones *See Osteogenesis imperfecta*

Calcification,   175, 276
Calcitonin,   81, 82, 226
  osteoclast activity reduced by,   264, 270
  osteoclastic bone resorption inhibited by,   94, 103
  receptors,   88, 98, 101, 109, 122, 235, 236, 270, 288
Calcium ATPases,   88, 89
Calcium binding, nucleation and,   196, 197
Calcium-binding glycoproteins,   33, 88, 89
Calcium, intracellular mobilization, growth factors and,   213
Calcium phosphate, MG-63.3A cells depositing,   134
Callus,   14
Calvaria(e),   13, 16, 17, 18, 63, 64, 80, 85, 95, 111, 114–117, 208, 263
  cells,   81, 82, 100
  osteocalcin production in,   82, 85
  precursor cell lines from,   43, 48, 54, 58
  resorption models,   102, 103
Calvaria-conditioned medium,   115, 117, 119
Calvaria-like cells, transformed,   82
Capillaries, bone formation, in,   6, 8, 9, 10
Cartilage,   6
  articular, resorption, 279
  avascularity of,   6, 21
  marrow and vascular cell invasion into,   11, 13–16, 18, 19
  osteogenesis and,   16–21
Cartilage model, bone formation, of,   6, 7, 8, 11, 15, 16, 17
Cartilage-derived factor (CDF), 208
Cartilage-derived growth factor (CDGF),   208, 210, 218
Cartilage induction factors,   208
Cartilage precursors,   44
Cathepsin B1,   127
Cell attachment determinants,   132, 137, 184, 203, 204
Cell cultures,   63, 64, 66, 69, 70, 72–76
  bone function *in vivo* and,   79, 80

Cell differentiation, tissue specific,   4, 5, 14, 19, 24, 32
Cell surface adhesion molecules (CAMs)
  bone development, in,   20
  tooth formation, in,   23, 24, 33
Cell surface antigens,   64, 66, 69, 70, 71, 75
Cell transmembrane linkage molecules,   23, 24, 32
Chick chorioallantoic membrane,   111, 117, 119, 120
Chicken
  adhesion receptor complex,   133
  bone phosphoprotein,   182, 191, 192
Chondroblasts,   221, 222
Chondrocytes,   221, 222
  articular,   242, 244, 254
  bone–cartilage interface, at,   11, 20
  calcification,   13, 16, 17, 75, 76
  hypertrophic,   7, 8, 11, 14, 17, 61, 75, 76
  survival in culture,   11, 18, 76, 254
Chondrogenic cells,   18, 20, 21, 73
Chondroid,   75
Chondroitin sulphate,   134, 136, 137
Collagen
  degradation,   194, 195, 196
  interactions with other matrix components,   146, 147
  mutant genes,   194
  synthesis,   245, 246
    growth factors and,   213
  pCcollagen,   145, 147
Collagen type I,   6, 7, 19, 25, 43–45, 50, 79, 80–82, 86, 88, 89, 119, 142
  biosynthesis,   143
    cysteine substitution in,   152, 159
  bone,   166, 174, 175, 176
  calcification,   175
  dentine, 166, 176
  fibril assembly,   142, 143, 145, 179
    thermodynamics of,   144, 145, 147
  purified (Vitrogen),   224
  receptor,   132, 140
  skin,   174, 175, 176
  synthesis,   208, 228, 230, 231, 245, 246
Collagen type II,   7, 17, 43, 44, 245, 246
Collagen type III,   25, 43, 44, 49, 50, 81, 245, 246
Collagen type IV,   23, 43, 49, 50, 57

Collagen type V, 23, 25, 49
Collagen type VII, 23, 33
Collagen type X, 11, 17
Collagenase
  cytokine-induced synthesis, 242,
    243, 244, 245, 247
  osteoblastic cells producing, 96,
    102, 228
  osteoclastic bone resorption
    and, 101, 125, 126
Colony-forming cells, marrow stroma
  (CFC-O), 76
Colony-forming unit-fibroblastic (CFU-
  F), 44, 45–47, 48, 49, 50, 54, 56
Colony-forming unit-granulocyte
  macrophage (CFU-GM), 109
Colony-forming unit-
  reticulofibroblastoid (CFU-
  RF), 48, 49, 50, 57
Colony-stimulating factors, 241, 257–
  261, 274
  haemopoietic, 48
Colony-stimulating factor-1 (CSF-1) See
  Macrophage-CSF
Connective tissue, molecular
  synthesis, 292, 293
Cyclic AMP, 79, 81, 88, 119, 122, 213,
  227, 228, 236, 237, 243
Cytokines, 239–256
  action of, 242, 243–247, 261
  gene transcription induced by, 244,
    245, 246
  osteoblast formation and function
    and, 266
  skeleton and, 247, 248
  sources and properties, 240–242
  See also Growth factors
Cytotactin See Tenascin
Dentine, 166
  biomineralization, 31, 32, 33, 35
  calcium hydroxyapatite crystal
    formation in, 25, 27, 29, 31, 36
  collagen type I of, 166, 176
  cultured osteoclasts and, 270, 271
  non-collagenous proteins of, 164–
    171, 176
  osteocalcin, 25
  phosphoprotein, 24, 25, 33, 166
    calcium binding by, 25, 171, 174,
      176, 177
    expression by odontoblasts, 25,
      38

ion diffusion by, 177
  polyclonal antibodies against, 25,
    38, 39
  phosphophoryns, 166, 167–171, 173
  secondary, 173, 174
  See also Enamel, Tooth
Dentinogenesis imperfecta, 33, 176
Diffusion chambers, 43, 44, 45, 53, 58,
  89
  bone formation in, 90, 139
  osteogenic tissues formed in, 54. 55,
    76, 81, 82
  UMR cell response in, 88
Dihydroxyvitamin D_3 See Vitamin D_3
Dunn mouse osteosarcoma, 90, 104

Ehlers-Danlos syndrome, 151, 152,
  153, 154, 287
Enamel
  biomineralization, 31, 32, 35
  maturation, 29
  See also Dentine, Tooth
Enamel proteins, 24, 26–29, 33, 35, 36
  See also Amelogenins, Enamelins
Enamelins, 26, 27
  See also Amelogenins, Enamel
    proteins
Enameloid, aprismatic, 36, 37
Endochondral sequence, classical, 14,
  19
Endothelial cell
  basement membrane, osteoblast
    proliferation and, 59
  response to BDGFs, 215, 216
Endothelial cell-derived growth factor
  (ECDGF), 209
Enzyme-linked immunosorbent assay
  (ELISA), 111, 113, 117, 122,
  123
Epidermal growth factor (EGF), 33,
  208, 211, 218
  CFU-F differentiation and, 46, 47
  mouse tooth morphogenesis and
    differentiation and, 36, 37
  response of osteoblast-like cells
    to, 81
Epigenesis, 39, 40
Erythropoietin, 258
Extracellular matrix, 1, 217, 223, 229,
  282
  bone, 119, 120, 178, 207, 208
  cartilage, 11

composition, 208
cytokine binding to, 245
degradation, 194, 242
dental, biomineralization, 22–41
enamel, proteins of, 29
Gla-protein, 182, 186, 223
polypeptide growth factors in, 207–225
proteins, polyclonal antibodies against, 75

Factor VIII, 49
Fc receptors, 263, 270, 272
Fibrinogen receptor, 132, 133
Fibroblast(s), 50, 239, 240, 244
    collagen type IV in, 57
    marrow, 48
    osteogenic tissue derived from, 45
    periosteal, 63
    synovial, 242, 243, 244, 246, 254
Fibroblast growth factor (FGF), 33, 180, 208, 210–212, 218
    osteoinduction of, 213, 215, 220–223
    receptors, 213, 215
    sensitivity to guanidine, 224
Fibroblastic colonies, 44, 45, 48, 54
Fibroblastic colony-forming cell (FCFC), 44, 45
Fibrogenic cells, 58
Fibronectin, 23, 25, 33, 35, 43, 57, 80, 131
    bone cell maturation and, 139
    receptor, 131–134, 136–140
    synthesis, IL-1 and, 246

Gap junctions, 140, 239, 293
Gelatinase, 244
Genes
    regulatory, 290
    structural, 142–160, 290
Gene transfer techniques, 51
Genomic repertoire, repair processes, in, 4, 5
Giant cells, marrow-derived, 117, 120
Gla-protein, 24, 25, 26, 38, 88
    bone See Osteocalcin
    matrix, 182, 186
Glucocorticoids
    bone growth regulation by, 221, 230, 231
    osteocalcin and osteopontin synthesis depressed by, 80

β-Glycerol phosphate, calcification and, 87, 89, 90
Glycoproteins, tooth development, in, 25, 40
Glycosaminoglycans
    bone, in, 185, 208
    tooth development, in, 25
Granulocyte colony-stimulating factor (G-CSF), 257, 258, 260
Granulocyte macrophage colony-stimulating factor (GM-CSF), 245, 257, 258, 260, 268, 274
Growth factors, polypeptide, 207–225, 258, 260, 261, 292
    dental biomineralization, in, 33, 36, 37
    pathways for mitogenic stimulation of cells, 217
    storage
        bone, in, 210, 211, 224
        extracellular matrix, in, 218, 220
    See also under specific names and under Cytokines
Growth hormone, bone growth regulation and, 227, 228
Growth plate, 16
Guanidine, growth factors and, 224

Haemopoiesis, 42, 43, 45, 48
    cell interactions in, 261, 262
Haemopoietic growth factors, 257–274
Haemopoietic tissue structure, 261, 262, 273, 274
Heparin, 208, 209, 210, 211, 212, 215, 220, 224
α₂-HS-glycoprotein, 180, 181
Human skeletal growth factor (hSGF), 208
Hydroxyapatite, 117, 120, 123, 124, 125, 166, 178, 179, 187, 188, 208

Immortalized cells, 82, 85, 86
Immunocytochemistry, 25, 26, 28
Indomethacin, 53, 95, 246, 252, 254
Insulin-like growth factor 1, 33, 180, 208, 211, 218, 222, 227, 228, 241
Integrin, 23, 32, 33
Intercellular adhesion molecule 1, 133
Interleukin 1 (IL-1), 180, 208, 227, 240, 242, 261, 272, 273
    α and β types, 243

Interleukin 1 (IL-1) (*contd.*)
   bone resorption and,   83, 85, 96, 98,
      103, 237, 247
   collagen types I and III synthesis
      and,   246, 247, 248, 252, 253,
      254
   collagen type II synthesis and,   246,
      247, 248, 253
   collagenase synthesis and,   244, 245
   mechanism of action,   243, 244
   osteoclast differentiation and,   98,
      124
   PGE$_2$ production and,   232
   receptors,   243, 245
   response of osteoblast-like cells
      to,   81
   TGF-β activity and,   247
Interleukin 2,   261
Interleukin 3 (Multi-CSF),   257, 258,
   260, 273
Interferon γ
   bone resorption inhibited by,   247
   collagen synthesis and,   245, 253
   nuclear factor 1 and,   255
   osteoclast regulation by,   288

Laminin,   23, 33, 43, 49, 50, 57
Leucocyte adhesion molecules,   133
Leukotrienes, bone resorption and,   95
Lining cells,   62, 81, 95, 290
Lymphocyte(s),   82, 239, 240, 241
Lymphokine *See Cytokine*
Lymphotoxin *See Tumour necrosis
   factor* β
Lysine hydroxylase deficiency,   287
Lysine oxidases,   57, 58

Macrophage(s)
   bone marrow cultures, in,   98, 101,
      269, 270, 271, 272
   lineage,   101, 272
Macrophage-CSF (M-CSF),   257, 258,
   260, 268, 272, 273, 274
Macrophage-derived growth
   factors,   208
Malignant cells, differentiation level
   of,   88
Manganese, osteoclast,   126
Marrow cells,   5, 11, 13, 14, 20
   cell lines derived from,   43
   co-cultures, 111, 114–117
   cultures,   59, 96, 101

fibroblasts,   48
mononuclear cells,   110, 111, 117,
   120
osteogenesis, in,   292
osteogenic precursors in,   43, 44
single cell suspensions, 43, 44
trypsinization,   56
Marrow stroma colony-forming cells
   (CFC-O),   75, 76
Matricrine concept,   217, 223, 229, 282
Matrix *See Extracellular matrix*
Mesenchymal stem cells,   5, 13, 14, 20,
   50, 55, 57, 58
   See also Stromal stem cells
Metalloproteinase-3 *See Stromelysin*
MG-63 human osteosarcoma
   cells,   133, 134, 139, 211
MG-63.3A cells,   134, 136, 139
β$_2$-microglobulin,   208, 222, 227
Monoclonal antibodies
   osteoclast-specific,   96, 98, 109–111,
      113, 270
   osteogenic cells, against,   63–75, 289
Monocytes,   239, 240, 241, 242, 243
Monocyte-conditioned medium,   242,
   243, 246, 254
Morphogens,   39
Multi-CSF *See Interleukin 3*
Multinucleated cells,   263, 264, 269,
   270, 272, 289
Myogenic cells,   54, 58
Myotendinous antigen *See Cytotactin*

Na$^+$/K$^+$-ATPase, osteoclasts, in,   109,
   123, 126
Non-collagenous proteins
   bone,   164–167, 178–202, 213, 282,
      283, 284
   dentine,   166, 176
   matrix,   291
Nuclear factor 1,   246, 255
Nucleation, crystal,   196, 197

Odontoblasts,   25, 35, 36, 37, 38, 173
Odontogenesis,   289
Oestradiol, bone metabolism
   and,   229, 230
Oestrogen,   83, 88, 227, 229, 229, 230
Osteoblasts,   5–10, 13–16, 38, 50, 57,
   79, 179, 180, 208, 239, 240
   atrial natriuretic peptide receptors
      on,   55

bone repair, in,   277, 278, 279
bone tissue function and,   83
cultured, bone formation in,   89
dedifferentiation,   62
differentiation,   4, 5, 20, 33, 61–77,
    81, 83, 216
genesis,   19
hormonal regulation,   63
LDL receptors on,   53
lineage cells,   61, 63, 87
–matrix interactions,   290
maturation-related changes in,   81
–osteoclast co-cultures, bone
    resorption in,   94, 95, 100, 102,
    103, 104
–osteoclast interactions,   227, 228,
    231, 290
polypeptide growth factor
    biosynthesis and,   212
proliferation,   62, 216
response
    bone growth factors, to,   212, 213,
        215
    in culture, diversity of,   81, 82
Osteoblastic cells
committed progenitors,   289
osteoclastic bone resorption regulated
    by,   92, 94, 95, 96
Osteoblastic phenotype,   62, 70, 71
diversity of,   78–91
Osteoblast-like cells,   63–69, 71
Osteoblast-like colonies, osteogenic
    activity of,   55, 56
Osteocalcin,   26, 40, 50, 55, 111, 117,
    124, 179, 182, 186, 187, 208
bone turnover and metabolism,
    in,   187, 192, 193
calcium binding and,   198, 199, 200,
    201
secretion, growth factors and,   213
synthesis,   79, 80, 81, 82, 85, 89
Osteoclasts,   5, 8, 13, 18, 57, 61, 94,
    179, 208, 216, 217, 239
activation,   288
activity,   125–130
binding to bone,   123, 125
bone marrow co-cultures,   111, 114–
    117, 119
bone repair, in,   277, 278, 279
bone resorption and,   94, 95, 96, 98,
    100, 101, 127, 247, 270, 271, 272
calcitonin and,   264

cell cultures,   110
cell surface antigens,   109, 111–120,
    122, 123, 126
cell surface ruffling,   124, 125
characteristics,   123
conditioned medium
    experiments,   113, 114–117,
    119
cultures, 110, 262–268, 269, 270, 271,
    273
'cutting cone',   217
development,   108–124
    CSFs inducing,   247, 266
    regulation of,   92–107
differentiation,   241, 289
function, regulation of,   92–107, 216
haemopoietic origin,   96, 104–106,
    128, 129, 208, 289, 291, 293
lead poisoning and,   129
lysosomal enzymes,   124, 125, 126
–matrix interactions,   290
multinuclearity,   98, 100, 101, 123
–osteoblast interactions,   290
pH,   126, 127
plasma membrane
    components,   108, 109
prostaglandins inhibiting,   95, 96,
    100, 231
PTH and,   227, 228, 235
recruitment,   123, 124
regulation,   124, 129
reverse differentiation,   126, 127
–specific monoclonal antibodies,   96,
    98, 109–113, 118, 270
terminal differentiation,   109, 110,
    117, 119, 120, 122
TGF-β release by,   217
Osteoclast resorption stimulating
    activity (ORSA),   94, 95, 96, 98,
    101
Osteocytes,   5, 62, 70, 71, 81, 289
cell cultures,   69, 71
cell surface antigens,   73, 74, 75
osteoid,   62
predetermined differentiation,   72,
    73
PTH binding to,   69, 71
'surface',   96
Osteocytic osteoblasts,   62
Osteogenesis See Bone formation
Osteogenesis imperfecta,   57, 147, 158,
    176, 178, 194, 248, 285–288

Osteogenesis imperfecta (*contd.*)
  heterogeneity of defect, 154, 156, 157, 287
  lethal variants, 148–151, 154
  non-collagen proteins in, 179
  non-lethal variants, 151–154
Osteogenic cells
  characteristics, 79
  differentiation, 45–47, 288, 289
  lineage, 61, 70
  monoclonal antibodies against, 63–75
  precursors, 6, 13, 14, 19, 61, 62, 70
    marrow-derived, 42–60
    See also Preosteoblasts
  tissue, fibroblast-derived, 45
  types, reversibility of, 75
Osteogenin, 208, 282, 292
Osteoinductive proteins, 281, 282
Osteonecrosis, 275, 277, 279
Osteonectin, 80, 81, 88, 111, 117, 182, 187–189, 192, 193, 199, 208
Osteopetrosis, 128, 129
Osteopontin, 75, 80, 81, 88, 89, 111, 117, 123, 133, 166, 171, 175, 182, 184, 191, 192, 203–206
Osteoporosis
  adipogenesis in, 53
  fractures in, 276
  gene defect in, 153, 158, 159
  glucocorticoid excess and, 230
  heterogeneity of, 159
  mechanisms of, 289, 292, 293, 294
  PTH and, 229, 235

Paget's disease, 178, 288
Paracrine factors, dental
    biomineralization, in, 32, 33
Parathyroid hormone (PTH)
  alkaline phosphatase levels and, 81
  anabolic effect, 228, 229, 235
  binding to osteoblast-like cells and osteocytes, 69, 71
  bone formation and, 79, 81, 83, 86, 95, 226, 228, 229, 234, 235
  bone resorption and, 94, 96, 98, 103, 115, 227–229, 234, 235
  cyclic AMP production and, 81, 82, 88, 213
  osteoclast differentiation and, 98, 100, 115
  osteoclast regulation by, 124

receptors, 6, 50, 127, 128
  -related peptide, 228, 236, 237
  TGF-β activity stimulated by, 247
Periodontal disease, 245
Phorbol ester-inducible genes, 244, 245, 247
Phosphophoryns, 164, 165, 167–170, 171, 173, 174, 177
Phosphoproteins, 166, 175, 191
Plasminogen activator, 228
Platelet-derived growth factor, 180, 208, 209, 210, 211, 212, 213, 215, 218, 241
Polyclonal antibodies, 28, 75
Preameloblasts, 36
Predentine, 36, 38
Preosteoblast, 62, 73, 76
Preosteoclast, 101, 104, 127, 128, 129, 235, 236
Procollagen type I
  biosynthesis, 143
  conversion to collagen, 144, 145, 146
  gene amplification, 157
  gene mutations, bone diseases and, 147–154, 287, 293
  proα1(I) chain, for, 142, 143, 151, 152, 153, 154, 159, 181, 192
    amino propeptide of, 181, 192
  proα2(I) chain, for, 142, 143, 158, 159, 176, 246
    gene mutations, 148, 149, 151–154
  C-proteinase, 145, 146, 147
  N-proteinase, 145, 146, 147, 151, 152, 153
  mRNA, 80
  structural gene expression, 142–160
    retroviruses and, 157
    tissue specific, 157, 158
Procollagenase, 244
Proenkephalin gene, 245, 247
Prostaglandins, 81, 83, 95, 96, 100
Prostaglandin E, response of osteoblast-like cells to, 81, 82, 83
Prostaglandin E$_2$
  bone growth regulation and, 227, 228, 230, 231, 232, 234, 236
  bone resorption and, 103, 213
  cytokine-stimulated collagen synthesis and, 246, 252, 254
  cytokine-stimulated release, 242, 243, 247, 253
  osteoclast regulation by, 124

PTH-stimulated synthesis, 228
IIb/IIIa Protein complex, 132
Protein kinase activity, 173, 192, 213, 243
'Protein suicide', 150, 151, 157, 194
Proteoglycans
  bone, 179, 182, 184–186, 208
  bone formation, in, 11, 17, 81, 137, 179, 184
  cartilage, 111, 185
  chondroitin sulphate, 134, 137, 185, 186
  dermatan sulphate, 186
  keratan sulphate, 184, 185
  tooth development, in, 23, 25
  types I and II, 185, 186

Rat
  osteoblast-like cells, 64, 66
  osteosarcoma 17/2 cells, 79, 80, 81, 82, 86, 87, 88, 139, 204
  44 kDa phosphoprotein, 182, 191
  RCT3 cells, 82, 85, 86, 88
  Restriction fragment length polymorphisms, 33
  Reticular cells, 50, 59
  Retinoic acid, osteoblast response to, 81, 86, 88
  Rheumatoid arthritis, 245
  Ridgeway mouse osteosarcoma, 90
  RGD receptor superfamily, 132, 139
  RGD sequence, 131, 132, 133, 137, 138, 139, 184, 203

Skeletal mass, 179
Skeletal muscle, repair cells, 5
Skeletal remodelling, cytokines and, 247, 248
Skin collagen, 174, 175, 176
Somatomedin C See Insulin-derived growth factor
SPARC, 182, 188, 192
Stacked cell layer, 6, 9, 10, 11, 13, 14, 15, 17, 19, 20, 21, 73
Stromal stem cells, 42–60
Stromelysin, 244, 245, 247
Substrate adhesion molecules (SAMs), 20
  tooth formation, in, 23, 24

Talin, 32
Target mesenchymal cells, 242

Tartrate-resistant acid phosphatase (TRAP), 123, 124, 236, 269
Tenascin, 23, 24, 25, 33, 40
TGF See Transforming growth factor
Thyroid hormones, bone growth regulation and, 227
Tissue plasminogen activator, osteoblastic cells producing, 96
Tooth
  biomineralization, 22–41
  morphogenesis, 22, 23–25
  See also Dentine, Enamel
Toxin gene expression, 248
Transcription factor AP-1, 245
Transferrin, 33
Transformed cells, 82, 88, 90
Transforming growth factor, 35, 208
Transforming growth factor β (TGF-β)
  bone formation and, 79, 80, 180, 208, 218, 222, 227, 240, 247
  collagen synthesis and, 246, 247, 255
  PGE$_2$ production stimulated by, 232
  release by osteoblasts, 217, 228
  response of osteoblast-like cells to, 81
Transgenic mice, 193, 248, 255, 292
Tumour necrosis factor, 227, 261
  α and β types, 241, 242, 243
  bone resorption and, 95
  collagen synthesis and, 246
  collagenase synthesis and, 244
  effect on target cell populations, 261, 274

U 937 cells, receptor regulation in, 140
UMR cells, 81, 88, 100, 236

Vasculature
  bone repair, in, 14, 290
  long bone formation, in, 9, 10, 13, 18, 19, 20, 21
Very late activation antigens, 133
1,25-(OH)$_2$ Vitamin D$_3$, 16, 79, 80, 81, 82, 83, 86, 88, 89, 94, 96, 98, 103, 124, 140, 187, 204, 226, 229, 247, 266, 271, 272
  binding proteins, 33
Vitrogen, 244
Vitronectin receptors, 132, 133, 134, 137
von Willebrand factor, 132, 133

Zinc collagenase, 57